DROUGHTS, FOOD
AND CULTURE

DROUGHTS, FOOD AND CULTURE

Ecological Change and Food Security in Africa's Later Prehistory

Edited by

Fekri A. Hassan

University College London
London, England

Kluwer Academic / Plenum Publishers
New York, Boston, Dordrecht, London, Moscow

Library of Congress Cataloging-in-Publication Data

Droughts, food, and culture: ecological change and food security in Africa's later
prehistory/edited by Fekri A. Hassan
    p. cm.
   Includes bibliographical references and index.
   ISBN 0-306-46755-0
    1. Prehistoric peoples—Food—Africa. 2. Agriculture, Prehistoric—Africa. 3. Climatic
changes—Africa. 4. Paleoclimatology—Africa. 5. Africa—Antiquities. I. Hassan, Fekri A.

GN861 .D76 2002
960'.1—dc21

2002022218

ISBN: 0-306-46755-0

©2002 Kluwer Academic / Plenum Publishers, New York
233 Spring Street, New York, New York 10013

http://www.wkap.nl/

10 9 8 7 6 5 4 3 2 1

A C.I.P. record for this book is available from the Library of Congress

Printed in the United States of America

CONTRIBUTORS

R. Abu-Zied, Southampton Oceanography Centre, Southampton, SO14 3ZH, UK.

Hala N. Barakat, Cairo University Herbarium, Faculty of Science, Cairo University, Giza, Egypt.

Barbara E. Barich, Dipartimento di Scienze Storiche, Archeologiche e Antropologiche dell'Antichità, University of Rome "La Sapienza", Via Palestro 63, 00185 Rome, Italy.

R. Bonnefille, CNRS, CEREGE, bp 80, 13545, Aix-en-Provence, Cedex 04, France.

Peter Breunig, Johann Wolfgang Goethe-Universität Frankfurt am Main, Seminar für Vor- und Frühgeschichte, Archäologie und Archäobotanik Afrikas, Grüneburgplatz 1, D - 60323 Frankfurt am Main, Germany.

H. Brinkhuis, Laboratory of Palynology and Paleobotany, Utrecht University, The Netherlands.

Ann Butler, Institute of Archaeology, University College London, 31-34 Gordon Square, London, WC1 0PY, UK.

J. Casford, Southampton Oceanography Centre, Southampton, SO14 3ZH, UK.

Mauro Cremaschi, CNR Centro di Geodinamica Alpina e Quaternaria, Dipartimento di Scienze della Terra, University of Milan, Italy.

I. Croudace, Southampton Oceanography Centre, Southampton, SO14 3ZH, UK.

Savino di Lernia, Centro Interuniversitario di Ricerca sulle Civiltà e l'Ambiente del Sahara Antico, University of Rome "La Sapienza", Via Palestro 63, 00185 Rome, Italy.

Achilles Gautier, Laboratorium voor Paleontologie, Vakgroep Geologie en Bodemkunde, Universiteit Gent, Krijgslaan 281/S8, B-9000 Gent, Belgium.

Fekri A. Hassan, Institute of Archaeology, University College London, 31-34 Gordon Square, London, WC1 0PY, UK.

Stan Hendrickx, Katholieke Universiteit Leuven, Laboratorium voor Prehistoire, Redingenstraat 16 bis, B-3000 Leuven, Belgium.

F. J. Jorissen, Department of Oceanography and Geology, University Bordeaux I, CNRS URA 197, France.

J. Kallmeyer, Department of Earth Sciences, Bremen University, Germany.

D. Mercone, Southampton Oceanography Centre, Southampton, SO14 3ZH, UK.

M. U. Mohammed, Department of Geology and Geophysics, Addis Ababa University, c/o P.O.Box 3434, Addis Ababa, Ethiopia.

Katharina Neumann, Johann Wolfgang Goethe-Universität Frankfurt am Main, Seminar für Vor- und Frühgeschichte, Archäologie und Archäobotanik Afrikas, Grüneburgplatz 1, D - 60323 Frankfurt am Main, Germany.

E. J. Rohling, School of Ocean and Earth Science, Southampton University, Southampton Oceanography Centre, Waterfront Campus, European Way, Southampton, SO14 3ZH, UK.

Martine Rossignol-Strick, Laboratoire de Paléobiologie et Palynologie, Boite 106, Université Pierre et Marie Curie, 4 Place Jussieu, 75252 Paris cedex 05, France.

M. Adebisi Sowunmi, Department of Archaeology and Anthropology, University of Ibadan, U. I. P. O. Box 20204, Ibadan, Nigeria.

J. Thomson, Southampton Oceanography Centre, Southampton, SO14 3ZH, UK.

Wim Van Neer, IUAP, Royal Museum of Central Africa, 3080 Tervuren, Belgium.

Robert Vernet, Département d'Histoire, Faculté des Lettres, BP 396, Nouakchott, Mauritanie.

G. Wefer, Department of Earth Sciences, Bremen University, Germany.

PREFACE

Recent droughts in Africa and Europe have serious implications for food security and grave consequences for local and international politics. The issues do not just concern the plight of African peoples or Europe's role in alleviating catastrophic conditions, but also Europe's own ecological future. Africa's Sahel zone is one of the most sensitive climatic regions in the world, and the events that gripped that region in the 1970s were the first indicators of a significant shift in global climatic conditions. Deterioration of living standards in Africa due to adverse climatic conditions is also likely to involve the world community in various dimensions.

With the realization that contemporary climatic events and their consequences could not be properly interpreted without an adequate knowledge of long-term climatic variability and associated cultural developments in Africa, a workshop was organized, from 15 to 18 September 1998 in London (UK), on 'Ecological Change and Food Security in Africa's Later Prehistory', with a generous grant from the European Science Foundation. In addition to substantive contributions to our knowledge of African prehistory, the workshop was also a venue to explore the potentials for a comprehensive, integrated program of future research with the following goals in mind:

(1) to develop *a common strategy for documenting and interpreting local responses to global ecological events and their influence on intercultural contacts*;
(2) to forge available data into a coherent ecological-cultural framework;
(3) to explore means by which data can be processed in *a dynamic model that can be used for interpretation of the long-term consequences of ecological events*; and
(4) to establish *a data bank of ecological and archaeological data to assist researchers and policy makers in assessing long-term ecological processes.*

The workshop was thus an opportunity to overcome the limitations imposed by the lack of coordination among research teams working in Africa and the absence of a concerted strategy of archaeological inquiry on a continental scale. Although the scope of this volume was limited by design and by practical necessities to a select group of eminent archaeologists and palaeoclimatologists, it is hoped that future meetings will integrate the results from this workshop within other regional contexts and specific topical domains.

It is also hoped that this volume will alert archaeologists to the effects of severe, abrupt, and closely-spaced climatic events on the course of cultural developments. Such events, which are now becoming quite evident thanks to high-resolution palaeoclimatic data, are likely to cause significant ruptures in the cultural fabric of societies on various scales. Some of the events discussed in this volume have proven to be of a global, trans-continental nature. This global interconnection is due to the response of the shift of the Intertropical Convergence Zone across an area that extends from parts of South and Mesoamerica to China. The contributions in this volume should be seen as parts of a whole, which is in turn a part of trans-continental cultural developments in response to climatic crises that punctuated the millennial change in climate. Archaeologists stand to gain a deeper understanding of the past by focusing on specific climatic events that set in motion cultural responses, that in turn have cultural consequence under similar or pre-existing long-term climatic conditions.

It would be a grave error to resurrect the ghost of environmental determinism in order to undermine current efforts to clarify the intricate link between climate and society. It is now abundantly clear that the impact of any climatic change is mediated by human perceptions and social action, and that many cultural developments are predicated upon previous cultural choices and norms. Cultural developments are not to be explained in terms of a single paramount cultural mechanism, such as diffusion or the banal "evolution", but in terms of a model of cultural innovations (that may range from prayers to migration) that are maintained or rejected by the local community, that may or may not be adopted (often with significant modifications) by neighboring groups, and that may or may not survive from one generation to the next.

In this volume, the appearance of cattle keeping in Africa and its impact on African societies in different ecological contexts, in response to adverse abrupt climatic crises, reveals the poverty of deterministic thinking, the shortcomings of simplistic diffusionary models, and the deficiency of linear evolutionary models. The origins and dispersal of cattle keeping in Africa, by contrast, reveal that societies chose from a set of probable responses, that they accepted, assimilated and modified innovations (local or borrowed) within the context of their field of knowledge and action, governed by norms, mores, and organizational parameters. The emergence of cattle pastoralism, for example, in Central, West and East Africa were independent local responses taking advantage of local ecological conditions and within a strategy that maintained earlier traditions. The disparity between Egypt and other neighboring African countries was a result of a different timing of events, different ecological settings (the Lower Nile as a significant factor), and the cultural geography of Egypt within North Africa and in proximity to Southwest Asia. Travelling on different tracks in response to the same climatic events, different groups responded, at a later time, to the next climatic event in a very different manner from before. Developments in a cultural region may also have an impact on neighboring regions.

A more profound understanding of how societies dealt with climatic change in the past is one of the means by which we can cope with our changing climate today. This volume not only highlights the importance of detecting the timing and severity of past abrupt climatic events, but also the role of archaeological investigations in teasing out such events and clarifying their impact on human societies. The palaeoclimatological community is also now challenged to explain the causes of such abrupt events and to provide measures of climatic change that are significant in the realm of human affairs.

This volume is thus an attempt to bring archaeology within the domain of contemporary human affairs, and to forge a new methodology for coping with environmental problems from an archaeological perspective.

Participants in the London workshop included: Dr. Hala Barakat (Cairo), Prof. Barbara Barich (Rome), Mr. Obare Bogado (Benin), Prof. Peter Breunig (Frankfurt), Dr. Ann Butler (London), Dr. R. Chedadi (Marseilles), Prof. Mauro Cremaschi (Milan), Dr. Savino di Lernia (Rome), Dr. Françoise Gasse (Aix-en-Provence), Prof. Achilles Gautier (Gent), Dr. Stan Hendrickx (Brussels), Dr. Steve Juggins (Newcastle-Upon-Tyne), Ms. Dagmar Kleinsgutl (London/Vienna), Dr. Henry Lamb (Aberstwyth, Wales), Dr. Kevin MacDonald (London), Dr. James McGlade (London), Dr. Katharina Neumann (Frankfurt), Dr. François Paris (Paris/Cairo), Dr. E. J. Rohling (Southampton), Dr. Martine Rossignol-Strick (Paris), Prof. Paul Sinclair (Uppsala), Prof. Adebisi Sowunmi (Ibadan, Nigeria), Prof. A. C. Stevenson (Newcastle-Upon-Tyne), Dr. Mohammed Umer (Addis Ababa), Prof. T. H. van Andel (Cambridge), Dr. Wim Van Neer (Tervuren), and Dr. Robert Vernet (Nouakchott, Mauritania). Mrs. Marianne Yagoubi, represented the European Science Foundation at the meeting.

I am most thankful to all who attended for the lively and stimulating discussions that have opened up new vistas for investigation and new possibilities for cooperation between scholars from different regions and disciplines. I am also particularly grateful to those who contributed to this volume. The workshop was held at the Institute of Archaeology, University College London, and I am indebted to Prof. Peter Ucko, Director of the Institute, for making the facilities of the Institute available for the workshop. The workshop has signaled for me a new direction in African archaeology, with theoretical implications for archaeology as a whole, and I wish to thank the European Science Foundation for its support of that event. Partial financial support through Prof. Paul Sinclair is also gratefully acknowledged. I wish also to acknowledge logistic support by Dr. Hala Barakat, Geoffrey Tassie, Joanne Rowland, Dagmar Kleinsgutl, and Janet Picton. Special thanks go to Julie Wilson who copy edited this work and prepared it for publication.

Cherish the land from the spoiling drought, from the raging wind, the dust-laden storm...and the beating rain.

CONTENTS

List of Figures... xiii

List of Tables... xvii

1.   Introduction...   1
     F. A. Hassan

2.   Palaeoclimate, Food and Culture Change in Africa: An Overview   11
     F. A. Hassan

Section I: Climatic Change..   27

3.   Rapid Holocene Climate Changes in the Eastern Mediterranean............   35
     E. J. Rohling, J. Casford, R. Abu-Zied, S. Cooke,
     D. Mercone, J. Thomson, I. Croudace, F. J.
     Jorissen, H. Brinkhuis, J. Kallmeyer and G. Wefer

4.   Climate During the Late Holocene in the Sahara and the Sahel: Evolution   47
     and Consequences on Human Settlement..............................
     R. Vernet

5.   Late Pleistocene and Holocene Climatic Changes in the Central Sahara:
     The Case Study of the Southwestern Fezzan, Libya.....................   65
     M. Cremaschi

6.   Late Holocene Climatic Fluctuations and Historical Records of Famine
     in Ethiopia...   83
     M.U. Mohammed and R. Bonnefille

7.   Environmental and Human Responses to Climatic Events in West and
     West Central Africa During the Late Holocene......................   95
     M. A. Sowunmi

Section II: Plant Cultivation.. 105

8.   Regional Pathways to Agriculture in Northeast Africa...................... 111
     H. N. Barakat

9.   From Hunters and Gatherers to Food Producers: New Archaeological
     and Archaeobotanical Evidence from the West African Sahel....... 123

P. Breunig and K. Neumann

10. Holocene Climatic Changes in the Eastern Mediterranean and the Spread
     of Food Production from Southwest Asia to Egypt............................. 157
     M. Rossignol-Strick

11. Sustainable Agriculture in a Harsh Environment: An Ethiopian
         Perspective... 171
     A. Butler

Section III: Pastoralism... 189

12. The Evidence for the Earliest Livestock in North Africa: Or Adventures
         with Large Bovids, Ovicaprids, Dogs and Pigs........................... 195
     A. Gautier

13. Cultural Responses to Climatic Changes in North Africa: Beginning and
         Spread of Pastoralism in the Sahara....................................... 209
     B. E. Barich

14. Dry Climatic Events and Cultural Trajectories: Adjusting Middle
         Holocene Pastoral Economy of the Libyan Sahara...................... 225
     S. di Lernia

15. Food Security in Western and Central Africa During the Late Holocene:
         The Role of Domestic Stock Keeping, Hunting and Fishing.......... 251
     W. Van Neer

16. Bovines in Egyptian Predynastic and Early Dynastic Iconography........... 275
     S. Hendrickx

Conclusion..

17. Ecological Changes and Food Security in the Later Prehistory of North
         Africa: Looking Forward.. 321
     F. A. Hassan

Index.. 335

LIST OF FIGURES

2.1   Distribution of earliest cattle in Africa.................................... 15

2.2   Draft sketch of Holocene climatic variability in Africa..................... 19

3.1   Warm versus cold planktonic foraminiferal percentages for cores IN68-9, LC21, and LC31, plotted against core-depth in cm as corrected for thicknesses of ash-layers and turbidites............................. 36

3.2   Key benthic foraminiferal species in cores LC31 and LC21.............. 38

3.3   Warm-cold record and stable oxygen isotope record (*G. ruber*) for LC21, versus age in calibrated yrs BP.................................... 41

4.1   Evolution of the number of palaeoenvironmental age determinations in the Sahara... 48

4.2   Radiocarbon and animal occurrences in the Holocene Sahara............ 53

4.3   Diachronic evolution of economic stock species in Kerma (Sudan): bovines and ovicaprides.. 56

4.4   Human occupation in the south Sahara and north Sahel at the end of the Neolithic and at the beginning of history, according to radiocarbon determinations (494 dates).. 58

4.5   Palaeoenvironments and human occupation during the Holocene in the southwest Sahara.. 59

5.1   Location of the area studied in Chapter 5................................. 66

5.2   The stratigraphic sequence of Uan Afuda................................. 67

5.3   The stratigraphic sequence of the site MT21 in the Messak Sattafet...... 68

5.4   U/Th dates of the travertine and anthropogenic deposits at site 96/50, Wadi Tanshalt.. 69

5.5   The first cycle of the rock shelter's fill: the Uan Tabu sequence.......... 71

5.6   The second cycle of the rock shelter's fill: the stratigraphic sequence of Uan Muhuggiag.. 72

5.7   Schematic stratigraphic sequence of a lacustrine basin in the Murzuq Edeyen.. 74

5.8   Schematic stratigraphic sequence of a lacustrine basin in the Erg Uan Kasa.. 75

5.9   Stratigraphy of a phytogenic dune at the site TAH4 along Wadi Tanezzuft at Tahala.. 76

5.10   The climate changes as reconstructed by geological proxy................ 77

5.11   Radiocarbon age determinations discussed in Chapter 5.................. 79

6.1   Location map of the core sites discussed in Chapter 6.................... 84

6.2a   Pollen diagram from the Orgoba 4 core (O4), 3880 m, Bale
        Mountains, Ethiopia, arboreal pollen taxa............................... 86
6.2b   Pollen diagram from the Orgoba 4 core (O4), 3880 m, Bale Mountains,
        Ethiopia, non-arboreal pollen taxa...................................... 87
6.3a   Pollen diagram from the Dega Sala core (D1), 3600 m, Arsi,
        Mountains, Ethiopia. Arboreal (AP) pollen taxa......................... 90
6.3b   Pollen diagram from the Dega Sala core (D1), 3600 m, Arsi Mountains,
        Ethiopia. Non-arboreal pollen taxa...................................... 91
7.1   Map of West and West Central Africa showing the vegetation zones of
        the Guineo-Congolian region and pollen core sites of Chapter
        7.. 96
8.1   Location map showing sites mentioned in Chapter 8...................... 112
8.2   Palaeovegetation map of Nabta area ca. 8000 yr bp...................... 114
9.1   Study areas under consideration in Chapter 9.......................... 124
9.2   Map of northern Burkina Faso with major excavation sites.............. 125
9.3   Radiocarbon chronology of the later prehistory of the Sahel Zone of
        Burkina Faso based on calibrated dates................................ 127
9.4   Cultural material from different Final Stone Age dune sites in the Sahel
        Zone of Burkina Faso.. 130
9.5   Map of the Chad Basin of Northeast Nigeria with archaeological sites... 133
9.6   Radiocarbon chronology of the Holocene prehistory of Northeast
        Nigeria... 134
9.7   Distribution of firki mounds and of sites of the Gajiganna Culture,
        Northeast Nigeria (ca. 1800 cal BC – 800 cal BC) divided into
        pastoral (phase I) and agropastoral (phase II) stages.................. 136
9.8   Pottery of phases I and II of the Gajiganna Culture.................... 139
9.9   Economic and cultural appearance of the Gajiganna Culture............. 141
9.10   Gajiganna: Percentage values of plant impressions in potsherds from
        phase I to phase IIc.. 142
9.11   Stone artifacts of the Gajiganna Culture.............................. 144
9.12   Bone tools of the Gajiganna Culture................................... 145
9.13   Map of sites with Gajiganna related pottery........................... 147
10.1   The Eastern Mediterranean (Levantine Basin) Ghab Valley and Lake
        Hula: pollen records, Merimde and the Fayum: first agricultural
        sites in Africa... 159
11.1   Treeless landscape with stone terracing near Adi Ainawalid, Tigray.... 174
11.2   Minimum tillage: 2-oxen scratch plow, leaving stones on the fields,
        Adi Ainawalid, Tigray... 175
11.3   Sorghum crop with mixed varieties. Adi Bakel, Tigray.................. 177
12.1   Location map of sites in Egypt and Sudan referred to in Chapter 12.... 199
13.1   Map of the Libyan Sahara.. 211
13.2   Pottery with rocker impressions from Uan Muhuggiag, Tadrart Acacus,
        Libya... 214

13.3   Map of the Egyptian Western Desert, showing directions of the
        dispersal of domestic cattle.. 216
13.4   Mid-late Holocene cultural sequence in the Sahara and the Nile
        Valley... 218
14.1   Recent production of the main alimentary sources in North Africa....... 226
14.2   The area licensed to the Italo-Libyan Joint Mission.................... 227
14.3   Dating of Holocene archaeological sites in the Acacus and
        surroundings, in radiocarbon uncalibrated years before present........ 230
14.4   A Middle Pastoral site in the erg Uan Kasa. The site 94/63 shows an
        articulated intrasite organization and evidence of re-occupations..... 233
14.5   A model of the Middle Pastoral settlement pattern in the Acacus and
        surroundings, based on a seasonal vertical transhumance between
        lowlands and mountains.. 234
14.6   The cranium of the mummified infant from Uan Muhuggiag.............. 236
14.7   A rock shelter in the Acacus used as sheep/goat dwelling during the
        Late Pastoral... 239
14.8   A model of the Late Pastoral settlement pattern of nomadic groups in
        the Acacus and surroundings, characterized by a high mobility and
        large-scale movements... 241
14.9   Examples of 'exotic' tools found in Late Pastoral sites..................... 242
14.10   The settlement pattern of nomadic and semi-nomadic pastoral groups in
        the area of the Acacus Mountains at the beginning of the century.... 243
14.11   The semi-sedentary groups of the Late Pastoral in the Wadi Tanezzuft
        Valley. Sites feature a high density and are concentrated near the   244
        paleo-oasis of Tahala...
14.12   Circular tumuli in the eastern slopes of the Acacus Mountain, facing
        the erg Uan Kasa.. 245
15.1   Map of the major localities mentioned in Chapter 15 for West and
        Central Africa.. 252
15.2   Relative importance of fishing, hunting and herding on sites in the
        region considered in Chapter 15 (calculated on the basis of number
        of identified specimens).. 266
15.3   Dietary contribution of fishing, hunting and herding on sites in the
        considered region (calculated on the basis of number of identified
        specimens multiplied by an average total weight)..................... 268
16.1   'Double bull's head amulets'. Abydos (?), formerly Hilton-Price
        collection (Brussels E.3381a-c)... 281
16.2   Bull's leg amulet. Provenance not recorded, formerly MacGregor
        collection (Brussels E.6154)... 282
16.3   Figurative flint. Nagada, 'royal tomb', formerly MacGregor collection
        (Brussels E.6185a)... 284
16.4   'Bull's head' amulet. Provenance not recorded (Brussels E.2335)....... 286

16.5   'Bull's head' amulet, late type. Provenance not recorded, formerly
        Scheurleer collection (Brussels E.7126)................................. 286
16.6   Palette in the shape of bovid with bird amulet as leg. Provenance not
        recorded (Brussels E.4992).. 290
16.7   Bird amulet. Provenance not recorded (Brussels E.2179)................. 290
16.8   'Pelta' palette with two bird heads. Provenance not recorded (Brussels
        E.421)... 291
16.9   'Pelta' palette with one bird head. Provenance not recorded (Brussels
        E.422)... 291
16.10   Rhomboidal palette decorated with double (?) bull's head. Provenance
        not recorded (Brussels E.2182).. 293
16.11   Greywacke needle. Provenance not recorded (Brussels E.2187).......... 294
16.12   Palette with simplified Bat emblem. Provenance not recorded,
        Scheurleer collection (Brussels E.7129)................................. 294
16.13   Amulets: a. Provenance not recorded (Brussels E.2882); b. Provenance
        not recorded (Brussels E.2880); c. Ballas-Zawaida (bought),
        formerly MacGregor collection (Brussels E.6188b)..................... 295
16.14   Amulets: a. Provenance not recorded (Brussels E.2881); b. Provenance
        not recorded, formerly MacGregor collection (Brussels E.6188e); c.
        Ballas-Zawaida (bought), formerly MacGregor collection (Brussels
        E.6188c); d. Unidentified Petrie excavation (Brussels E.1231)........ 296
16.15   'Double bird head' palette. Provenance not recorded (Brussels E.2886). 297

LIST OF TABLES

1.1 Frequency of climatic events of different duration during the Holocene of Africa... 5

1.2 Conversion table of radiocarbon age estimates............................... 8

2.1 Calibrated radiocarbon age determinations of oldest domestic or putatively domestic cattle in Africa....................................... 14

3.1 Uncorrected radiocarbon ages (bp) for biozonal boundaries in Central Mediterranean cores studied by Jorissen *et al.* (1993)................. 39

6.1 Uncalibrated ^{14}C dates in yr bp of the studied cores (D1 and O4)......... 85

6.2 Synchronous pollen events in cores Orgoba 4, Bale Mountains and Dega Sala 1, Arsi Mountains... 88

6.3 Climatic phases inferred from the sunchronous pollen zones of cores Orgoba 4, Bale Mountains and Dega Sala 1, Arsi Mountains......... 89

7.1 Summary of late Holocene environmental changes in West and West Central Africa... 98

7.2 Some of the uses of the oil palm tree... 102

8.1 Presence/absence of a variety of grass types at Nabta, Hidden Valley and Abu Ballas... 115

9.1 Radiocarbon dates from archaeological sites in the Sahel zone of Burkina Faso... 128

9.2 Radiocarbon dates of the Gajiganna complex in the Chad Basin of Northeast Nigeria... 138

11.1 Strategies for sustainable subsistence at Adi Ainwalid, Tigray............ 183

12.1 Some Latin labels applied to animal domesticates......................... 197

13.1 Libya: synthetic radiocarbon chronology of deposits at Uan Muhuggiag... 212

14.1 Domestic animals from Holocene archaeological sites in the Acacus and surroundings... 235

15.1 Radiocarbon dates discussed in Chapter 15................................. 254

17.1 Chronology of climatic and cultural events discussed in this work....... 322

DROUGHTS, FOOD
AND CULTURE

1
INTRODUCTION

F. A. Hassan

This book is an attempt to situate archaeology within the domain of contemporary human affairs, and to forge a new methodology for coping with environmental problems from an archaeological perspective. From this perspective, the papers included in this volume highlight the aspects of the historical relationships between people and climatic change that are potentially useful in coping with current climatic fluctuations. Those aspects include:

(1) The ubiquity and prominence of abrupt, severe climatic events as triggers of human responses.
(2) The importance of regional variability in response to global climatic phenomena as a result of historical antecedents and local ecological settings. The response of plants, animals, and water resources to climatic events are particularly significant as a stimulus to human actions.
(3) The indeterminacy of human responses and the importance of environmental perception and decision making, given the options and opportunities available to communities at times of repeated harsh weather conditions. These responses may include: population dispersal, modifications of social organization, technological or economic innovations, and initiation of ritual, religious, or ideological practices.
(4) The role of population movements and dispersal, demographic flux, information networks, exchanges of food and goods and, in certain contexts, trade, make it necessary to consider major cultural transformations in terms of trans-regional and long-distance interactions.

Within the scope of contemporary archaeology, as represented by several contributions in this volume from current archaeological projects in Africa, the interactions between culture and the environment are not a matter of polemical debate, but of continued research to improve our understanding of the nature of that relationship. In this context, the investigation of cultural-environmental relationships on the Colorado Plateau (Gumerman, 1988) is particularly useful because of its theoretical perspicacity and methodological rigor. The study, which focused on the impact of changing climate for the 600-1300 AD interval, also a period of special interest in Africa's climatic history, concluded that responses to frequent droughts included:

(1) abandonment of settlements;
(2) movement to lowlands; and

(3) initiation of intensified subsistence practices.

The study also revealed a direct correlation between settlement behavior and interaction among groups, and spatial variability in precipitation.

Although there are archaeologists who would rather not concern themselves with environmental issues, not to mention those who think that cultures are entirely arbitrary subjective constructs, they might have found it disturbing that water had to be carried by trucks to fill the water reservoirs in England in 1995.

The impact of droughts since the 1970s on vegetation, food production and water resources in the African Sahel and all over the world, from China to Brazil, including Spain and the UK, for example, clearly demonstrate that a change in climate could have had an impact on cultures in the past. The current debate on global warming, its causes and consequences, has also focused both scholarly investigations and public attention on the relationship between climate change and the future of our food security and political stability. In the 1980s, it was estimated that roughly 30 billion dollars are lost annually due to changes in weather (Riebsame, 1989). Perhaps more significant than economic loss is the degradation of human life and the proliferation of violent conflict. Homer-Dixon, Boutwell, and Rathjens (1993, p. 38), who co-directed a project on environmental change and acute conflict, concluded that "scarcities of renewable resources are already contributing to violent conflicts in many parts of the developing world". These conflicts may foreshadow a surge of similar violence in coming decades, particularly in poor countries, where shortages of water, forests, and especially fertile land, coupled with rapidly expanding populations, already cause great hardship.

Within a few decades, the output from palaeoclimatic investigations has radically modified our traditional concepts of climate and climatic change. Archaeological investigations, which have pioneered the study of the relationship between climatic-environmental fluctuations and cultural change, are also beginning to clarify the complex connections that link cultures to their environmental setting. Gone is the naive presumption that climate is the prime mover of change, and gone also is the simplistic correlation of long-term climatic periods with significant cultural innovations, and the belief that such correlations reflect a deterministic, causal link; with climate as a cause and cultural developments as an effect (Sprout and Sprout, 1965).

The dynamic interactions between climate and culture are still far from clear. We are now in need of a genuine collaboration of archaeologists, palaeoclimatologists and palaeoecologists within the framework of a research program with explicit foci of joint investigations. Such a program of research will not only provide us with much needed basic information (for example, high resolution chronology of climatic events), but will also bring about a new perspective on the methodology and theoretical aspects of the relationships between climate and culture.

One of the fundamental notions that we may consider is that of the mechanisms by which climatic events may cause a cultural response (Kates, 1985). Perhaps central to these mechanisms is the way by which a climatic event influences resources of economic, social, or ideological relevance to a group of people. For example, a farming community may react to a climatic event that may influence the growing season, the area irrigated, or ground moisture. Pastoralists may react to changes in the location and permanency of water holes and the quality and extent of pastures. Inland populations may not worry about coastal changes as a result of a sea transgression, but coastal groups depending on fishing and collecting mussels may respond effectively to the changing circumstances.

It is also perhaps prudent not to ignore the mechanisms by which an environmental change is perceived (Sonnenfeld, 1972; Whyte, 1985), or the chain that leads from the perception of environmental change to making certain decisions (Craik, 1972) and then implementing them. It would appear that environmental perception will depend on the rapidity of a change, its intensity and its frequency (Hassan, 2000). Events that happen in a person's lifetime (or that of his/her immediate kin and social group), ranging from those that may cause a deterioration in the quality of life to life-threatening famines or disasters, are most likely to be perceived. Measures to alleviate, side-step or overcome such events are likely to be adopted, or not, depending on the social milieu of innovation and the criteria set by society for acceptable conduct (for example, remedying food shortages through cannibalism or infanticide, or switching from a meat-based diet to the ingestion of cereal grains). In general, food shortages may be met by: (1) modifying diet by exploiting less desirable resources to expand the width of the food niche; (2) developing innovative technological, economic, social or ideological means to enhance the productivity or quality of new or pre-existing resources (for example, developing the plow to replace the hoe, using grinding stones, engaging in food exchange networks, or changing the size and/or organization of the labor force, instituting or abandoning certain beliefs concerning the harvesting or consumption of certain resources); (3) re-organizing the demographic regime through changes in local densities, demographic flux, emigration, seasonal aggregation/dispersal, altering the dependency ratio, and manipulating fertility or mortality practices (from abortion to abstinence). Among the responses to drought years in selected villages within the arid zone of India (where a normal year is characterized by 377 mm of rain over 21 days compared with 159 mm over 8 days), recorded by Riebsame (1989, p. 14) were:

(1) collecting weeds as fodder;
(2) harvesting field borders for fodder;
(3) harvesting premature crops;
(4) harvesting crop by-products;
(5) more weeding; and
(6) lopping trees for fodder.

In Tanzania, reactions to droughts by at least fifty per cent of farmers in the Usambara mountains included (Riebsame, 1989, p. 14):

(1) selling cattle to buy food;
(2) storing more than one season's food when crop is good;
(3) moving;
(4) asking for help from friends and relatives;
(5) weeding plots;

(6) not planting when rain is not enough;
(7) planting drought resistant crops;
(8) paying for a rain maker;
(9) praying; and
(10) irrigation.

The likelihood of the sustainability, collapse or expansion of a certain cultural mode depends thus not only on the rapidity, intensity and frequency of climatic-environmental signals, but also on the elements and structural organization of the particular culture involved. In general, we may also make a distinction between cultures with different organizational structures. Such structures are, in essence, modes of management, and vary from those where decision making is left to individual choice to the other extreme, where decision-making and action are monopolized by an elite. Such an elite, who are common in complex state societies, acting for their own good or that of the larger society in as much as it enhances their own, may take decisions that may be harmful to certain local communities or individuals. It may be beneficial for the common good, if by enhancing innovative developments it may succeed in sustaining a quality of life or a group size that otherwise would not be possible. However, the elite may make and implement decisions that may be short-sighted or partial to their interests. In addition, they may sustain a population size and a high living standard that places society at greater risk of climatic-environmental anomalies. Moreover, such an elite institute ideologies and practices (for example, ideologies in favor of large farming families, acquisition of resources by force, or consumption of luxury goods) that may increase the vulnerability of the cultural regime to environmental disasters. In fact, the success of such cultures may lead to environmental stress or adverse environmental consequences (for example, as a result of deforestation, overgrazing or monocropping).

The ability to monitor and predict climatic change is predicated upon the time scale of human perception and the scale of environmental change. The earliest civilizations have used writing to preserve historical knowledge of astronomical and environmental events (for example, droughts or Nile flood levels). Before that, oral traditions constituted the repository of environmental knowledge. Until recently, when long-term records of climatic history extending as far back as millions of years became known, with great details of events over the last 100,000 years, perception of environmental events was extremely limited and patchy. The scale of human perception was incapable of construing climatic history in terms of events that differed in scale from variations over intervals of a few millennia, a few hundreds of years or a few decades, and those less than two decades (for time-scale and environmental change, see Driver and Chapman, 1996).

Such millennial variations are suggested by recent palaeoclimatic investigations of sea surface temperature (presented here by Rohling and colleagues) and ice-rafted debris in the North Atlantic (Bond *et al.*, 1997). From my analysis of climatic variability during the post-glacial period from 13 kyr cal BC to 1250 AD in Africa, I detect thirty events ranging in duration from a millennium to a century (Table 1.1). Shorter events are masked by the lack of

Table 1.1. Frequency of climatic events of different duration during the Holocene of Africa

Duration in years	Number of events	Frequency
100-200	10	33.3%
300-400	10	33.3%
500-600	4	13.3%
700-800	3	10%
1000	3	10%

high resolution and all estimates are subject to significant errors as a result of the lack of precision that accompanies our radiocarbon age estimates. We may also recognize periods that vary in length from 1700 to 4000 years in duration. Unlike abrupt events, the shift from the wet conditions of the early Holocene to the dry climate of the late Holocene was fairly gradual and might have spanned several millennia (Haynes, 1987).

From a human perspective, the duration of events is not only important because of its relevance to the ability to perceive or predict an event, but also because the persistence of similar weather conditions over a relatively long period of time permits certain cultural traditions to develop and persist. For example, the prevalence of relatively warm and wet conditions in the Sahara from 8400 to 7600 cal BC (9.4 to 7.7 kyr bp), and again from 7600 to 6800 cal BC (8.6 to 7.9 kyr bp), for periods of 800 and 700 years respectively, allowed gathering and, where lakes developed, fishing (see Phillipson, 1993 for a general review). By contrast, the period from 6100 to 4800 cal BC (7.25 to 5.9 kyr bp) was characterized by shorter intervals of approximately 300, 100, 200, and 100 years respectively. The period from 6100 to 4800 cal BC included episodes of severe cold and pronounced aridity (Street-Perrott and Perrott, 1990; Sirocko et al., 1996; Gasse and Van Campo, 1994; Roberts et al., 1994; Lamb et al., 1995; Hassan, 1996, 1997). Moreover, it also involved occasional switches in seasonality from monsoon-related summer rain to autumn and perhaps spring rain, especially in the northern part of the Sahara. The southern limit of the Mediterranean vegetation in the Sahara has been explored by Schultz (1994).

In modeling the relationship between climate and people, the dynamics of the landscape are of the utmost significance (see, for example, McGlade, 1995). In North Africa, the impact of cold, reduction in rainfall and seasonal variability on aeolian activity, surface water in lakes, runoff, waterflow in wadis, the density and distribution of trees and grasses were compounded by an increase in interannual variability that accompanies a reduction in rainfall in arid and sub-arid climatic belts. Of particular importance is the change in the geomorphological processes accompanying climatic events. For example, a study of the drought-related geomorphological processes in the Inland Niger Delta, central Mali (Jacobberger, 1988), revealed that a reduction in precipitation and fluvial discharge led to reduced frequency of overbank deposition. Reduction in floodplain vegetation density, damage to floodplain soil integrity, and confined flow in main channels caused a transition from vertical to lateral accretion and a corresponding increase in the meandering tendency of main river channels. A return to high flow rates would result in catastrophic flooding and floodplain stripping. The droughts are also causing a remobilization of dunes.

The impact of such changes on the landscape and food as well as water resources depends, in part, on the local setting and the position of the region in its climatic belt. In the Sahara, we must not only distinguish the plains from the massifs, the coastal fringe and the Atlas from the desert interior, but we must also recognize the importance of basins and the variations in the size and location of their catchment areas. We may, for example, contrast Lake Chad with the much more ephemeral and dramatically smaller Nabta Playa.

Furthermore, we cannot ignore the peculiarities of river valleys and deltas, whether it is the Niger or the Nile. Moreover, clinal differences in rainfall and seasonality differentiate the northern part of the Sahara from its southern fringe. At Siwa, the ephemeral lakes were much smaller than those that formed at Nabta and Bir Kiseiba, or at Kharga farther to the south. There is also a marked decrease in rainfall from the west of the Sahara eastward. These overall regional patterns, as well as the intra-regional differences, were crucial in shaping the cultural developments in Africa over the last 13,000 years.

In this volume, the contributors trace both the vagaries of climatic change in Africa, the role of local factors, and the means by which people developed subsistence and cultural strategies to maintain food security. The contributors provide the most-up-to-date information on the most significant climatic changes in various parts of Africa, and examine in specific contexts the accompanying cultural events. Their contributions deal primarily with three major themes in the later prehistory of Africa, namely: (1) climatic change; (2) the beginnings of plant cultivation; and (3) the origins and dispersal of cattle keeping and pastoralism.

REFERENCES

Bond, G., Showers, W., Cheseby, M., Lotti, R., Almasi, P., de Menocal, P., Priore, P., Cullen, H., Hajdas, I., and Bonani, G. (1997). A pervasive millenial-scale cycle in North Atlantic Holocene and Glacial climates. *Science* **278**: 1256-1266.

Craik, K. H. (1972). An ecological perspective on environmental decision making. *Human Ecology* **1**: 69-80.

Driver, T. S., and Chapman, G. P. (eds.) (1996). *Time Scale and Environmental Change*, Routledge, London.

Gasse, F., and Van Campo, E. (1994). Abrupt post-glacial climate events in West Asia and North Africa monsoon domains. *Earth and Planetary Science Letters* **126**: 435-456.

Gumerman, G. J. (ed.) (1988). *The Anasazi in a Changing Environment*, Cambridge University Press, Cambridge.

Hassan F.A. (1996). Abrupt Holocene climatic events in Africa. In Pwiti, G., and Soper, R. (eds.), *Aspects of African Archaeology. Papers from the 10th Congress of the PanAfrican Association for Prehistory and Related Studies*, University of Zimbabwe Publications, Harare, pp. 83-89.

Hassan F.A. (1997). Holocene Palaeoclimates of Africa. *African Archaeological Review* **14**(4): 213-231.

Hassan F.A. (2000). Holocene environmental change and the origins and spread of food production in the Middle East. *Adumatu* **1**: 7-28.

Haynes, C.V. Jr. (1987). Holocene migration rates of the Sudano-Sahelian wetting from the Arba'in Desert, Eastern Sahara. In Close, A.E. (ed.), *Prehistory of Arid North Africa, Essays in Honor of Fred Wendorf*, Southern Methodist University Press, Dallas, pp. 69-84.

Homer-Dixon, T.F., Boutwell, J.H., and Rathjens, G.W. (1993). Environmental change and violent conflict. *Scientific American* **February**: 38-45.

Jacobberger, P.A. (1988). Drought-related changes to geomorphological processes in central Mali. *Geologic Society of America Bulletin* **100**: 351-361.

Kates, R.W. (1985). The interaction of climate and society. In Kates, R.W., Ausubel, J.H., and Berberian, M. (eds.), *Climate Impact Assessment*, Wiley, Chichester, pp. 3-36.

Lamb, H.F., Gasse, F., Benkaddour, A., El Hamouti, N., van der Kaars, S., Perkins, W.T., Pearce, N. J., and Roberts, C.N. (1995). Relations between century-scale Holocene arid intervals in tropical and temperate zones. *Nature* **373**: 134-137.

McGlade, J. (1995). Archaeology and the ecodynamics of human-modified landscapes. *Antiquity* **69**: 113-132.

Phillipson, D.W. (1993). *African Archaeology*, Cambridge University Press, Cambridge. (2nd edition).

Riebsame, W.E. (1989). *Assessing the Social Implications of Climatic Fluctuations*, United Nations Environment Programme, Nairobi.

Roberts, N., Lamb, H.F., El Hamouti, N., and Barker, P. (1994). Abrupt Holocene hydro-climatic events: Palaeolimnological evidence from North-West Africa. In Millington, A.C., and Pye, K. (eds.), *Environmental Change in Drylands: Biogeographical and Geomorphological Perspectives*, Wiley, New York, pp. 163-175.

Schulz, E. (1994). The southern limit of the Mediterranean vegetation in the Sahara during the Holocene. *Historical Biology* **9**: 137-156.

Sirocko, F., Garbe-Schonberg, D., McIntyre, A., and Molfino, B. (1996). Teleconnections between the subtropical monsoons and high latitude climates during the last deglaciation. *Science* **272**: 526-529.

Sonnenfeld, J. (1972). Geography, perception and the behavioural environment. In English, P.W., and Mayfield R.C. (eds.), *Man, Space and Environment*, Oxford University Press, New York, pp. 244-251.

Sprout, H., and Sprout, M. (1965). *The Ecological Perspective on Human Affairs*, Princeton University Press, Princeton.

Street-Perrott F.A., and Perrott, R.A. (1990). Abrupt climate fluctuations in the tropics: the influence of Atlantic ocean circulation. *Nature* **343**: 607-612.

Whyte, A.V. T. (1985). Perception. In Kates, R.W., Austubel, J.H., and Berberian, M. (eds.), *Climate Impact Assessment*, Wiley, Chichester, pp. 403-436.

APPENDIX

One of the key methodological issues hampering interdisciplinary communication and correlation of environmental and cultural data is the lack of standardization in reporting radiocarbon dating. Accordingly, during the workshop and in this volume, we have opted for standardized abbreviations as follows:

- Year	yr
- 1000 years	kyr
- Uncalibrated radiocarbon years before present	yr bp
- Uncalibrated radiocarbon kyr before present	kyr bp
- Calibrated radiocarbon years before present	cal BP
- Calibrated radiocarbon kyr before present	kyr cal BP
- Calibrated radiocarbon kyr BC	kyr cal BC
- Calibrated radiocarbon kyr AD	kyr cal AD
- Historical years BC	BC
- Historical years AD	AD

In order to facilitate comparisons, a conversion table is provided to show approximate equivalent age estimates (Table 1.2).

Table 1.2. Conversion table of radiocarbon age estimates

Radiocarbon kyr bp	Radiocarbon kyr cal BP	Calibrated year
0.35		AD 1550
0.5		AD 1430
1.2-0.7		AD 780-1290
1.6-1.7		AD 400s
1.7-1.8		AD 300s
2.2-2	2.3-1.94	360/200 BC - AD 12
2.5-2.2	2.7-2.15	750-360/200 BC
2.5	2.7	775 BC
2.8	2.8	850 BC
3.0	3.1	1300 (1200) 1150 BC
3.3	2.55-2.45	1600-1500 BC
3.5	3.8	1880-1750 BC
3.7	4.05	2100 BC (2140-1980 BC)
4.0	4.4	2500 BC
4.4	5.0	3100-2920 BC
4.5	5.2	3335-3100 BC
4.9-4.5	5.65	3700-3300 BC
5.2-4.85	5.9-5.6	3950-3650 BC
5.4-4.5	6.25-5.25	4300-3300BC
5.5-5.4	6.35-6.25	4400-4300 BC
5.72	6.5	4500 BC
5.9-5.5	6.75-6.35	4800-4400 BC

cont...

6.0	6.8	4900 BC
6.5	7.35	5400 BC
6.6	7.45	5500-5400 BC
6.9-6.3	7.65-7.15	5700-5200 BC
6.9	7.65	5700 BC
7.0	7.85	5800 BC
7.26	8.0	6100 BC
7.9-7.7	8.75-8.45	6800-6500 BC
8.1	9.0	7000BC
8.6	9.6	7570 BC
9.4-8.6	10.5-9.6	8420-7570 BC
9.4	10.5	8420 BC
9.8	11.2	9040 BC
10	11.5	9250-9150 BC
10.6-10	12.55-11.15	10,600-9200 BC
11.7-10.7	13-12.65	11,100-10,700 BC
11.7-10.6	13.65-12.55	11,700-10,600 BC
12	14.0	12,000 BC
12.7	14.95	13,000 BC

2
PALAEOCLIMATE, FOOD AND CULTURE CHANGE IN AFRICA: AN OVERVIEW

F. A. Hassan

INTRODUCTION

Over the last 14,000 years, Africa has passed through no less than thirty major climatic events which varied in scale from 700-1000 years to 100-400 years, with low order variations in the range of 10-15 and 30-40 years. Among these events, several abrupt and severe dry episodes appear to have stimulated a variety of responses that include: (1) demographic shuffle, dispersal, or aggregation; (2) adoption or creation of a host of technological innovations; (3) acceptance and development of a variety of food producing modes; (4) evolution and transformation of social and political organizational arrangements; and (5) instigation of certain ideological beliefs and ritual practices.

The cultural changes resulted from the impact of the climatic events on the quality, amount, distribution, interannual variability and spatial unpredictability of water and food resources. Although certain events were localized, the impact of a change in one region inevitably spread, triggering a range of cultural developments in the whole continent. The social impact of a climatic-environmental event depended not only on the amplitude, frequency, duration, and rapidity of climatic signals and the modifications of the landscape, but also on the potential for perceiving environmental change, decision-making and group action.

The preservation and amplification of certain societal actions as cultural modes and traditions depends, in general, on the perceived utility and social matrix of prior actions and the opportunities presented by novel social or environmental situations. Like other cultures, those that emerged in Africa were flexible and responsive to change. The transition from hunting-gathering to pastoralism and farming, and subsequently the emergence of urban centers, complex political entities, from tribal chiefdoms to empires, constituted a sequence of developments chosen by certain communities as a means to overcome food insecurity. However, the rise of hierarchical and militarized societies to secure territorial ranges, enlarge the labor force, integrate regional resources, or extract revenues and tribute, marked a threshold of cultural evolution that aggravated the impact on the landscape and engendered modes of social display and consumption that precipitated gender and social inequalities, magnified population size, and triggered, in many cases, hostile inter-group contacts. In this paper, my aim is to provide an overall perspective on the role of climatic change as a major factor in the dynamics that shaped the core of African cultures before the impact of metals and the cultural developments over the last two millennia. This is a second approximation (cf. Hassan, 2000) of a model that attempts to provide a coherent framework for the later prehistory of Africa in which cattle and climate are two critical forces in African societies.

ORIGINS AND INITIAL DISPERSAL OF CATTLE-KEEPING

The communities in a region that includes Nabta, Bir Kiseiba and perhaps Uweinat and Kufra, with an annual rainfall of 100-200 mm per year, were more vulnerable to climatic oscillations, especially droughts, than other areas to the West (Hassan, 1986). It was also an area that supported more people than other regions to the north in Egypt. Recent investigations have not only shown that communities were well established there by 12,000 years ago (10 kyr bp), but that they were keeping cattle by 11,000 years ago, ca 9 kyr bp (Wendorf et al., 1987; Wendorf and Schild, 1994).

Following the abrupt and severe spell of aridity ca. 8.6 kyr bp (7.6 kyr cal BC), re-tooling of lithic artifacts and the establishment of slab-lined huts suggests the emergence of new organizational and technological developments. Farther to the west in the Adrar Acacus, the management of wild animals dates to ca. 9 kyr bp (10 kyr cal BP; Cremaschi and di Lernia, 1996; di Lernia and Cremaschi, 1996), as well as intensive utilization of grasses and collection of seeds and fruits at Uan Afuda at ca. 8.9-7.5 kyr bp (Castellatti et al., 1998), which leads me to suspect that the 8.6 kyr bp droughts might have been superimposed on a trend of drying conditions associated with an increase in interannual variability. Alternatively, the success of Saharan communities over a thousand years of occupation or more might have led to an increase in the size of local communities, placing them at the risk of unanticipated climatic upheavals.

The abrupt events at 8.6 kyr bp were sufficiently severe that the population might have dispersed before it returned to Nabta a century later when wetter conditions were re-established. According to Fred Wendorf (in lit.), that climatic event caused a drying of the water pools and ponds, a lowering of the water table, erosion by gulleying, and the deflation of older lacustrine sediments by wind erosion.

The chronology of the earliest remains of cattle in the Acacus, the Ennedi, and the Aïr is consistent with a point of origin in the Nabta-Bir Kiseiba region at 7600 cal BC or 8.6 kyr bp (Hassan, 2000).

In attempting to elucidate the relationship between climatic change and the origin and dispersal of cattle, we are immediately faced by some of the most intractable methodological issues that hamper and frustrate our efforts. First of all there is the difficulty in the identification and nomenclature of wild and domestic livestock (Gautier, 1989). There is also the ongoing debate concerning the domestic status of the Nabta cattle (Smith, 1986; Clutton-Brock, 1993). In accepting the judgment of Gautier for several reasons (see also DNA results by Loftus et al., 1994, and discussion by Gifford-Gonzalez, 1998), including those that have stemmed from my exploratory study of the distribution of putatively domestic cattle in Africa (Hassan, 2000), I wish to draw attention to the variety of ways in which livestock may constitute a part of the economy, which may include: (1) keeping a few cattle for social, ritual or ideological beliefs in an economy characterized primarily by hunting and foraging; (2) keeping a few cattle or livestock as a part of a generalized economic base characterized primarily by hunting and foraging; (3) the maintenance of herds of cattle within a short-range pastoral economy; (4) the maintenance of large herds of cattle by long-range transhumant pastoralism; (5) cattle herding as a supplement to cultivation, and vice versa, within a single community; (6) cattle keeping or herding by a community integrated with the economic pursuits of other communities of the same society; and (7) coexistence and symbiosis with other farming or foraging communities. The implications for the percentage and characteristics of the livestock in an archaeological assemblage are considerable. It is important, for example, to differentiate between the limited scatter of cattle bones at Nabta Playa and the richer assemblages from East Africa and, for that matter, between the later assemblages in East Africa, with a high frequency of the bones of domestic cattle, and earlier assemblages, where the frequency of domestic livestock is low. Compare here the assemblage from Ngamuriak, Kenya, dating to 2000 bp, where only 22 bones of wild animals are recovered from an assemblage of some 60,000 bones (Marshall, 1994, p. 24), with the faunal assemblage from Prolonged Drift, where domestic stock comprises 22% of the fauna (Ambrose, 1984).

The dating of the earliest cattle in Africa (see Gautier, 1987) depends in some cases on a single age determination (for example, at Meniet and Amekni), or on dubious associations (for example, at Haua Fteah). It is also common for the oldest date in a series to be singled out as the date of the earliest cattle (for example, at Grotte Capéletti). At Ti-n-Hanaketen in the Tassili n'Ajjaralso, an early introduction of cattle is based on a date of 7220 bp; predating the level where cattle appear (Aumassip, 1987). By eliminating, for the moment, occurrences with a single age determination, and averaging the two or three oldest consistent age measurements, the picture of the dispersal of cattle in Africa becomes much clearer. The picture also improves when calibrated radiocarbon average estimates are used (Table 2.1). Where neighboring sites show differences in age, the oldest site meeting the above criteria is taken as the earliest documented occurrence of cattle in that region.

TIMING AND DIRECTIONS OF THE SPREAD OF CATTLE-KEEPING 8600-6500 bp

In examining the distribution pattern of the earliest cattle in Africa it becomes clear that cattle do not cross south of latitude 15°N until after 2000 years cal BC (Figure 2.1). The oldest remains, from Délébo and Enneri Bardagué, date to 5900 and 5800 cal BC respectively. Occurrences farther west at Adrar Bous and Uan Muhuggiag date to 5000 and 4900 cal BC respectively (Barich, 1989; Paris, 1997). At the very same time, domestic cattle appear at Merimde Beni Salama in the eastern Delta, and Rabak, in the central Sudan. Using the most likely earliest

Table 2.1. Calibrated radiocarbon age determinations of oldest domestic or putatively domestic cattle[1] in Africa

Site	^{14}C year bp	cal year BC[2]	Av. cal year
Délébo, Chad	7180 ± 300	5990	5855 BC
	6900 ± 300	5720	
Enneri Bardagué, Chad	7450 ± 180	6230	5783 BC
	6930 ± 370	5740	
	6440 ± 225	5380	
Uan Muhuggiag, Libya	6035 ± 110	4930	4930/4830 BC
	6030 ± 80	4930	
	5950 ± 90	4830	
	5780 ± 80	4640	
Adrar Bous, Niger	6325 ± 300	5260	5180 BC
	6200 ± 250	5100	
Grotte Capalétti, Algeria	5940 ± 150	4780	4680 BC
	5750 ± 140	4580	
Merimde, Egypt	6130 ± 110	5060	4880 BC
	5970 ± 120	4870	
	5940 ± 100	4810	
	5890 ± 60	4780	
Rabak, Sudan	6050 ± 100	4940	4930 BC
	6020 ± 130	4920	
Tamaya Mellet, Niger	5245 ± 55	4040	4040 BC
	5230 ± 100	4020	
	5260 ± 120	4060	
Adrar Tiouyine, Algerian Sahara	5320 ± 130	4175	4020 BC
	5150 ± 140	3870	
Dongodien, Kenya	4100 ± 125	2610	2463 BC
	3860 ± 60	2320	
	3945 ± 130	2460	
Anezrouft, Northern Mali	4320 ± 100	2910	2670 BC
	3920 ± 199	2450/2410	
Karkarichinkat, Mali	4070 ± 90	2580	2520 BC
	3960 ± 160	2460	
Kintampo, Ghana	3550 ± 127	1890	1885 BC
	3560 ± 100	1890	
Dhar Tichitt, Mauritania	3465 ± 160	1750	1690 BC
	3425 ± 110	1630	
Khatt Lemaïteg	3350 ± 130	1630	1600 BC
	3310 ± 200	1570	
Gajiganna, Nigeria	3150 ± 70	1410	1410 BC
	3140 ± 110	1410	
Nkag, Cameroon	2580 ± 70	790	725 BC
	2490 ± 90	760,680,650,550	
Tongo, Zaire	1620 ± 90	AD 430	AD 405
	1690 ± 80	AD 380	
Cape, South Africa	2105 ±65	110 BC	AD 475
	1630 ± 60	AD 420	
	1430 ± 55	AD 640	

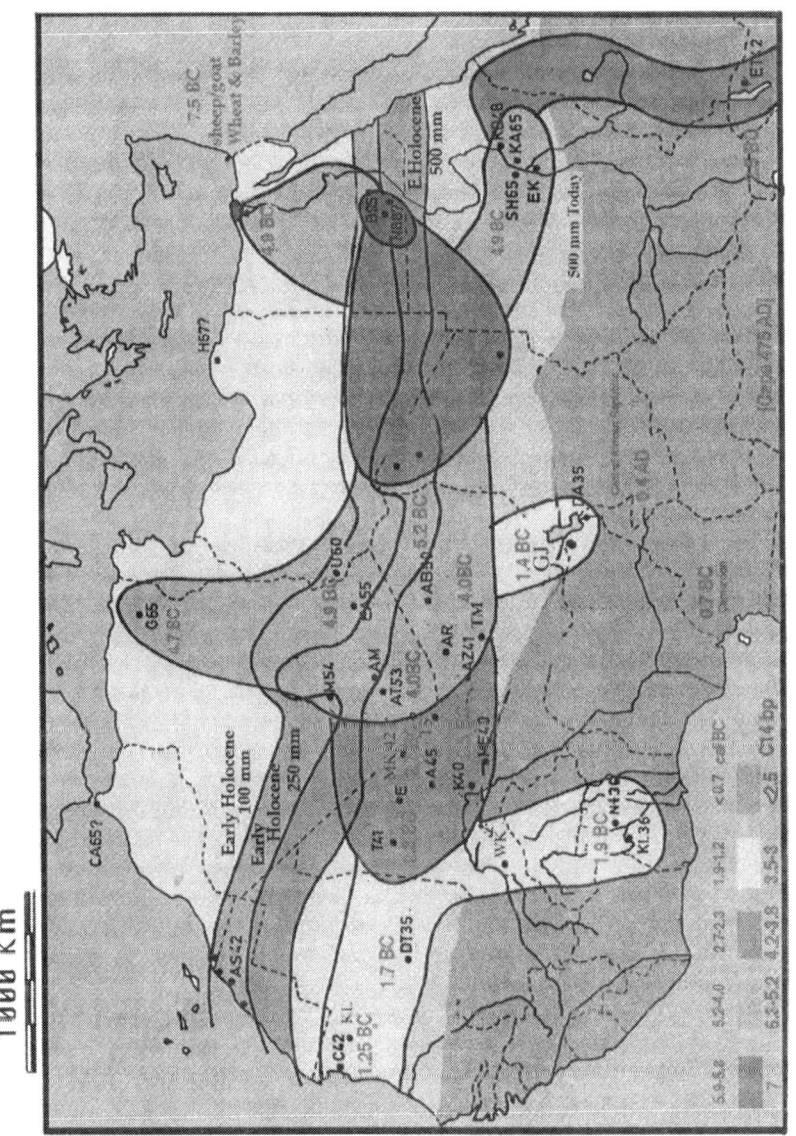

Figure 2.1. Distribution of earliest cattle in Africa.

dates at Grotte Capéletti, on the Mediterranean littoral of northern Algeria, the early cattle at Haua Fteah (coastal Cyrenaica), initially dated to 6800 bp, are now in doubt (Klein and Scott, 1986).

The initial movements westwards across the Sahara and, almost a millennium later, eastwards towards the Nile Valley, are likely to have been caused by the succession of drought episodes at 7600, 6800-6500, 6100, 5800 and 5500-5400 cal BC (8.6, 7.9-7.7, 7.26, 7, 6.6-6.5 kyr bp). Domestic African cattle were thus introduced by successive moves to the Central Sahara, the Nile Delta and the Sudan within a span of a millennium. The link between Saharan droughts and the origins, as well as the spread, of cattle in North Africa have been previously tackled by Clark (1962, 1967, 1980, 1984), Mauny (1967), Shaw (1976, 1977), Bower (1991, 1996), Hassan (1996, 2000), and Gifford-Gonzalez (2000). However, the emphasis in this contribution is on: (1) refinement of radiocarbon age chronology; (2) a consideration of the mode of dispersal not as a wave, but as a series of moves and developments that varied depending on the ecological opportunities, and the preceding cultural situation; and (3) a model of dispersal not as a result of a general trend of desertification, but as a response to a repetition of abrupt spells of droughts superimposed on a sequence of long-term climatic-ecological regimes.

The appearance of sheep and goat along the Red Sea coast and in the Eastern Sahara by 5800 cal BC, ca. 7 kyr bp (Vermeersch *et al.*, 1996) suggests that droughts in the southern Levant were also responsible for the emigration of nomads from that region, already in possession of Levantine wheat and barley as well as sheep and goat, into Africa. Sheep and goat, more adapted to desert conditions than cattle, were readily adopted by the Saharan groups. Successive droughts, as well as the beginning of a shift toward desertification, stimulated intensive collecting, storing and processing of local cereal grasses, particularly the drought resistant sorghum, among Saharan groups, as indicated by the wild sorghum and millet found in storage pits at Nabta ca. 7000 cal BC, 8.1-7.9 kyr bp (Wendorf *et al.*, 1998). Aridity and a lowering of the water table is indicated by the excavation of deep walk-in wells at that time. Intensive exploitation and processing of wild grasses including sorghum is also documented in the Central Sudan at 7.5 to 6.1 kyr bp or 6.3-5 cal BC (Magid and Caneva, 1998). This change at Nabta seems to have been associated with changes in social organization, as indicated by the remains of a village community.

In the Nile Delta, farming communities cultivating wheat and barley appear ca. 4900 cal BC at Merimde Beni Salama, on the western edge of the Delta (Hassan, 1998, 1997a). The cultivation of these cereals eventually replaces animal husbandry and other subsistence activities as the primary source of staple food. Within five hundred years the cultivation of cereals spread to southern Egypt and, five hundred years later, by 3900 cal BC, farming communities became a diagnostic feature of the Nile Valley in Egypt. Within the span of almost a millennium from this economic transformation, Egypt emerges as a unified state. Its political evolution was in part based on the regional integration of food products to counteract the effect of food shortages from a succession of low Nile floods (Hassan, 1988; 1997b).

Although there is evidence of domesticated cattle, sheep, and livestock in the Central Sudan by 4900 cal BC (Gautier, 1989), evidence for the cultivation of either sorghum or wheat and barley is lacking, perhaps because of a climatic regime that still allowed the raising of livestock in the low desert adjacent to the Nile River (Magid and Caneva, 1998). Reliance on livestock and pastoral nomadic subsistence explains the divergence in the cultural trajectory of Egypt and the Sudan thereafter. By 3500 years cal BC, Egypt and the Sudan were probably still on a par as far as cultural complexity and power are concerned. But by 3000 cal BC, increasing aridity in the central Sudan partially, and at least temporarily, undermined the resource base of the nomadic tribes and tipped the balance in favor of the Egyptian monarchs, who could count on a plentiful supply of grain. The viability of the Sudanese communities resided in the development of chiefdoms and petty states with leaders who depended on a cadre of young warriors for raids and defense.

SPECIALIZED PASTORALISM IN NORTH AFRICA

In the Central Sahara, pastoralism developed by ca. 5200-4800 BC (6300-6000 bp; Barich, this volume), giving rise to a rich Saharan art, depicting pastoral scenes and subjects of ideological significance (Muzzolini, 1995). The pastoral economy, under conditions of occasional droughts, probably fostered the emergence of differential wealth, territoriality, and raids. During periods of sustained optimal conditions, large herds and powerful leaders emerged. The earliest human burials with monumental superstructures in the Nigerian Sahara appear ca. 4700-4200 cal BC (Paris, 1996) and could have been linked to territorial claims and tribal chiefdoms.

The appearance of cow burials, dated at Nabta to 6450 ± 270 uncalibrated radiocarbon years bp (5400 cal BC) (Wendorf et al., 1997, p. 96) and almost at same time at Adrar Bous (6350 ± 260 uncalibrated radiocarbon years bp) and later at Chin Tafidet and In Tuduf, dating to 2400-2000 cal BC (Paris, 1997), also suggests that cows (and bulls) were already paramount in the myths and rituals of Saharan groups. The iconography of rock art along the Nile Valley in Nubia ca. 3500 BC, and that of ceremonial palettes dating from 3300 to 3000 BC, clearly links males with bulls, violence, hunting and attack dogs (Hassan, 1993). Cows, on the other hand, are linked with women, birth, nursing and possibly life after death. A distinct difference between the sexes was already evident in the preference to bury males with monumental tombs in the Sahara (Paris, 1996).

The appearance of megalithic monumental tombs and tumuli in the Nigerian Sahara as early as 4700 BC, and presumably much earlier in the Eastern Sahara (Wendorf et al., 1992/3), clearly predates the construction of Egyptian mastabas and pyramids by a very long time. Such structures might have been the precursors of Egyptian pyramids and monumental architecture. It is noteworthy that tumuli appear in Egypt in several localities, such as at Nagada and Helwan, during Predynastic times (Hassan, 1988).

The continuity and progressive development of culture in Egypt was undoubtedly made possible by the flow of the waters of the Nile, fed by distant tributaries in Equatorial Africa and Ethiopia, at a time when North Africa was facing increasing aridity and desertification (Hassan, 1993).

CATTLE IN WEST AFRICA

Cattle do not appear in Northeastern Mali until ca. 2910-2450 cal BC (Commelin *et al.*, 1993; Raimbault, 1995) and perhaps earlier (a single date from Village de la Frontière, of 3590 cal BC, awaits further substantiation). They are recorded farther south in Mali, at Karkarichinkat, at 2500 cal BC (Smith, 1974). Links between climatic oscillations and cultural developments in this region have been discussed by McIntosh (1993). At the same time (2500 BC), the earliest livestock is recorded from East Africa (Bower, 1996). The adoption and dispersal of livestock from the Sudan to Kenya and from Niger to Mali was perhaps a result of the droughts of 3100-2900 BC (4.5-4 kyr bp). From East Africa, livestock (sheep and goats) then dispersed to southern Africa. The mode by which livestock were incorporated into the economy of various communities in southern Africa has been a subject of debate (see for example, Kinahan, 1996). Dwarf sheep and dwarf goat do not appear at Nkang in Cameroon until 700 BC, and are not known from the inter-lacustrine region before 300 AD (Van Neer, 2000). The earliest secure dates on ovicaprids from southern Africa are 110 BC at Spoegrivier, and AD 420 and AD 640 at Kasteelberg A (Bousman, 1998; Sadr, 1998).

The dispersal of livestock, which included cattle, sheep, and goat, after the droughts of 2100 BC (3.7 kyr bp) in the Western Sahara and West Africa was rapid and widespread. Occurrences of domestic cattle were reported from Dhar Tichitt at 1700 cal BC, Kintampo at 1900 cal BC, at Khatt Lemaïteg at 1600 BC (Bathily *et al.*, 1998), and at Chami on the Atlantic Coast at 1250 cal BC (Gautier, 1987). The Sahara was becoming, by that time, hyperarid, and pastoralism with long-distance transhumance became a necessity (Figure 2.2). We have a glimpse of the local developments from that time on in the Nigerian Sahara thanks to the recent well-dated investigations reported by François Paris (1996), where changes in funerary architecture reflect cultural transitions at 2300/1750 cal BC, 1700/1000, and 800/750 BC, with marked construction of funerary monuments between ca. 800/750 and 550/200 cal BC. Later occupations are rather insignificant, suggesting a diminution of cultural activities that accompanied the desertification of the Central Sahara by 2200 bp (360/200 cal BC to 12 AD).

One of the most outstanding developments, coincident with the dispersal of livestock westward, was the penetration of the Sahel southwards, breaking for the first time the divide between Sahara and Sahel. The earliest livestock are reported from Gajiganna on the shores of Lake Chad in Nigeria at 1400 cal BC (Breunig *et al.*, 1993, 1996). At that time small-scale farming among hunter-gatherers is also reported both at Burkina Faso and in Northeast Nigeria, (Neumann *et al.*, 1996). It probably resulted from increasing pressure on populations squeezed between the desiccated Sahara and the edge of the forest. The period from 2100 BC to 1300 BC (3700 to 3000 bp) requires special attention as a critical interval in the relationship between climatic change and culture: this period witnessed the culmination of dry conditions in the Sahel; hyperarid conditions in the Egyptian Sahara; the retreat of Mega-Lake Chad; and included the collapse of the Old Kingdom at 2180 BC (Hassan, 1997c) followed by a period of political instability in Egypt, as well as, as some argue, upheavals

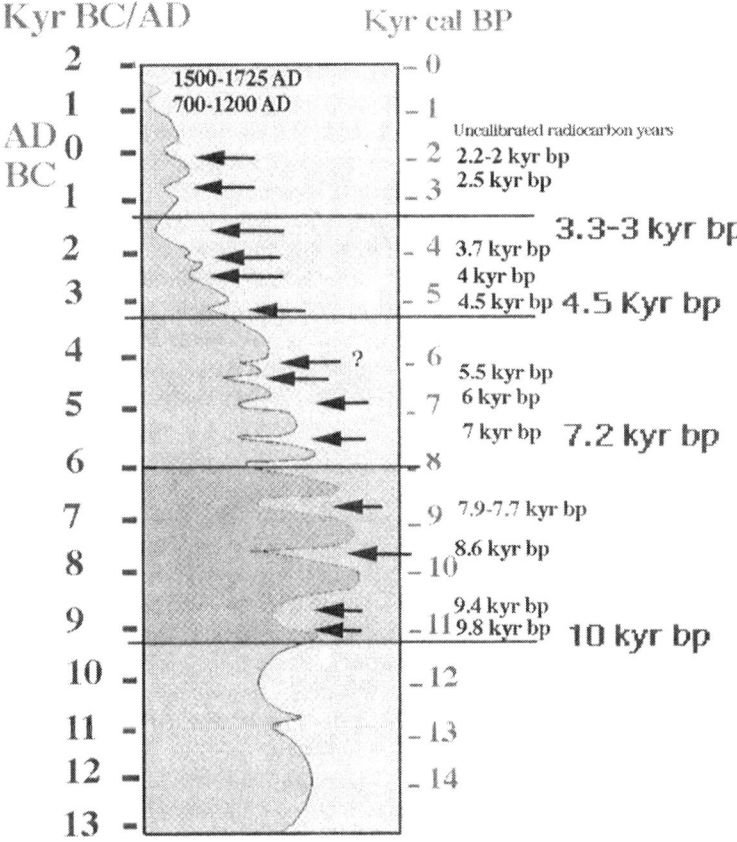

Figure 2.2. Draft sketch of Holocene climatic variability in Africa.

in Mesopotamia, Anatolia, and India (Bowden *et al.*, 1981; Weiss, 1997). In addition I will submit here that it was the combination of climatic events and the cultural antecedents of a pastoralist economy in the Sahara/Sahel setting of the western Sahel that stimulated indigenous domestication of sorghum and millet.

EARLY AGRICULTURE IN SUB-SAHARAN AFRICA

In a fairly recent review of indigenous African agriculture, Harlan (1992) notes that, with the exception of cultivated pearl millet at Ntereso at 1250 BC, and a shift to the cultivation of pearl millet in the eleventh century BC in southern Mauritania, nearly all finds are dated after the first millennium AD. (He questions the two pollen grains presumed to be of pearl millet at Amekni and the Brachiaria at Adrar Bous dated to 4000 BC. He also mentions that Clark reports sorghum at 2000 BC at Adrar Bous, however, in *The Prehistory of Africa* (1970, p. 202), Clark makes no mention of domesticated cereals before those at Dhar Tichitt). The recent claim for domesticated pearl millet on the basis of impressions on potsherds dating to 3500 ± 100 bp, 1936-1683 cal BC (Pa-1157) and 3420 ± 120 bp (1881-1527 cal BC; Pa-1299), from villages 72 and 149, on the Dhars Tichitt and Oualata (Amblard, 1996), fits chronologically with the present model, but further confirmation of the status of the grains is required.

The domestication or incipient domestication of oil palm and yam might have also been stimulated by the retreat of the forest margin in West Africa around 2200/2100 BC. However, it is after 2800 bp (850 BC) that oil palm suddenly increases in association with weeds of cultivated fields (Andah, 1993). The decisive climatic factors for the appearance of cultivation in West Africa are clearly related to the arid conditions during the period from 2200 to 1300 cal BC. Breunig and Neumann (this volume) note that food production appears in the West African Sahel after 1200 cal BC. They also postulate that agriculture was introduced from the Sahara further northeast from the hypothetical area supposed by Harlan (1992).

It is perhaps no coincidence that it is at that time that specialized pastoralism appears in East Africa, and it is also no coincidence that between 2500-1800 BC (4000 and 3500 bp) cool-season crops such as barley, emmer, and other tetraploid wheats, chickpeas, lentil, flax, fava bean and pea (Harlan, 1992), which were already established in Egypt and the Sudan (dating to 3500-3400 cal BC, 4700 bp, in the A-Group and late third millennium BC at Kerma, according to Shinnie, 1996), were introduced into Ethiopia. The cultivation of these crops perhaps stimulated the cultivation of local summer grasses and native crops to supplement herding, fishing, hunting, and foraging.

It does appear that cattle herding, rather than farming, was the main subsistence base for the people of Punt/Gash and that fishing and hunting, as well as foraging, supplemented the diet. The transition to arid climate 2100 cal BC (ca. 3700 bp), and again during the first millennium BC (1600-1300 cal BC), probably led to an emphasis on the cultivation of Near Eastern crops, and later the domestication and cultivation of the indigenous crops such as Teff, the most important native crop and the principal cereal of Ethiopia. Teff (*Eergostis tef*) is labor intensive; the seed bed must be carefully prepared and the fields must be meticulously weeded. A prior knowledge of cultivation, particularly of barley, was probably helpful in the domestication process of Teff (Harlan, 1992, p. 68). An economic model may not explain, however, the impetus for the initial cultivation of Qaat, *Catha edulis* (a mild narcotic) or *Coffea Arabica*, both of which are now major export crops.

Domesticated rice appears first at Jenné-Jeno in the first century AD (McIntosch and McIntosh, 1993), perhaps associated with the exploitation of the seasonal flooding of the Niger, and might have been developed in the inner Delta ca. 1800 cal BC or 3500 years bp (Portères, 1976). It subsequently spread to other areas via the valleys of the upper Niger and Senegal, where non-floating and upland races adapted to dry farming were developed.

COPING WITH DROUGHTS: THE AFRICAN LEGACY

During the third and second millennia BC, Egypt maintained its civilization in the face of drought by various administrative and organizational devices, as well as by embarking on the first major hydraulic project in history, in the Fayum (Hassan, 1998). At the same time farming and herding were spreading to southern Africa, with various types of indigenous innovations; West Africa was enhancing its food security by developing lines of interregional exchanges that were inevitably accompanied by the rise of important exchange centers and hierarchical societies. McIntosh and McIntosh (1993), for example, hypothesize that Jenné-jeno was well situated to serve as a center for the exchange of foods produced in the Inland Delta (with access to rice, dried fish, and fish oil), for Saharan copper, salt, and savanna products, including iron from Benedougu.

It is interesting that Saharan communities facing food scarcities resorted to mining salt and developing metallurgical industries to secure food from their southern neighbors. It is perhaps with the advent of trade, industry, and hierarchical societies following the 2100-1300 cal BC climatic crises that Africa entered a new stage in its history. Nevertheless, the impact of droughts since then, especially between 700 and 1200 AD, was crucial in shaping the quasi-present cultural layout of the Sahara. The contemporary crisis since 1972 is alarming because it occurs at a time when national boundaries are established and schemes to settle the nomads for national security and other reasons are underway.

The legacy of Africa is that climate is not destiny. It reveals human ingenuity and creativity, and highlights the need to investigate in more detail the factors responsible for ethnic fluidity, demographic flexibility and cultural malleability that have been in the past an element in the cultural transformation of Africa. It also calls for a closer look at the factors that led to conflict, warfare and gender inequalities as societies changed their social organization and ideology. Freed from the blindfolds of taxonomic and typological research, and eschewing imaginary unilinear modes of evolutionary change, we are capable of examining the richness and dynamism of African cultures. For a long time, Africa was conceived in terms of, and I quote from Bogado (1995, p. 116), "primitive, and static (non-evolutive) and a-historical peoples". Africans do not need to seek self-confirmation in the construction of historically flawed models of identity and heritage. Cultural developments in different continents were constrained by ecological parameters, historical antecedents and geographic proximity to other cultures. Neither history nor prehistory, not to mention genetics and bioanthropology, provide any evidence for the inherent superior genius of any people. The westward spread of farming into Europe spanned more than 2500 years from 7000 to 4500 BC (Whittle, 1997; Sherratt, 1997; van Andel and Runnels, 1995). The spread was selective and opportunistic and appears to have been influenced by the same abrupt climatic events that played a role in the similar westward spread of food production into Africa. As in Africa, the farming communities initially occupied a tiny part of Europe and similarly neither the spread of farming nor the local developments were uniform. Within two millennia after the initial phase of dispersal, farming became widespread and was followed by the emergence of regional cultures, some with megalithic tombs and tumuli recalling those of the Nigerian Sahara. At that time Europe did not have an irrigation technology, did not use the plow, and was not characterized by urban centers. It was not until Europe began to experience the influence of the expanding trade networks of Mesopotamian cities that significant social disparities relating to status and consumption became evident. From that time, as Sherratt (1997, p. 201) has eloquently put it, speaking for Europe, but also in my opinion for elsewhere, "control of men [and I should add women], animals, and the powers of nature became the paths to social and material success".

The myth of exclusive indigenous developments anywhere in the world is one of the most vicious myths of the past and the present. It is as dangerous in Africa as it is in Europe. In an editorial to the second issue of the ESF Standing Committee for the Humanities (May 1998), the Chair of the Committee, Professor Wim Blockmans, warns against radical nationalism and against a European identity that simply replaces previous nationalist visions. He affirms that "...cultural diversity has been the best guarantee of development in the past. At the very least, correct information is required at all levels of society in and around Europe and their interactions in the shaping of a European heritage". Not only has Europe been enriched in the past by what it inherited from Anatolia and Mesopotamia, but also in historical times from what it learned from the Arabs, who in turn were the beneficiaries of Indian, Chinese, Egyptian and Levantine heritage.

In concluding his book on *Migrations and Cultures: A World View,* Thomas Sowell (1996, pp. 379-391) states that cultures do not exist simply as 'static' differences, or as sentimental badges of identity that permit their carriers to engage in breast-beating, but are in actuality the manifestations of a cultural 'capital' (using the favorite trope in European thought) that accumulates as people develop, borrow or adapt ideas, manners, or objects. The guarantee for cultural resilience is definitely not chauvinism or insularity, but creativity, selectivity, and maximization of the sources of innovation by allowing unobstructed intermingling of cultures. As people from different continents face the threat not only of global climatic fluctuations but also the common threats of consumerism and militant fanaticism, there is a need to move beyond subjects of scholarly discourse entrenched in nationalist ideologies to a whole range of humanistic issues that can only flourish when our topics of discussion are of concern to individuals in every corner of the planet.

I aimed, in this brief sketch, to integrate the snippets of information we have in a coherent synthesis in order to pinpoint specific areas where our data are blatantly incomplete or pathetically inadequate, and to flag some topics of theoretical and methodological relevance. I hope that this introduction will serve as a means to create a dialog between participants from different research communities and facilitate the identification of future research projects.

Africans have responded to the recurrence of abrupt, severe droughts in Africa over the last 12,000 years with ingenuity; developing a variety of social and technical innovations. One of the key developments consisted of keeping cattle, a practice first documented in the Egyptian Sahara. Cattle-keeping spread subsequently to the Central Sahara, where specialized pastoralism developed. Later, it also spread to the Nile Valley in Egypt and the Sudan, and further west at a time when North Africa was drying up. It also spread from Sudan to East Africa. In West Africa, cultivation appeared ca. 3700 in response to a severe episode of droughts that also stimulated the emergence of pastoralism in East Africa. The shift in subsistence to agriculture in the Nile Valley as early as 6300 bp was followed with sociopolitical change that culminated in the rise of the state. Pastoralist societies outside Egypt formed chiefdoms and other political organizations, setting the stage for the recent history of Africa.

NOTES

(1)   Except for the Cape (S. Africa) and Nkag (Cameroon) where the earliest livestock consists of ovicaprids.

(2)   Using Calib 3.0.3c program by Stuiver, M., and Reimer, P.J. (1993) *Radiocarbon* 35: 215-230.

REFERENCES

Amblard, S. (1996). Agricultural evidence and its interpretation on the Dhars Tichitt and Oulata, south-eastern Mauritania. In Pwiti, G., and Soper, R. (eds.), *Aspects of African Archaeology. Papers from the 10th Congress of the PanAfrican Association for Prehistory and Related Studies*, University of Zimbabwe Publications, Harare, pp. 421-427.

Ambrose, S.H. (1984). The introduction of pastoral adaptation to the highlands of East Africa. In Clark, J.D., and Brandt, S.A. (eds.), *From Hunters to Farmers: Causes and Consequences of Food-Production in Africa*, University of California Press, Berkeley, pp. 212-239.

Andah, B. (1993). Identifying early farming traditions of West Africa. In Shaw, T., Sinclair, P., Andah, B., and Okpoko, A. (eds.), *The Archaeology of Africa: Food, Metals and Towns*, Routledge, London, pp. 240-254.

Aumassip, G. (1987). Le Néolithique en Algérie: etat de la question, *L'Anthropologie* 91(2): 585-622.

Barich, B. (1989). Uan Muhuggiag rock shelter (Tadrart Acacus) and the late prehistory of the Libyan Sahara. In Krzyzaniak, L., and Kobusiewicz, M (eds.), *Late Prehistory of the Nile Basin and the Sahara*, Polish Academy of Sciences, Poznan, pp. 499-505.

Bathily, M., Khattar, M.O., and Vernet, R. (1998). *Les Sites Néolithiques de Khatt Lemaëteg (Amatlich) en Mauritanie Occidentale*, Centre Culturel Francais de Nouakchott, Meudon.

Bogado, O. (1995). Terminology, typology and nomenclature in African prehistory: scientific and ethical considerations for the 21st century and onwards. In Andah, B.W. (ed.), *Rethinking the African Cultural Script: An Overview of African Historiography*, West African Journal of Archaeology and Anthropology, Ibadan, pp. 112-125.

Bousman, C.B. (1998). The chronological evidence for the introduction of domestic stock into southern Africa. *African Archaeological Review* 15: 133-151.

Bowden, M.J., Kates, R.W., Kay, P.A., Riebsame, W.E., Warrick, R.A., Johnson, D.L., Gould, H.A., and Weiner, D. (1981). The effect of climatic fluctuations on human populations: two hypotheses. In Wigley, T.M.L., Ingram, M.J., and Farmer, G. (eds.), *Climate and History*, Cambridge University Press, Cambridge, pp. 479-513.

Bower, J. (1991). The pastoral Neolithic of East Africa. *Journal of World Prehistory* 5: 49-82.

Bower, J. (1996). Early food production in Africa. *Evolutionary Ecology* 4: 130-139.

Breunig, P., Ballouche, A., Neumann, K., Rosing, F., Thiemeyer, J., Wendt, P., and Van Neer, W. (1993). Gajiganna - new data on early settlement and environment in the Chad Basin. *Berichte des Sonderforschungsbereichs* 268, Bd. 2. Frankfurt/Main, pp. 51-74.

Breunig, P., Neumann, K., and Van Neer, W. (1996). New Research on the Holocene settlement and environment in Nigeria. *African Archaeological Review* 13: 111-146.

Castellatti, L., Cottini, M., and Rotolli, M. (1998). Early Holocene plant remains from Uan Afuda Cave, Tadrart Acacus (Libyan Sahara). In di Lernia, S., and Manzi, G. (eds.), *Before Food Production in North Africa. Proceedings of the Homonymous Workshop held in Forli, September 1996, within the XIII World Congress of the International Union of the Prehistoric and Protohistoric Sciences*,ABACO Edizioni, Rome, pp. 91-102.

Clark, J.D. (1962). The spread of food production in sub-Saharan Africa. *Journal of African History* 3: 211-228.

Clark, J.D. (1967). The problem of Neolithic culture in sub-Saharan Africa. In Bishop, W.W., and Clark, J.D. (eds.), *Background to Evolution in Africa*, The University of Chicago Press, Chicago and London, pp. 601-627.

Clark, J.D. (1970). *The Prehistory of Africa*, Praeger, New York.

Clark, J.D. (1980). Human populations and cultural adaptations in the Sahara and Nile during prehistoric times. In Williams, M.A.J., and Faure, H. (eds.), *Late Prehistory of the Nile Basin and the Sahara*, Balkema, Rotterdam, pp. 387-410.

Clark, J.D. (1984). The domestication process in Northeast Africa: ecological change and adaptive strategies. In Krzyzaniak, L., and Kobusiewicz, M. (eds.), *Origin and Early Development of Food-Producing Cultures in North-Eastern Africa*, Polish Academy of Sciences, Poznan, pp. 23-41.

Clutton-Brock, J. (1993). The spread of domestic animals in Africa. In Shaw, T., Sinclair, P., Andah, B., and Okpoko, A. (eds.), *The Archaeology of Africa: Food, Metals and Towns*, Routledge, London, pp. 61-70.

Commelin, D., Raimbault, M., and Saliège, J.-F. (1993). Nouvelles données sur la chronologie du Néolithique au Sahara malien. *Académie des Sciences* C.R. 2 (T.317) Série II **41**: 543-550.

Cremaschi, M., and di Lernia, S. (1996). Climatic changes and human adaptive strategies in the Central Saharan Massifs: the Tadrart Acacus and Messak Settafet perspective (Fezzan, Libya). In Pwiti, G., and Soper R. (eds.), *Aspects of African Archaeology. Papers from the 10th Congress of the PanAfrican Association for Prehistory and Related Studies*, University of Zimbabwe Publications, Harare, pp. 39-51.

di Lernia, S., and Cremaschi, M. (1996). Analysis of the Pleistocene-Holocene transition in the Central Sahara: culture and environment in the Uan Afuda Cave (Tadrart Acacus, Libya). In Pwiti, G., and Soper R. (eds.), *Aspects of African Archaeology. Papers from the 10th Congress of the PanAfrican Association for Prehistory and Related Studies*, University of Zimbabwe Publications, Harare, pp. 429-440.

Gautier, A. (1987). Prehistoric men and cattle in North Africa: a dearth of data and a surfeit of models. In Close, A.E. (ed.), *Prehistory of North Africa*, Southern Methodist Press, Dallas, pp. 163-187.

Gautier, A. (1989). A general review of the known prehistoric faunas of the Central Sudanese Nile Valley. In Krzyzaniak, L., and Kobusiewicz, M. (eds.), *Origin and Early Development of Food-Producing Cultures in North-Eastern Africa*, Polish Academy of Sciences, Poznan, pp. 353-357.

Gifford-Gonzalez, D. (1998). Early Pastoralists in East Africa: ecological and social dimensions. *Journal of Anthropological Archaeology* **17**: 166-200.

Gifford-Gonzalez, D. (2000). Animal disease challenges to the emergence of pastoralism in Sub-Saharan Africa. *African Archaeological Review* **17**(3): 95-139.

Harlan, J.R. (1992). Indigenous African Agriculture. In Cowan, C.W., and Watson, P.J. (eds.), *The Origins of Agriculture: An International Perspective*, Smithsonian Institution Press, Washington, D.C., pp. 59-70.

Hassan F.A. (1986). Holocene lakes and prehistoric settlements of the Western Faiyum. *Journal of Archaeological Science* **13**: 483-501.

Hassan F.A. (1988). The Predynastic of Egypt. *Journal of World Prehistory* **2**: 135-185.

Hassan F.A. (1993). Population ecology and civilization in Ancient Egypt. In Crumley, C.L. (ed.), *Historical Ecology: Cultural Knowledge and Changing Landscapes*, School of American Research, Santa Fe, pp. 155-181.

Hassan F.A. (1996). Abrupt Holocene climatic events in Africa. In Pwiti, G., and Soper, R. (eds.), *Aspects of African Archaeology. Papers from the 10th Congress of the PanAfrican Association for Prehistory and Related Studies*, University of Zimbabwe Publications, Harare, pp. 83-89.

Hassan F.A. (1997a). Egypt: beginnings of agriculture. In Vogel, J.O. (ed.), *Encyclopedia of Precolonial Africa*, Altmira, Walnut Creek and London, pp. 472-479.

Hassan F.A. (1997b). Egypt: the emergence of state society, In Vogel, J.O. (ed.), *Encyclopedia of Precolonial Africa*, Altmira, Walnut Creek and London, pp. 405-409.

Hassan F.A. (1997c). Nile floods and political disorder in early Egypt. In Dalfes, H.N., Kukla, G., and Weiss, H. (eds.), *Third Millennium BC Climate Change and Old World Collapse*, NATO ASI Series, 149, Springer-Verlag, Berlin / Heidelberg, pp. 1-23.

Hassan F.A. (1998). Climatic change, Nile floods and civilization. UNESCO *Nature and Resources* **342**: 34-40.

Hassan F.A. (2000). Climate and cattle in North Africa. In Blench, R., and MacDonald, K. C. (eds.), *The Origins and Development of African Livestock: Archaeology, Genetics, Linguistics, and Ethnography*, University College London Press, London, pp. 61-86.

Kinahan, J. (1996). A new archaeological perspective on nomadic pastoralist expansion in southwestern Africa. In Sutton, J.E.G. (ed.), The Growth of Farming Communities in Africa from the Equator Southwards. *Azania* 29-30: 211-226.

Klein, R., and Scott, K. (1986). Re-analysis of faunal assemblages from the Haua Fteah and other late Quaternary archaeological sites in Cyrenaican Libya. *Journal of Archaeological Science* 13: 515-42.

Loftus, R.T., David, E.M., Bradley, D.G., Sharp, P.M., and Cunningham, P. (1994). Evidence for two independent domestications of cattle. *Proceedings of the National Academy of Sciences of the USA* 91: 2757-2761.

Magid, A.A., and Caneva, I. (1998). Exploitation of food plants in the early Holocene central Sudan: a reconsideration. In di Lernia, S., and. Manzi, G. (eds.), *Before Food Production in North Africa. Proceedings of the Homonymous Workshop held in Forlì, September 1996, within the XIII World Congress of the International Union of the Prehistoric and Protohistoric Sciences*, ABACO Edizioni, Rome, pp. 79-89.

Marshall, F. (1994). Archaeological perspectives on East African pastoralism. In Fratkin, E., Galvin, K., and Roth, E. (eds.), *African Pastoralist Systems*, Lynne Rienner, Boulder, pp. 17-43.

Mauny R. (1967). L'Afrique et les origines de la domestication. In Bishop, W., and Desmond-Clark, J. (eds.), *Background to Evolution in Africa*, The University of Chicago Press, Chicago and London, pp. 583-599.

McIntosh, R.J. (1993). The pulse model: genesis and accommodation of specialization in the Middle Niger. *Journal of African History* 34: 181-220.

McIntosh, S.K., and McIntosh, R.J. (1993). Cities without citadel: understanding urban origins along middle Niger. In Shaw, T., Sinclair, P., Andah, P., and Okpoko, A., (eds.), *The Archaeology of Africa: Food, Metals and Towns*, Routledge, London, pp. 622-641.

Muzzolini, A. (1995). *Les Images Rupestres du Sahara*, Toulouse.

Neumann, K., Ballouche, and Klee, M. (1996). The emergence of plant food production in the West African Sahel: new evidence from northeast Nigeria and northern Burkina Faso. In Shaw, T., Sinclair, P., Andah, B., and Okpoko, A., (eds.), *The Archaeology of Africa: Food, Metals and Towns*, Routledge, London, pp. 440-448.

Paris, F. (1996). *Les Sépultures du Sahara Nigérien du Néolithique à l'Islamisation*, 2 vols. ORSTOM, Paris.

Paris, F. (1997). Les inhumations de bos au Sahara méridional au Néolithique. *Archaeozoologia* 9: 113-122.

Portères, R. (1976). African cereals. In Harlan, J.R., de Wit, J.M.J., and Stemler, A.B.L. (eds.), *The Origins Of African Plant Domestication*, Mouton, The Hague, pp. 409-452.

Raimbault, M. (1995). *La Culture Néolithique des 'Villages à Enceinte' Dans la Région de Tessalit, au Nord-Est du Sahara Malien. L'Homme Mediterranéen*, Publications de l'Université de Provence, Aix-en-Provence.

Sadr, K. (1998). The first herders at the Cape of Good Hope. *African Archaeological Review* 15(2): 101-132.

Shaw, T. (1976). Early crops in Africa: a review of the evidence. In Harlan, J.R., de Wit, J.M.J., and Stemler, A.B.L. (eds.), *The Origins Of African Plant Domestication*, Mouton, The Hague, pp. 107-153.

Shaw, T. (1977). Hunters, gatherers and first farmers in West Africa. In Megaw, J.V.S. (ed.), *Hunters, Gatherers and First Farmers Beyond Europe: An Archaeological Survey*, Leicester University Press, Leicester, pp. 69-125.

Sherrat, A. (1997). The transformation of early agrarian Europe: the later Neolithic and Copper Ages 4500-2500 BC. In Cunliffe, B. (ed.), *Prehistoric Europe: An Illustrated History*, Oxford University Press, Oxford, pp. 167-201.

Shinnie, P.L. (1996). *Ancient Nubia*, Kegan Paul International, London.

Smith A.B. (1974). Preliminary report of excavations at Karkarichinkat Nord and Karkarichinkat Sud, Tilemsi Valley, Republic of Mali, Spring 1972. *West African Journal of Archaeology* 4: 33-55.

Smith A.B. (1986). Cattle domestication in North Africa. *African Archaeological Review* 4: 197-203.

Sowell, T. (1996). *Migrations and Cultures: A World View*, Harper Collins, New York.

van Andel, T. H., and Runnels, C. N. (1995). The earliest farmers in Europe. *Antiquity* 69(264): 481-500.

Van Neer, W. (2000). Domestic animals from archaeological sites in Central and West-central Africa. In Blench, R and MacDonald, K (eds.), *Origins and Development of African Livestock.* University College London Press, London, pp. 163-190.

Vermeersch, P., Van Peer, P., Moeyersons, J., and Van Neer, W. (1996). Sodmein Cave site, Red Sea Mountains (Egypt). *Sahara* 6: 31-40.

Weiss, H. (1997). Late Third Millennium abrupt climate change and social collapse in West Asia and Egypt. In Dalfes, H.N., Kukula, G., and Weiss, H. (eds.), *Third Millennium BC Climate Change And Old World Collapse,* Springer-Verlag, Berlin, pp. 711-722.

Wendorf, F., and Schild, R. (1994). Are the early Holocene cattle in the eastern Sahara domestic or wild? *Evolutionary Anthropology* 3: 118-128.

Wendorf, F., Close, A.E., and Schild, R. (1987). Early domestic cattle in the eastern Sahara. *Palaeoecology of Africa* 18: 441-448.

Wendorf, F., Close, A.E., and Schild, R. (1992/3). Megaliths in the Egyptian Sahara. *Sahara* 5: 7-16.

Wendorf, F., Schild, R., Appelgate, A., and Gautier, A. (1997). Tumuli, cattle burials and society in the Eastern Sahara. In DBrich, B.A., and Gatto, M.C. (eds.), *Dynamics of Movements and Responses to Climate Change in Africa,* Bonsignori Editore, Rome, pp. 90-101.

Wendorf, F., Schild, R., Wasylikowa, K., Dahlberg, J., Evans, J., and Biehl, E. (1998). The use of plants during the early Holocene in the Egyptian Sahara; Early Neolithic food economies. In di Lernia, S and Manzi, G. (eds.), *Before Food Production in North Africa. Proceedings of the Homonymous Workshop held in Forli, September 1996, within the XIII World Congress of the International Union of the Prehistoric and Protohistoric Sciences,* ABACO Edizioni, Rome, pp. 71-89.

Whittle, A. (1997). The first farmers. In Cunliffe, B. (ed.), *Prehistoric Europe: An illustrated history,* Oxford University Press, Oxford, pp. 136-166.

SECTION I

CLIMATIC CHANGE

The old nazir, Sheikh Hassan's father – may God have mercy on him – used to pitch his *dikka* at the watering-place here in winter. There was grass everywhere then, and it grew tall and green. There was game of all sorts. By God, you could find oryx even in the wadi in those days! There were no bore-wells then; no one had even heard of them. The Arabs watered from the hand-wells in summer and the rainwater pools in winter. The rain was more plentiful in those times. Even if the rains failed, as they did in some years, there was enough grazing left from previous years for the animals to eat.

Michael Asher, *A Desert Dies*, Viking, London, 1986, p. 85.

In general, some of the key cultural developments in the later prehistory of Africa were, in part, ultimately controlled by severe abrupt climatic oscillations accompanying post-glacial warming and subsequent events. In general, the period from 14,000 years ago to 9500 years ago was a period of oscillating conditions marking the initial phase of post-glacial warming. In North Africa, global warming was characterized by wetter conditions.

An account of the timing and duration of some of the critical climatic events of global significance is given in this section by **Rohling** *et al.* (Chapter 3), **Vernet** (Chapter 4), **Cremaschi** (Chapter 5), **Mohammed and Bonnefille** (Chapter 6), and **Sowunmi** (Chapter 7). **Rohling** and co-workers, using marine records, detect events from indicators of the change in sea surface temperature (SST). According to **Rohling** and his team, preliminary results from three high resolution marine records from the South Adriatic Sea, Southeast Aegean Sea, and North Levantine Sea show marked SST reductions around 18-16, 13.4-11, 10-9.5, 8.6-7.9, 6.7-5.8, and 3.8-3 cal kyr BP. The SST fluctuations were determined from planktonic foraminiferal ratios of warm, oligotrophic mixed layer species relative to cool, more eutrophic deep mixing indicator species. These records portray the prevalence of the seasonal thermocline, and modern distribution patterns suggest that a reasonable approximation of SST is possible. The timing of the cooling events correspond with Holocene climate fluctuations recognized previously in the North Atlantic and the Greenland Ice Sheet. The 8.6-7.9 kyr cal BP event has been investigated in special detail. A temperature reduction of a few • C appears to have taken place abruptly within 50 years. The peak cool conditions appear to have prevailed over a period less than two centuries, after which a more gradual return to warmer conditions is found. Re-establishment of warm conditions more or less similar to those preceding the cooling was completed on average two to three centuries after the onset of the cooling. Oxygen isotope data suggest an increase in humidity from ~9.5 kyr cal BP and an onset of gradual aridification around 7.4 kyr cal BP to reach present-day conditions for the first time ~5.5 kyr cal BP, with a slight amelioration ~3.5 kyr cal BP. This is in reasonable agreement with abundance variations of radiocarbon dated Saharan humidity markers.

The history of climatic change during the Holocene in the Sahara and Sahel is reviewed in this section by **Vernet**, who recognizes an early Holocene humid period that began ca. 10 kyr bp. This wet period ended by a short arid episode ca. 7.9 yr cal BP (around 7.2-7 kyr bp), which had considerable consequences for human occupation. The middle Holocene witnessed the progressive onset of desert conditions in the Central Sahara as well as the maintenance of a relatively humid climate in the south. An arid crisis by 5 kyr cal

BP (4.2-4 kyr bp) led to a marked separation between the Sahara in the north (by then a desert) and the south, where a Sahelian climate lasted for another millennium or more. The Sahel, to the south of 22/21• N (but with considerable latitudinal range from Ténéré to Niger), was dominated by the summer monsoon regime. In the Sahel, people survived wherever water emerged at the surface (or was just below it). The plains with thick sandy cover were excluded since water disappeared rapidly and deeply into the ground, and was thus inaccessible (this would explain why the Ténéré was uninhabited after 5 kyr cal BP).

The Sahelian zone was still moist during the third millennium bp and wadi watercourses were often active. At that time, Lake Chad rose again. From Mauritania to the Sahelian Sudan the most dense occupation occurs during this millennium. Activities during this phase included hunting, fishing, herding, sometimes agriculture, and the beginning of metallurgy. By the second millennium bp, the Sahara became entirely desert, except at its southern margins. However, conditions worsened during the second half of the first millennium AD, which led the Sahelian groups to settle in the Central Saharan mountains, from Adrar in Mauritania to Tibesti. They were eventually replaced by the ancestors of the Maures and the Tuaregs (in Mali and in the Niger) by the eleventh and twelfth centuries AD.

Geological indicators of climatic change during the late Pleistocene and Holocene in the Western Fezzan (Libya, Central Sahara) are examined in this section by **Cremaschi** (Chapter 5). The area includes the Central Tadrart Acacus Massif, the Messak Settafet Plateau, the sand seas of Uan Kasa and Murzuq, and the Tanezzuft Valley. In this area, wetter phases since the late Pleistocene are indicated by a wide range of proxies: alluvial deposits, fill of caves and rock shelters, travertine, and palaeosols in the mountains; as well as by lacustrine and marsh deposits, archaeological sites, and palaeosols in the ergs. Dating and correlation of proxy data from these different domains allow a palaeoclimatic reconstruction which shows a slow progressive change, from a very wet period at the boundary between late Pleistocene and Holocene toward progressively drier conditions. Travertine sedimentation in the mountain ranges dates from 14-9 cal kyr BP as a consequence of a sharp increase of precipitation at the end of the late Pleistocene and at the very beginning of the Holocene. During this phase, enhanced fluvial activity, rubification and clay translocation in soils were also recorded. Lakes and ponds covered a very large surface from 9.6 cal kyr BP to about 6.3 cal kyr BP, with a first cycle of high stand dated to the tenth millennium bp. This phase is contemporaneous with the older phase of deposition of the anthropogenic fill in the caves of the Tadrart Acacus, which is also indicative of wet conditions. Both dried lakes and shelter fills display a sedimentary gap, related to a dry environment, at about the middle of the ninth millennium bp. It may be interpreted as the local effect of the middle Holocene dry period, showing some overlap with the dry event recorded by windblown chemical indicators and a decrease in methane in Greenland ice-core proxy. The second half of the seventh millennium bp is one more period of high stand of the lakes and of wet climate in the mountains, as indicated by the humified deposits inside caves and shelters and further supported by wet savanna pollen record. A short arid spell also initiated at 7.2-6.9 cal kyr BP.

The link between palynological data of late Holocene climatic change and historical records of famines in Ethiopia is examined in this section by **Mohammed and Bonnefille** (Chapter 6) on the basis of palynological data used to detect significant changes in vegetation during the late Holocene. Mohammed and Bonnefille detected an abrupt dry event ca. 4.4 kyr cal BP, which was followed by a moist episode indicated by an increase in tree pollen. This moist phase coincided with a phase of low amplitude lake level rises in the Ethiopian Rift Valley. During the same time interval, upland swamps, which provided organic sediment for pollen analysis, showed a very important development. From ca. AD 1450 (about 0.5 kyr bp) to the last few centuries, a fall in tree pollen taxa is documented on a South Ethiopian highland site at the edge of the forest. This event marks the onset of relatively colder and perhaps drier conditions. Frequent drought and famine events were recorded in historical archives during the same time interval.

On the other side of the continent, the changes in vegetation, as a proxy to climatic change, in the Guineao-Congolian region are examined in the final chapter of this section by **Sowunmi** (Chapter 7), who recognizes a marked dry phase about 3800 cal BP. The rapid onset of aridity led to a reduction of the extent of the rain forest and its edaphic variability, with an expansion of woodland/grassy savanna. The arid spells promoted slash-and-burn farming and an increase in the distribution of oil palm and its exploitation.

3

RAPID HOLOCENE CLIMATE CHANGES IN THE EASTERN MEDITERRANEAN

E. J. Rohling, J. Casford, R. Abu-Zied, S. Cooke, D. Mercone, J. Thomson, I. Croudace, F. J. Jorissen, H. Brinkhuis, J. Kallmeyer and G. Wefer

INTRODUCTION

Rapid climatic fluctuations over the Eastern Mediterranean are recognized in three high resolution marine sediment cores—IN68-9 for the south Adriatic Sea, LC21 from the southeast Aegean Sea, and LC31 from west of Cyprus. Although investigations are not yet completed, the preliminary results demonstrate that unprecedentedly high quality information may be gained from these records because of the high accumulation rates.

LITHOLOGY

Lithological changes within core IN68-9, showing several distinct turbidites and ash-layers (Figure 3.1), have been described previously, and detailed age assessment suggests that no significant turbidites or hiatuses were overlooked (Jorissen *et al.*, 1993; Rohling *et al.*, 1993, 1997). Core IN68-9 contains a composite dark-colored sapropelic unit that is 'interrupted' by lighter homogeneous sediments. Lamination denoting bottom water anoxia was observed in the lower of the two sapropelic sub-units, but not in the upper sub-unit.

The results for core LC21 presented here concern the upper two sections of a longer piston core. In these upper sections, a composite dark-colored sapropelic unit is again found, with intercalated lighter colored homogeneous sediments. Higher up in the core, a 10 cm thick ash layer relates to the Santorini eruption of 1628 BC (Kuniholm *et al.*, 1996).

The record for core LC31 also represents the upper two sections of a longer core. As in the other two cores, a composite sapropelic unit is observed, separated by lighter-colored homogeneous sediments. Some 50 to 60 cm below the sapropel, a level was found with significant volcanic ash admixture. As it does not constitute a discrete ash-layer, however, it has not been used as a datum for correlation to the Mediterranean tephrachronology.

The high accumulation rates in the cores (see below) immediately bring

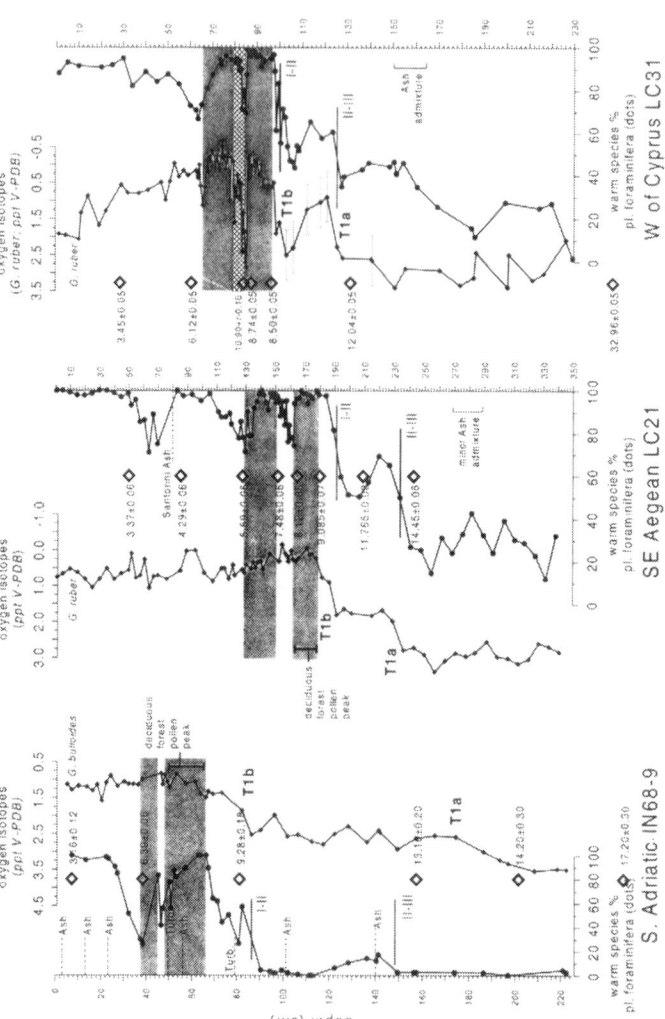

Figure 3.1. Warm versus cold planktonic foraminiferal percentages (dots) for cores IN68-9 (41.75.5 N, 17.54.5 E; 1234 m), LC21 (35.66 N. 26.58E; 1522 m), and LC31 (34.59.76 N, 31.09.81 E; 2298 m), plotted against core-depth in cm as corrected for thicknesses of ash-layers and turbidites. Grey bands indicate dark colored sapropelic sediments. Positions where intercalated ash-layers and turbidites were observed are indicated with dashed lines. Open diamonds and numbers indicate raw AMS radiocarbon results in radiocarbon years uncorrected for the 400 yr reservoir age. Bold codes T1a and T1b indicate the positions of glacial terminations 1a and 1b. Biozonal boundaries I-II and II-III as defined by Jorissen *et al.* (1993). Cross-hatched interval in LC31 represents suspected slump as described in the present paper. Oxygen isotope records (filled diamonds) are in ‰ relative to PDB standard, for *Globigerinoides ruber* in the 250-350 μm size range (LC21 and LC31) and *G. bulloides* > 150 μm (IN68-9).

the nature of the sapropel 'interruptions' into question. The intervals might be genuine, or a result of redeposition. In core IN68-9, abundance variations of benthic foraminifera (Rohling *et al.*, 1997) and benthic molluscs (Van Straaten, 1966, 1972) indicate that the interval is genuine.

In LC21 we observe that the lighter coloured interval between the two sapropel sub-units is also characterized by specific benthic foraminiferal assemblage changes that argue against redeposition (Figure 3.2). Following negligible faunal abundances in the lower sapropel sub-unit, we find repopulation in the 'interruption' by opportunistic species which were hardly present in the area before sapropel formation started (Figure 3.2). If the sediments of the 'interruption' were to represent a mass deposition event, then its faunal content would be expected to bear some similarity to pre-sapropelic sediments.

The benthic foraminiferal record is much less decisive in LC31 than in LC21 (Figure 3.2). In addition, inorganic geochemical data demonstrate conspicuous returns to pre-sapropelic values within the 'interruption' in LC31. A similar drastic shift is seen in the oxygen isotope record (Figure 3.1). These changes suggest that the 'interruption' of the sapropel in LC31 resulted from a redeposition event. This interpretation is corroborated by radiocarbon dating, which returned an anomalous dating profile through the sapropel with the sediments of the 'interruption' dated at 10,900 uncorrected radiocarbon years bp (Figure 3.1). The complete lack of obvious size/shape sorting of fauna, and good preservation of even very fragile foraminiferal shells within the interval, indicate that redeposition took place in the form of a minor slump. On the basis of Ba/Al and Ti/Al profiles (not shown), we infer that the anomaly extends between 85 and 80 cm in the core.

In LC31, as in LC21 and IN68-9, the upper sapropel unit is found to be faunistically distinct from the lower unit, which excludes the possibility that the upper unit could represent a redeposited part of the lower unit. Hence, we conclude that the double appearance of S1 in core LC21 reflects genuine changes in the sedimentary environment, and has not resulted from redeposition processes, while there are strong indications that the double aspect of the sapropel in LC31 is an artifact resulting from intercalation of 5 to 6 cm of pre-sapropelic sediments due to slumping.

TIME-STRATIGRAPHIC FRAMEWORK

The framework for IN68-9 has been previously discussed by Rohling *et al.* (1997). The framework for cores LC21 and LC31 is based on AMS radiocarbon age determinations (Figure 3.1), supported by dated planktonic foraminiferal biozonal boundaries that were previously described for the Central Mediterranean region (Jorissen *et al.*, 1993). Two biozonal boundaries were distinguished within the last 18,000 years: the zone I/II boundary, with a mean uncorrected radiocarbon age range of 9850 ± 470 bp and the zone II/III boundary around 13250 ± 380 bp (Table 3.1; ages in radiocarbon years uncorrected for reservoir age).

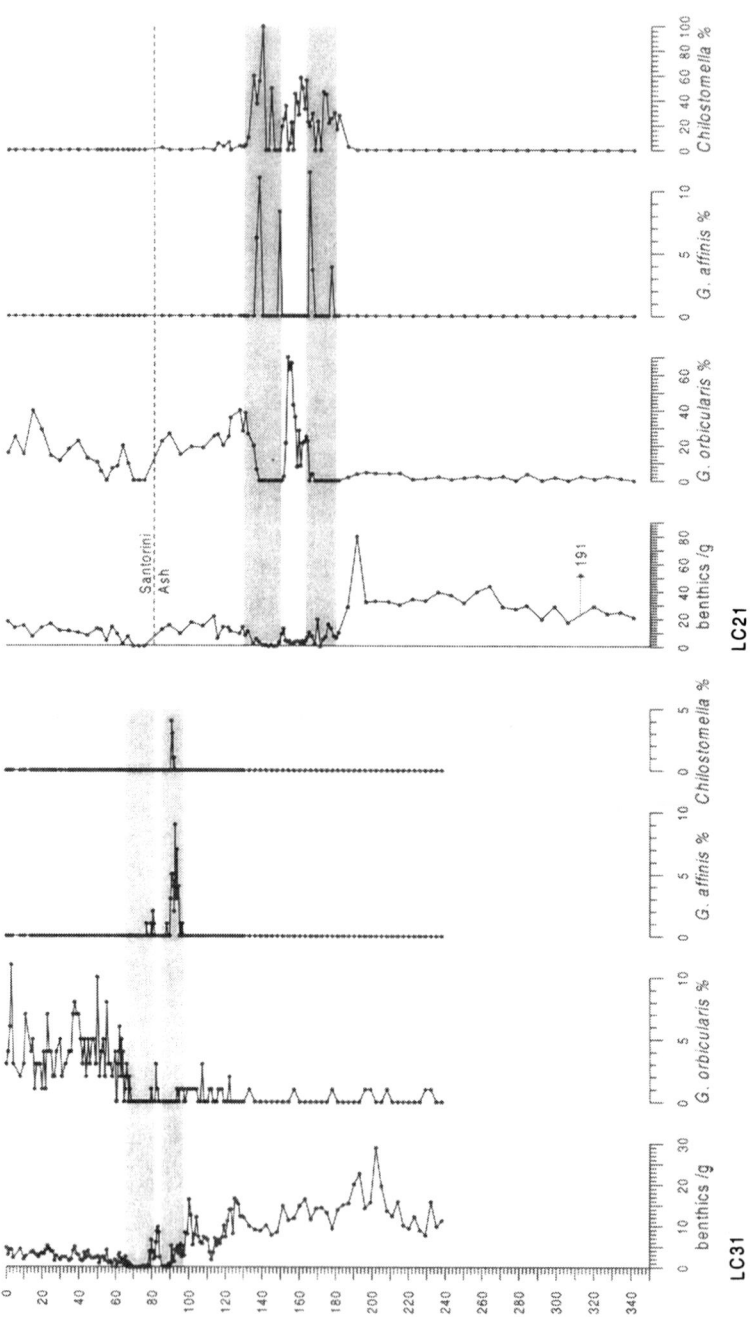

Figure 3.2. Key benthic foraminiferal species in cores LC31 and LC21, as examples to illustrate faunal differences between non-sapropelic and sapropelic sediments, between upper and lower sapropelic units, between pre- and post-sapropelic sediments, and of >interruption= sediments versus other intervals. Grey bands indicate sapropelic intervals.

Table 3.1. Uncorrected radiocarbon ages (bp) for biozonal boundaries in Central Mediterranean cores studied by Jorissen *et al.* (1993)

I/II boundary	II/III boundary
10,180	12,870
9620	13,590
10,100	12,680
9060	13,110
9780	13,610
10,330	13,630
$\bar{x} = 9850 \pm 470$	$\bar{x} = 13,250 \pm 380$

Age determinations on a West Aegean core by Zachariasse *et al.* (1997) range near to the values reported in Table 3.1, suggesting that the biozonal boundaries are more-or-less synchronous between the Central Mediterranean and the Aegean Sea. A recent study on a core with a 'double' Holocene sapropel from the Southwest Aegean Sea also confirms the validity of this extrapolation of the biozones (Geraga *et al.*, 1997, 2000). The positions of the planktonic foraminiferal biozonal boundaries I/II and II/III in the three cores presented here are indicated in Figure 3.1.

Further age-control was obtained for LC21 from a distinct ash layer related to the Santorini eruption of 1628 BC (Kuniholm *et al.*, 1996). Since 'before present' is measured by convention relative to 1950, this implies an age of 3578 cal BP for that eruption.

DISCUSSION AND CONCLUSIONS

Sea Surface Temperature Fluctuations

Sea surface temperature (SST) fluctuations in absolute terms are a contentious issue in palaeoceanography. A common tool for first-order assessment uses ratios of known warm (oligotrophic) versus cool (eutrophic) planktonic foraminiferal species, an approach that has been found to provide very similar results to more sophisticated statistical methods in the Eastern Mediterranean (cf. Thunell *et al.*, 1977; Rohling *et al.*, 1993, 1997). Whatever technique based on faunal assemblage data is used, there is always an overprint of the degree of eutrophication prevailing in the study area today, and in the past. The two influences—SST and eutrophication—are intricately linked and as yet there are no sound methods for their deconvolution. Here, we follow the simple *a priori* method of Rohling *et al.* (1997), plotting abundances of the warm (oligotrophic) species *Globigerinoides ruber* and the SPRUDTS-group (*Globigerinoides sacculifer, Hastigerina pelagica, Globoturborotalita rubescens, Orbulina universa, Globigerinella digitata, Globoturborotalita tenella,* and *Globigerinella siphonifera*) versus the cool (eutrophic) species *Globorotalia scitula*, *Turborotalita quinqueloba*, *Globorotalia inflata*, and *Neogloboquadrina pachyderma* (Figure 3.1).

The warm (oligotrophic) species are those that dominate in subtropical waters of the present-day ocean, within the shallow (upper ~50 m), warm, and nutrient-depleted surface mixed-layer. Many of these species are spinose and contain photosynthetic symbionts. In subtropical waters, the cool (eutrophic) species are today mostly found in winter, when SST is reduced and the mixed layer has considerably deepened (>100 m) and becomes enriched in nutrients. Several of these species have distinct preferences for specific habitats, for example in association with upwelling, subsurface chlorophyll maxima and/or deep frontal mixing (for summaries of habitat characteristics, see Rohling *et al.*, 1993, 1997). Despite the imprint of food-related habitat preferences, a basic warm/cold subdivision remains possible. Recent multivariate statistical studies have highlighted that temperature is one of the main controls on global planktonic foraminiferal distribution patterns (Pflaumann *et al.*, 1996; Kallel *et al.*, 1997).

Using our simple *a priori* grouping in 'warm' and 'cold' faunal categories and plotting these in relative abundance ratios, rapid and repeated SST fluctuations are recognized in all three cores investigated (Figure 3.1). The Adriatic record is corroborated by good agreement with an SST record based on dinoflagellates (Targarona *et al.*, 1997), while that data had also been successfully related to local continental records (Zonneveld, 1995, 1996; Zonneveld and Boessenkool, 1996). In Figure 3.3, LC21 is used as a master record, and cooling events have been identified that can be recognized at time-equivalent intervals in the other cores. Cooling events are recognized around 3.0-3.8 (LC21 only), 5.8-6.7, 7.9-8.6, (9.5-10.0, Adriatic only), 11.0-13.4, and (poorly defined) 16-18 kyr cal BP.

Comparison With Extra-Mediterranean Climate Fluctuations

Millennial-scale climatic cycles have been described for study areas in Greenland, Canada, and the northwest Indian Ocean (Campbell *et al.*, 1998), corroborating the apparent climatic variability we infer from the Eastern Mediterranean records. Bond *et al.* (1997) provide age determinations for ice-rafted debris (IRD) events in the North Atlantic, around 1.4, 2.8, 4.2, 5.9, 8.1, 9.4, 10.3, and 11.1 kyr cal BP. Several of these correlate reasonably well with cooling events in our Eastern Mediterranean records.

We contend that the IRD events in the North Atlantic likely are contemporaneous to the cooling events in the Eastern Mediterranean, within the resolution of the time-stratigraphic framework. Our oldest cooling event has a similar age as the youngest peak of left-coiling *Neogloboquadrina pachyderma* observed in the Gulf of Lions (Western Mediterranean), which correlates with Heinrich Event 1 in the North Atlantic (Rohling *et al.*, 1998). The main cooling event spanning 11.0-13.4 kyr cal BP correlates with the Younger Dryas expressions around the Eastern Mediterranean (Rossignol-Strick, 1995, 1997).

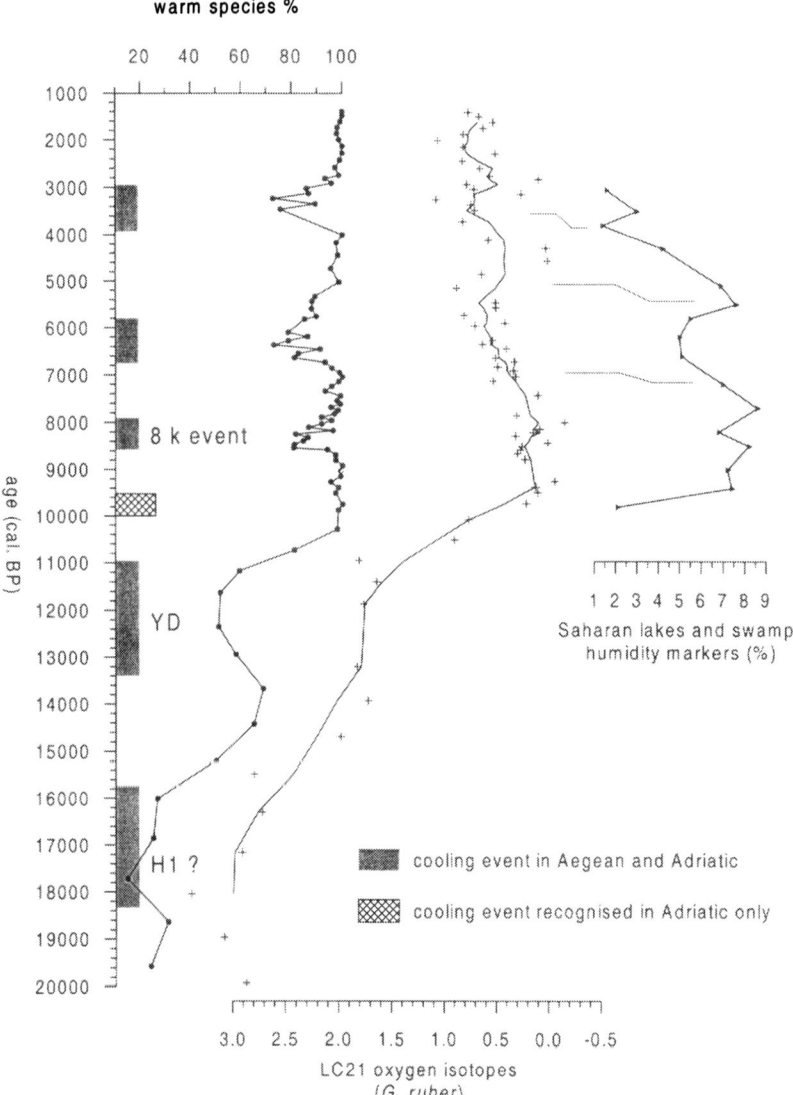

Figure 3.3. Warm-cold record and stable oxygen isotope record (*G. ruber*) for LC21, versus age in calibrated yrs BP, based on the time-stratigraphic framework explained in the text and Figure 3.1. The solid line in the isotope record represents a 5-point moving average, to highlight fundamental trends. Radiocarbon dated Saharan humidity marker abundances are plotted for comparison (Petit-Maire and Guo, 1997). Grey bands represent cool intervals, and cross-hatched band indicates a local (?) Adriatic cool event, noted only in IN68-9. YD stands for Younger Dryas, H1 (?) for a possible Heinrich Layer 1 equivalent cooling.

The 7.9-8.6 kyr cal BP cooling event 'splits' the anaerobic sapropel sediments into two sub-units in the two cores from the Adriatic and Aegean seas, which are both areas of active deep water ventilation in the present-day climate. We are studying more cores from the Eastern Mediterranean to verify whether this event can be recognized only in the northern satellite basins of the Eastern Mediterranean, or indeed is present throughout the basin. This event appears to correlate approximately with the well documented '8 kyr cal BP' event, the major Holocene cooling episode recognized beyond the North Atlantic basin (Alley *et al.*, 1997; DeVernal *et al.*, 1997). Because of the slump in LC31, we can not confirm or exclude whether the 8 kyr cal BP event is present in that core as well. The oxygen isotope excursion to high values does continue for 1 or 2 cm above 80 cm core depth, but at this stage we do not know whether this is an expression of the event, or a residual effect of the slump. Certainly the reconstructed low SST values do not extend outside the suspect 80-85 cm interval, suggesting that no cooling influences are recorded throughout the intact segments of S1.

Comparison With North African Climate Changes

The temporal distribution of various types of humidity indicators in the Sahara region shows a distinct wet phase between ~9.5 and ~4.0 kyr cal BP (Petit-Maire and Guo, 1997). Around 8.0 kyr cal BP, a brief aridity crisis is suggested, and a major aridity crisis is indicated between 7.0 and 5.5 kyr cal BP (Figure 3.3). We find a correlation between the onset of the wet phase ~9.5 kyr cal BP and the strong warming into the early Holocene parts of our records, which in the Adriatic notably ends a local cool phase. In addition, we also find good correspondence with the onset of the main early Holocene $\delta^{18}O$ depletion in our records (Figure 3.3).

The thermal maximum and strongest $\delta^{18}O$ depletions between ~9.0 and 6.0 kyr cal BP in our records correspond to the maximum humidity period in the Sahara record, while the 8 kyr cal BP cooling event is noted as a brief aridity crisis in the record of Saharan humidity markers. Our main cooling event of 6.7-5.8 kyr cal BP coincides closely with a prolonged Saharan aridity crisis dated at 7.0-5.5 kyr cal BP (Figure 3.3).

The 13.4-11.0 kyr cal BP cooling event in LC21 corresponds well in time with the major peak in *Chenopodiaceae* pollen found in sections around the Eastern Mediterranean and Northwest Indian Ocean, which is interpreted as a cold and arid phase corresponding to the Younger Dryas (Rossignol-Strick, 1995, 1997). Those studies also suggest warm and wet conditions, between about 10.0 and 6.5 kyr cal BP, based on peak abundances of *Pistacea* pollen, which would correspond to the Eastern Mediterranean thermal and Saharan humidity maxima shown in Figures 3.1 and 3.3. Brief intervals of increased *Chenopodiaceae* and *Artemisia* within the *Pistacea* zone, that may signal brief relapses to cooler and drier conditions (Rossignol-Strick, 1995), may relate to the 8 kyr cal BP cooling event observed in our records.

The overall waning of the humid period from 5.0 to 4.0 kyr cal BP shows good agreement with the stable oxygen isotope records (Figures 3.1 and

3.3). For LC21 and LC31, these are based on the shallow dwelling planktonic foraminiferal species *Globigerinoides ruber* that predominates today in the late summer to fall (August-October) season (Pujol and Vergnaud-Grazzini, 1995), which coincided with the maximum monsoonal discharge from the Nile River in pre-Aswan Dam times. For ease of comparison with the continental humidity record, the oxygen isotope record of LC21 has been plotted versus time in Figure 3.3.

To understand the isotopic changes, several major influences may be separated. Oxygen isotope records are influenced by: (1) sea level change as a function of global ice-volume, as preferential sequestration of ^{16}O in continental ice-sheets leaves the global ocean enriched in ^{18}O; (2) sea water temperature, as the $^{16}O:^{18}O$ ratio during uptake into carbonate changes with the temperature of the ambient water; and (3) the relative importance of evaporation—which preferentially removes ^{16}O—and freshwater input into the basin, both volumetrically and in terms of oxygen isotope composition.

In the conventional notation of oxygen isotope ratios, the 'glacial' effect comprises a maximum $\delta^{18}O$ difference of 1.2‰ between more positive Last Glacial Maximum values and the present. The general trends of the glacial effect reflect the history of deglacial melt water addition to the global ocean/sea level rise (Fairbanks, 1989, 1990). The 'temperature' effect entails enrichment/ depletion of about 0.2‰ per °C cooling/warming. The total glacial-interglacial range of temperature change in the area midway between the Levantine Sea and Aegean Sea is thought to have been of order 5°C (see overview in Rohling and De Rijk, 1999a,b), so that this would have a maximum influence of roughly 1‰ on the $\delta^{18}O$ record. Note that the main trends and fluctuations within the warming since the Last Glacial Maximum are reflected in the warm species records (Figure 3.3). The $\delta^{18}O$ record of LC21 shows greater amplitudes of change than may be explained purely from the glacial and temperature effects. The deviations from the expected fluctuations are likely caused by surface water $\delta^{18}O$ enrichment due to evaporation, and/or $\delta^{18}O$ depletion due to precipitation and run-off (cf. Rohling, 1999). By approximation, persistent low values (plotted according to convention to the right) suggest an overall more humid climate, while maintained enrichments suggest that aridity prevailed. Hence, isotope residual fluctuations demonstrate general agreement with the Saharan humidity markers. Humidity increased around 9.5-9.0 kyr cal BP. From about 7.5-7.0 kyr cal BP a trend started towards the present aridity that was firmly established around 3.5 kyr cal BP. The timing of the onset of this trend matches the observations of Grove (1993), as summarized also in Hassan (1996).

The 7.9-8.6 kyr cal BP Cooling Event

This cooling event (Figures 3.1 and 3.3) was also highlighted independently for IN68-9 by Targarona *et al.* (1997), and merits some further attention. Its magnitude may be estimated from the total glacial to interglacial temperature difference of about 5°C. Assuming that the deglacial temperature rise was distributed equally over the two 'terminations' (which approximately coincide with the two biozonal boundaries; Jorissen *et al.*, 1993), then the 100% change in the SST record of LC21 might be linearly calibrated to temperature change. This would imply that a 10% increase in this record corresponds roughly to a 0.5°C temperature rise. As a first order approximation, the temperature drop over the Eastern Mediterranean associated with the onset of the 8 kyr cal BP event would have been of order 1-2°C. Rohling *et al.* (1997) argued that calibration for the record of IN68-9 is more complicated (a logarithmic calibration seemed more appropriate), and tentatively concluded that the temperature drop was of the order of 2°C. Myers and Rohling (2000) used a numerical circulation model to demonstrate that the 'sapropel' mode may indeed be interrupted in the Adriatic and Aegean basins for such a cooling.

The rapidity of this temperature change is remarkable (Figure 3.3). The change takes place from one sample to the next (i.e. over 1 cm) in LC21, i.e. over a period of ~50 years. The LC21 record then suggests that peak cool conditions were maintained over a period of 150-250 years, after which a more gradual return to warm conditions occurred over a period of several centuries.

ACKNOWLEDGMENTS

Thanks are due to M. Segl at the isotope laboratory at Bremen. Support for JC was obtained from the NERC, JK from EC-SOCRATES exchange funds, and for RA-Z from the Egyptian government. This study is part of EC MAST-3 programmes CLIVAMP (MAS-CT95-0043) and SAP (MAS3-CT98-0137). Cores LC21 and LC31 were recovered by *R.V. Marion Dufresne*, chief scientist G. Rothwell, within the framework of EC-MAST2 programme PALEOFLUX, and are stored at the SOC-BOSCOR repository. This study contributes to UNESCO-IUGS project Climates in the Past (CLIP). Thanks are due to ESF for sponsoring the workshop in London at which this paper was discussed within the context of the Saharan climate developments.

REFERENCES

Alley, R.B., Mayewski, P.A., Sowers, T., Stuiver, M., Taylor, K.C., and Clarck, P.U. (1997). Holocene climatic instability: a prominent, widespread event 8200 yr ago. *Geology* **25**: 483-486.

Bond, G., Showers, W., Cheseby, M., Lotti, R., Almasi, P., deMenocal, P., Priore, P., Cullen, H., Hajdas, I., and Bonani, G. (1997). A pervasive millenial-scale cycle in North Atlantic Holocene and Glacial climates. *Science* **278**: 1257-1266.

Campbell, I.D., Campbell, C., Apps, M.J., Rutter, N.W., and Bush, A.B.G. (1998). Late Holocene ~1500 yr climatic periodicities and their implications. *Geology* **26**: 471-473.

DeVernal, A., Von Grafenstein, U., and Barber, D. (1997). Researchers look for links among paleoclimate events. *EOS: Transactions of the American Geophysical Union* **June**: 247-249.

Fairbanks, R.G. (1989). A 17,000 year glacio-eustatic sea level record: influence of glacial melting rates on the Younger Dryas event and deep-ocean circulation. *Nature* **342**: 637-642.

Fairbanks, R.G. (1990). The age and origin of the "Younger Dryas Climate Event" in Greenland Ice Cores. *Paleoceanography* **5**: 937-948.

Geraga, M., Tsaila-Monopolis, S., Tripsanas, E., Papatheodorou, G., Ferentinos, G., Hasiotis, T., Seymour, K.S., and Ioakim, C. (1997). Paleoceanographic and paleoclimatic changes in the central Aegean Sea (Myrtoon basin) during Holocene-Late Pleistocene. Preliminary results, *Proceedings 5th Hellenic Symposium on Oceanography and Fisheries, Kavalla, Greece, 15-18 April, 1997, Vol. 1*, National Centre for Maritime Research, Athens, pp. 213-216.

Geraga, M., Tsaila-Monopolis, S., Ioakim, C., Papatheodorou, G., and Ferentinos, G. (2000). Evaluation of palaeoenvironmental changes during the last 18,000 years in the Myrtoon basin, SW Aegean Sea. *Paleogeography, Palaeoclimatology, Palaeoecology* **156**: 1-17.

Grove, A.T. (1993). Africa's climate in the Holocene. In Shaw, T., Sinclair, P., Andah, B., and Okpoko, A. (eds.), *The Archaeology of Africa: Food, Metals and Towns*, Routledge, London, pp. 32-42.

Hassan, F.A. (1996). Abrupt Holocene climatic events in Africa. In Pwiti, G., and Soper, R. (eds.), *Aspects of African Archaeology. Papers from the 10th Congress of the PanAfrican Association for Prehistory and Related Studies*, University of Zimbabwe Publications, Harare, pp. 83-89.

Jorissen, F.J., Asioli, A., Borsetti, A.M., Capotondi, De Visser, J.P., Hilgen, F.J., Rohling, E.J., Van der Borg, K., Vergnaud-Grazzini, C., and Zachariasse, W.J. (1993). Late Quaternary central Mediterranean biochronology. *Marine Micropaleontology* **21**: 169-189,.

Kallel, N., Paterne, M., Duplessy, J.C., Vergnaud-Grazzini, C., Pujol, C., Labeyrie, L.D., Arnold, M., Fontugne, M., and Pierre, C. (1997). Enhanced rainfall in the Mediterranean region during the last sapropel event. *Oceanology Acta* **20**: 697-712.

Kuniholm, P.I., Kromer, B., Manning, S.W., Newton, M., Latini, C.E., and Bruce, M.J. (1996). Anatolian tree rings and the absolute chronology of the Eastern Mediterranean, 2220-718 BP. *Nature* **381**: 780-783.

Myers, P.G., and Rohling, E.J. (2000). Modelling a 200 year interruption of the Holocene sapropel S1. *Quaternary Research* **53**: 98-104.

Petit-Maire, N., and Guo, Z. (1997). In-phase Holocene climate variations in the present-day desert areas of China and Northern Africa. In Meco, J., and Petit-Maire, N. (eds.), *Climates of the Past, Proceedings of the UNESCO-IUGS, June 2-7, 1995, CLIP meeting*, University de Las Palmas, Gran Canaria, pp. 137-140.

Pflaumann, U., Duprat, J., Pujol, C., and Labeyrie, L.D. (1996). SIMMAX: a modern analog technique to deduce Atlantic sea surface temperatures from planktonic foraminifera in deep-sea sediments. *Paleoceanography* **11**: 15-35.

Pujol, C., and Vergnaud-Grazzini, C. (1995). Distribution patterns of live planktic foraminifers as related to regional hydrography and productive systems of the Mediterranean Sea. *Marine Micropaleontology* **25**: 187-217.

Rohling, E.J. (1999). Environmental controls on salinity and $\delta^{18}O$ in the Mediterranean. *Paleoceanography* **14**: 706-715.

Rohling, E.J., and De Rijk. S. (1999a). The Holocene Climate Optimum and Last Glacial Maximum in the Mediterranean: the marine oxygen isotope record. *Marine Geology* **153**: 57-75.

Rohling, E.J., and De Rijk. S. (1999b). Erratum to "The Holocene Climate Optimum and Last Glacial Maximum in the Mediterranean: the marine oxygen isotope record". *Marine Geology* **161**: 385-387.

Rohling, E.J., Jorissen, F.J., Vergnaud-Grazzini, C., and Zachariasse, W.J. (1993). Northern Levantine and Adriatic Quaternary planktic foraminifera; reconstruction of paleoenvironmental gradients. *Marine Micropaleontology* **21**: 191-218.

Rohling, E.J., Jorissen, F.J., and De Stigter, H.C. (1997). 200 year interruption of Holocene sapropel formation in the Adriatic Sea. *Journal of Micropalaeontology* **16**: 97-108.

Rohling, E.J., Hayes, A., De Rijk, S., Kroon, D., Zachariasse, W.J., and Eisma, D. (1998). Abrupt cold spells in the northwest Mediterranean. *Paleoceanography* **13**: 316-322.

Rossignol-Strick, M. (1995). Sea-land correlation of pollen records in the Eastern Mediterranean for the glacial-interglacial transition: biostratigraphy versus radiometric time-scale. *Quaternary Science Reviews* **14**: 893-915.

Rossignol-Strick, M. (1997). Paléoclimat de la Méditerranée orientale et de l'Asie du Sud-Ouest de 15000 à 6000 BP, *Paléorient* **23**(2): 175-186.

Stuiver, M., and Reimer, P.J. (1993). Extended 14C data base and revised CALIB 3.0 14C age calibration program, *Radiocarbon* **35**: 215-230.

Targarona, J., Boessekool, K., Brinkhuis, H., Visscher, H., and Zonneveld, K. (1997). Land-Sea correlation of events in relation to the onset and ending of sapropel S1: a palynological approach. In Targarona, J. (ed.), *Climatic and Oceanographic Evolution of the*

     Mediterranean Region over the last Glacial-interglacial Transition: a Palynological Approach, LPP Contributions Series, 7, Febo, Utrecht, pp. 87-112.

Thunell, R.C., Williams, D.F., and Kennett, J.P. (1977). Late Quaternary paleoclimatology, stratigraphy and sapropel history in Eastern Mediterranean deep-sea sediments. *Marine Micropaleontology* **2**: 371-388.

Van Straaten, L.M.J.U. (1966). Micro-malacological investigation of cores from the southeastern Adriatic Sea. *Proc. Kon. Ned. Akad. Science* **69**: 429-445.

Van Straaten, L.M.J.U. (1972). Holocene stages of Oxygen depletion in deep waters of the Adriatic Sea. In Stanley, D.J. (ed.), *The Mediterranean Sea*, Dowden Hutchinson & Ross Inc., Stroudsburg Pa, pp. 631-643.

Zachariasse, W.J., Jorissen, F.J., Perissoratis, C., Rohling, E.J., and Tsapralis, V. (1997). Late Quaternary foraminiferal changes and the nature of sapropel S1 in Skopelos Basin. *Proceedings 5th Hellenic symposium on Oceanography and Fisheries, Kavalla, Greece, 15-18 April, 1997, Vol. 1*, National Centre for Maritime Research, Athens, pp. 391-394.

Zonneveld, K.A.F. (1995). Palaeoclimatic and palaeo-ecological changes during the last deglaciation in the Eastern Mediterranean: implications for dinoflagellate ecology. *Review of Palaeobotany and Palynology* **84**: 221-253.

Zonneveld, K.A.F. (1996). *Palaeoclimatic and Palaeo-ecologic Changes in the Eastern Mediterranean and Arabian Sea Regions During the Last Deglaciation: a Palynological Approach to Land-sea Correlation*, PhD thesis, Utrecht University, Utrecht.

Zonneveld, K.A.F., and Boessenkool, K.P. (1996). Palynology as a tool for land-sea correlation: and example from the eastern Mediterranean. In Andrews, J.T., Austin, W.E.N., Bergsten, H., and Jennings, A.E. (eds.), *Late Quaternary Palaeoceanography of the North Atlantic Margins*, The Geological Society, London, pp. 351-357.

4

CLIMATE DURING THE LATE HOLOCENE IN THE SAHARA AND THE SAHEL: EVOLUTION AND CONSEQUENCES ON HUMAN SETTLEMENT

R. Vernet

INTRODUCTION

The main stages of the Holocene climatic history are known (Figure 4.1). After a long dry period during the late Pleistocene, ending earlier in the central mountains than on the plains, a first humid sequence was followed around 10,700 bp by a brief dry episode, corresponding on the European continent to the Younger Dryas. The Holocene humid period extended from around 10,000 to 7500/7200 bp, with optima that may reasonably be dated around 9500 and 8500. It was followed by a sustained drought, lasting several centuries, and then by a long period (lasting up to 4200 bp) of globally warmer climate, marked by alternating humid peaks (around 6700 and 4500) and dry peaks (6400, 5500 and 4800). The extent of this second humid Holocene, however, is weaker.

A dry episode, around 4200/4000, was quite severe and marks, climatically speaking, the end of an era. It was followed, during the second millennium BC, by a third humid phase which, in the southern Sahara (and only there), can be viewed as a genuine optimum. Indeed, it is from 4000 bp onwards that the Sahara becomes divided into two great ensembles. North of the tropic, aridity became the norm and the Sahara once more became a desert, while in the south, the process took a slower pace, aridity progressing more gradually toward the Sahel. There were still a few climatic remissions during the first millennium BC. A final one occurred during the second half of the first millennium AD. These pauses in the desertification process, however, became shorter and less marked as time progressed.

Researchers have extracted this information from more than a thousand examples of carbon dating, as well as from other types of data (including thermoluminescence dating). Obviously, the data gathered varies in quality, its reliability being at times questioned, as in the case of water or shells. What is more, most of the age determinations available concern humid episodes, the only exceptions being ostrich eggshell and TL age determinations on sandy sediments.

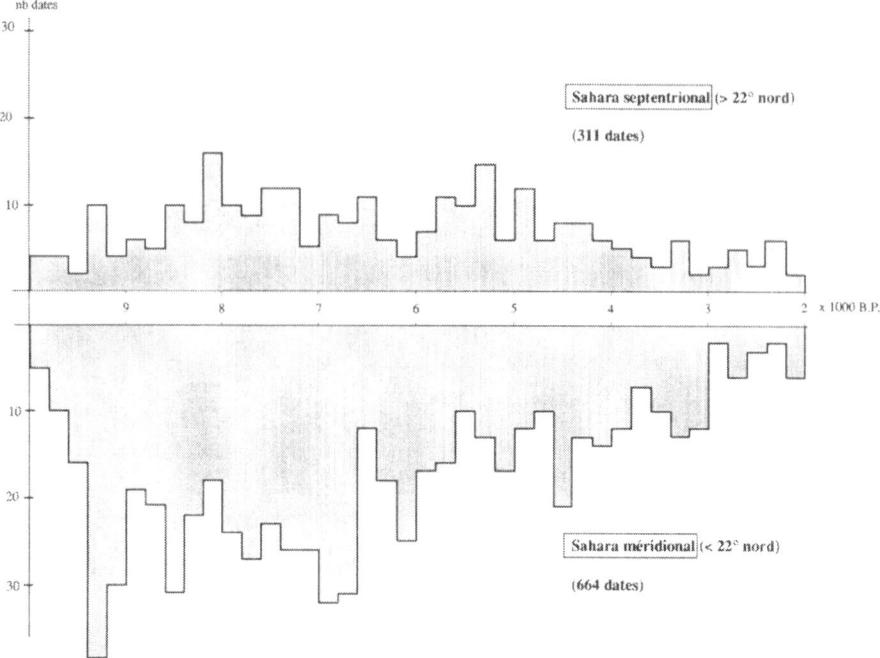

Figure 4.1. Evolution of the number of palaeoenvironmental age determinations in the Sahara.

Finally, these age determinations seldom give us much precision as to the impact of the event on the sequence. In order to gain this information, a good stratigraphy is needed. Since the effect of a climatic event on man varies throughout any given episode, little more precision can be reached until a sufficient volume of long stratigraphies is made available. In view of all this, the climatic evolution mentioned above may be somewhat oversimplified.

SAHARAN ENVIRONMENTAL VARIABILITY

The Sahara and the Sahel encompass several geographical zones. Latitude is not the only variable at work, for altitude, at times elevated throughout vast areas, may induce a 'water tower' effect. Proximity of the Atlantic or the Mediterranean coastlines, of allogen rivers or of Lake Chad, constitute yet other factors of differentiation, while the nature of the soil (whether it is rocky or sandy) will have a direct impact on the level of water conservation. For all these reasons, the Sahara should be viewed as a geographical mosaic. This observation is particularly true around mountain ranges or the great hydrological networks, where landscape units, at times, cover extremely small areas.

The scope of climatic events is not well known. More precise information is needed regarding their importance (volume and duration of precipitation), and even more so regarding the rhythm of any given climatic change, which is essential for a fuller understanding of its potential impact on man (to take an example from the twentieth century: precipitation in the Sahel in view of the average rainfall throughout the century: 1950-65 > 30%; 1970-90 < 30%). Moreover, researchers have accumulated a better knowledge of humid episodes than of arid ones, which in most cases are not dated directly. Indeed, while radiocarbon age determinations may provide satisfactory measures of acute dry episodes (ca. 7200/7000; ca. 4200/4000; ca. 3000 bp), they do not give much information about intermediary stages, tending to blend them into the more humid periods. As a result, the latter appear to have lasted up to one or even several thousand years, which everyone would agree is quite a doubtful result.

It seems clear that little progress will be made in our knowledge of the evolution of human settlements in the Sahara until we are able to reach a century-level order of precision. In fact, one should ideally be able to reach a quarter of a century level of precision—that is, the level of a human generation.

Climatic events hold consequences for ecosystems which, again, we do not measure with enough precision. Such factors as the type of precipitation (e.g., mild and long lasting; violent and brief; warm or cold), the type of soil concerned (e.g., permeable or not; flat or uneven; rocky or sandy; presence or absence of a vegetation cover), the length of the event, etc., will have a significant effect on whether the water infiltrates the soil or evaporates; on whether the vegetation prospers or wilts; on the evolution of the fauna, etc.

Furthermore, little is known about the way in which water tables evolve. Scholars are more and more inclined to believe that modest variations in the level of a water table may have a significant effect on existing open water fields (possibly making them disappear), or make new ones appear. In other words, changes in the ecosystem may well have occurred without any great variations in the level of precipitation. Consequences on the study of human settlements are evident: indeed, given the mosaic-effect mentioned above, a semi-arid Sahara could possibly have hosted prosperous human societies.

We also know little of the vegetation and fauna's capacity to resist a given dry episode and to regenerate itself. Here again, there are too many variables at work. Recent evolution, however, demonstrates that, in many cases, the flora of the southern Sahara, adapted to extreme conditions, has better resisted the droughts of the 1970s and 80s than the vegetation in the Sahel, which has at times been swept away, and suffers to this day from the (irreversible?) drop in the water table level. Where trees could reach the water table, the devastating effects of the drought were erased in a surprisingly short period of time.

In fact, we need to know the reaction time of water tables, and of the flora and fauna, to any given climatic event of average length (for example, no longer than 10 to 25 years). This information is crucial for a full understanding of the evolution of human settlements. A brief incident may well be absorbed with no major consequences, while a series of climatic crises spread over a generation is likely to push human groups into moving in search of less hostile conditions, or at least to modify significantly their mode of adaptation to the ecosystem.

THE EFFECT OF THE EVOLUTION OF THE SAHARAN MILIEU ON HUMAN SETTLEMENTS DURING THE NEOLITHIC

From 10,000 to 2000 bp, the Sahara and the Sahel hosted just about every type of milieu. The desert of Libya or the Egyptian Western Desert have never experienced the abundance once associated with the southern Sahara and, inversely, there was never a truly desert stage in the Sahel. Moreover, mountainous regions (the Ahaggar, Tibesti, Saharan foothills of the Atlas), the allogenous river valleys—perhaps even their southern tributaries, and Lake Chad have never really lacked water, and may have served as places of refuge—as remains the case to this day.

The Sahara as a mosaic consisted of cave paintings and traces of hippopotamus, of elephants, of rhinoceroses and fishes; of palaeolakes; of gazelles and ostriches; of empty ergs (the Tenerian culture is not rooted in the Ténéré *per se* but in its contours and the inselbergs scattering them—Kawar, Fachi, Termit); and, most of all, the Sahara of yellow and dusty grassy fields, of dried up wadis eleven months out of the year, of meager livestock and game animals haunting waterholes. A Sahara, then, comparable to the present-day Sahel, itself a mosaic of dry areas and green patches located around permanent sources of water, whether on the surface or just beneath.

It is in this milieu, oscillating between humidity and aridity, and most of all between semi-humidity and semi-aridity, that cultures have evolved throughout the eight centuries of the Holocene. Hunter-gatherers, fishermen, later herders, perhaps even farmers, still Epi-Palaeolithic or already Neolithic, inventing metallurgy, in Niger or elsewhere, during the second millennium BC. But also, leaving the Eastern Sahara, joining in the emergence of the Pharaonic civilization, from the fourth cataract to the delta of the Nile, and again, much later, creating the culture of Tichitt in Mauritania, and, during the following millennium, the historical kingdoms of Tekrour, Ghana and Kanem.

Very different peoples have participated in the adventure, coming from north and south, and most likely also from the east (the Sudanese Nile Valley). As suggested by the stone paintings of the Central Sahara, these populations succeeded one another for thousands of years, at times coming in contact with each other. They have, most of all, showed exceptional capacities of adaptation. It is thus crucial that we improve our understanding of their evolution. Indeed, just as we lack information on the duration and rhythm of climatic changes in the region, we remain unable to evaluate with satisfactory precision the ability demonstrated by Neolithic peoples to adapt to the environmental changes that took place.

Yet it is evident that humanity was subjected to environment constraints. In certain zones, particularly in the south, huge ensembles, at least partially sedentarized, have left traces of an intense occupation of space, sustained in some cases throughout the late Neolithic. Borders of mountain ranges and plateaus, river and lake shores, dune formations of the southern Sahara or the Atlantic coastline have, at times, been associated with enormous population densities, sharply contrasting with the present-day situation.

It seems, however, that the people of the Holocene had arrived in a Sahara virtually devoid of human life, while its border areas—North Africa, Nile, Sahel—were already inhabited. Will we be able to know with certainty how they arrived, and what modes of life they initially adopted?

As for later periods, one should seek to understand how people reacted to the aridity of 7000 bp, then to the progressive drying up of the Northern and Eastern Sahara—a process which took place during the centuries following 6000 bp, reaching an irreversible stage by the arid crisis of 4000 bp, and, finally, to the desertification of the southern Sahara.

Adaptation—with a certain lag—to the new conditions occurred in different ways, with a significant impact on population density:

(1) by sustaining activities while reducing their scope to the areas that remained humid (wells, seasonal nomadism, etc.; Trousset, 1985);
(2) by adopting a new calendar of activities, from hunting to herding, cattle to goats, herding to agriculture;
(3) by adding a new activity, aimed at increasing resources; and
(4) by operating a radical change in activity and therefore in modes of life: hunters becoming farmers, herders becoming warriors or caravaniers, herders or farmers returning to hunting, with all possible levels of overlap between nomadism and sedentary life.

Will we be able one day to do much more than observe the chronological coincidence, in the Central Sahara, between the development of herding and the arid phase of 7000 bp? Indeed, it is more or less with (or rather following?) this extremely dry event that man begins to conquer the Sahara and North Africa, in expanding circles. The same process seems to be at work, again, during the arid phase of 4000 bp, this time in the southern Sahara and in the Sahel.

Will we be able to make progress in our grasp of the relationship between the desertification of the northern and Eastern Sahara and the Palaeo-Berber progression toward the east (Nile Valley)? One thing that we do know, however, is that periods of major arid events, marking the boundaries of the middle Holocene, happen to coincide with the stages leading to Pharaonic Egypt (around 6000/5500 and 3500 bp; Hassan, 1988, this volume).

What were the consequences of the arid phase of 4000 bp on human settlement in the Sahara and the Sahel —and even in the Nile Valley? Here again, we do know at least that the wars of the pharaohs of the XVIII Dynasty against the 'Libyans' and the 'People of the Sea' (not to mention the Nubians and populations of the Eastern Sahara) coincided, around 3000 bp, with yet another climatic crisis.

Concerning the southern Sahara, stone paintings clearly indicate a process of cultural transition, with the evolution from a 'bovidean' imagery to an ensemble that is still pastoral, yet where cattle gradually become overshadowed by goats, as man draws well scenes and humans holding weapons—warriors as well as hunters, ancestors of the Toubous, of the Tuaregs and (in part) of the Maures. Ancient cultures disappear, new ones prosper. In this view, the example of the culture of Tichitt, in southeastern Mauritania, is remarkable. For some two thousand years, after 4000 bp, several hundred stone villages were built alongside a 500 km long cliff, around 18° latitude north—at the top or at the bottom, in a defensive or open position. The environment was rich: backwaters and waterholes at the bottom of the relief, pastures to the north as well as to the south. Man's activities included hunting, fishing, gathering, herding and agriculture, and the economy prospered. These populations had probably come from the north, pursued by the drought of 4000 bp.

They were forced to adapt once more, after 3000 bp, as the climate became dry again. In some cases, the process is defensive: control over narrows and springs, settlements around the main depressions, water stocking jars, fortified access to wells, etc. Other cases, which are not yet well enough documented (Ould Khattar, 1995), seem, however, to indicate the determination of villages to maintain agricultural activities despite increasing aridity: terraces are built, small dams and embankments aimed at containing water are erected. These techniques are used to this day by the Dogons, on the Malian cliff of Bandiagara, and there are other illustrations of such evolution, for instance in the Fayum, on the Saharan border of the Atlas chain, or in the Aïr.

By contrast, the last stages of the Tenerian culture were far more abrupt: the arid phase of 4000 bp had irreversibly broken the equilibrium and forced the populations of herders and hunters to flee the region, seeking refuge in the mountain ranges (Aïr, in particular), or pushing toward the south and the southwest (Termit and, most of all, Azawagh, where a hydrological network was maintained, mainly fed throughout the following millennium by the Ahaggar and Aïr ranges).

The end of the Neolithic in the Sahara which, despite the presence of metallurgy in certain regions, is at times effective around 2000 bp, is associated with the same mechanisms of adaptation by human groups: they either held on, resolving to modify the structure of their economy and thus their mode of life, or they chose to maintain their prior habits, thus migrating further south.

THE NORTHERN AND EASTERN SAHARA'S RETURN TO DESERT CONDITIONS

In the regions located north of the Tropic, the last Holocene moist climatic episode falls around 6500 bp. The period was favorable to human settlements, yet humanity had by now learned to live, even to prosper, under relatively hostile conditions. Given the technical means, people have an ability to practice herding, perhaps even agriculture, in certain zones; they have become able to live relatively at ease in hardly humid conditions (Figure 4.2). The notion of 'climatic optimum', used to refer to this period, is to be tempered, as shown by the examples of the Egyptian Western Desert, of Libya, or the Algerian Lower Sahara. Contrasts between north and south became acute, as the southern Sahara overcame the effects of several arid crises.

After 6000 bp, the levels of the Nile were no longer very high. After 5000 bp, flood levels dropped by a range of 25% to 30%, with yet another sharp decrease between 3900 and 3700 (see Hassan, 1986, 1997). There were still a few

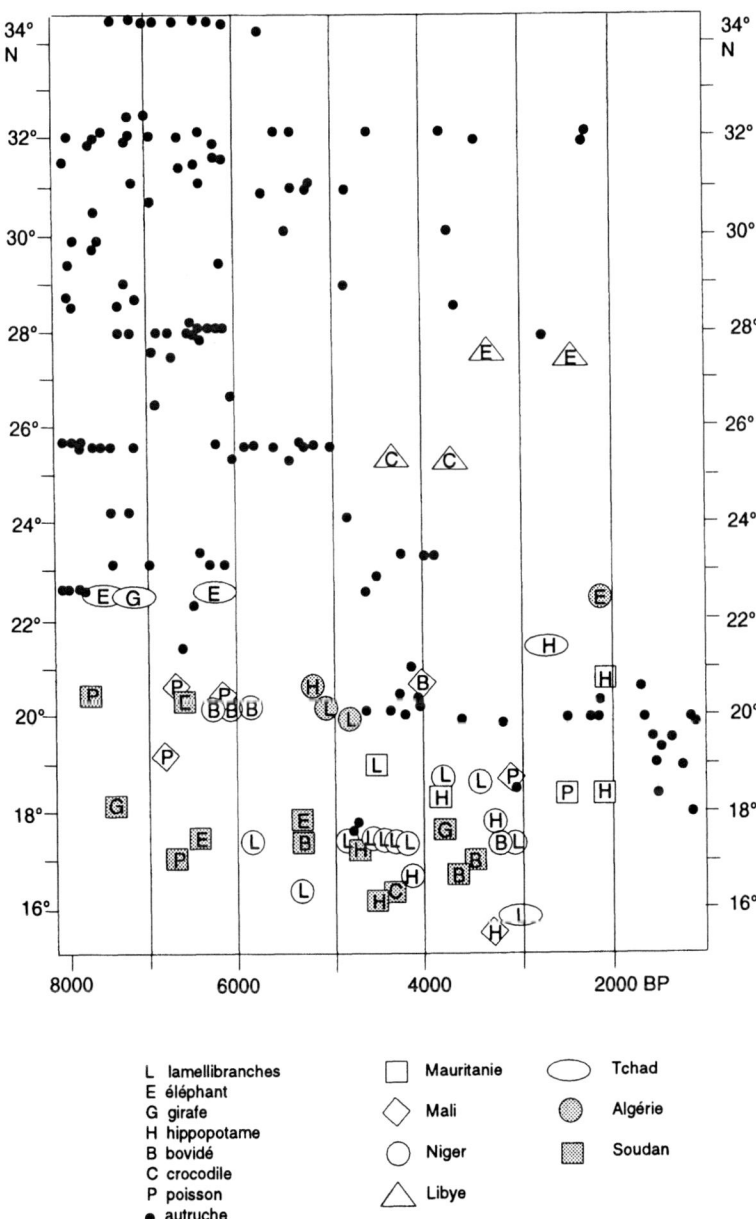

Figure 4.2. Radiocarbon and animal occurrences in the Holocene Sahara.

favorable periods during the second half of the millennium, and again around 2900 bp. Research in the Western Desert, and in particular the work of Neumann (1989, 1992), demonstrates that the region had not fallen below the level of semi-aridity (100 to 150 mm), while aridity never ceased to progress toward the south: 21°20' in 5000 bp; 20°30' in 3800; 19° in 3500; 17°30' in 1500 bp. The same observation is valid for the Libyan Sahara (see Barakat, this volume).

North of the Algerian Sahara, the middle Holocene Optimum was brief: already by 6000 bp climate deteriorated and surface water became saline (Callot, 1987). Aumassip's exhaustive survey of the region's northwest (1986) shows that, with the exception of a few privileged spots where surface water had not evaporated, the Holocene fauna remained for the most part limited to gazelles, ostriches and warthogs. Aridity was present as early as 4500 bp. From then on there were only short remissions, as in the case of the Great Western Erg around 3280-3080 and 2800 bp. These dates are consistent with what we know about regions of the Sahara located more to the south, where the phenomenon is significantly more marked.

The frontier seems to settle around 22°N, about 400 km north of the Sahara's present-day limit. The 'central mountain ranges' appear as areas of junction, northern slopes degrading at a higher pace. Rock art also indicates that, in high altitude and relief zones, relatively favorable conditions were maintained up to around 3500 bp in spite of the climatic crises.

EVOLUTION OF CLIMATE IN THE SAHARA AND THE SAHEL DURING THE LATE HOLOCENE

Prior to 4000 bp, rainfall conditions in the southern Sahara were globally favorable, in spite of a few drier episodes (around 6500 and 5500 bp, and especially after 4500 bp). Conditions may even have been 'too' favorable further south, as the main rivers and the Sahelian depressions were flooded under the abundance of water discharges coming at once from tropical zones such as the Fouta, and from the Sahara (Ahaggar, Adrar of the Ifoghas, Tibesti). Lake Chad, the Niger and the Senegal rivers and their tributaries reached dimensions and flood levels possibly unmatched since the appearance of man in the region. Lake Chad, at the time, filled not only its present-day bed, but also the huge depression of Borkou, at the foot of the Tibesti. The inner delta of the Niger, in Mali, presents itself as an immense basin of running water and swamps.

Man is not attracted to this overabundance of water, with its treacherous floods and dangerous fauna, including hippopotamuses, crocodiles, lions, elephants, as well as, insects and parasites carrying diseases (malaria, bilharzia, onchocercosis).

Much too humid, densely covered by trees and thus hard to penetrate, especially during the rainy season, infested with dangerous animals, these regions probably did not seem very attractive to man. The northern borders of the flooded plains, by contrast, appear favorable to human settlement. It is, however, the Saharan plains, still endowed with functional hydrological networks, that most attracted the people of the Neolithic.

The Arid Crisis around 4000 bp

Around this period, there was an acute dry episode in the Sahara. It was particularly marked in its southern part and in the Sahel, as suggested in a number of stratigraphies by the presence of a sandy layer. What is more, listings of radiocarbon age determinations invariably show a hiatus around this period. Finally, in most cases, northern Sahelian regions do not hold archaeological artifacts until the end of this arid period, which seems to have provoked radical changes in the organization of human settlement.

The Humid Period of the Fourth Millennium BC

This was the last Holocene Optimum. It was associated with an exceptionally high population density in the southern Sahara and the northern Sahel. For the last time, hydrological networks expanding from Saharan basins were functional (Gorgol, Karakoro, Azawad, Tilemsi, Azawagh, Wadi Howar). This optimum did not, however, reach the scope of those that preceded it, and climatic and ecological conditions progressively deteriorated.

Though illustrations may be extracted from other regions as well, cited here are a few examples from Sudan. Between 3800 and 3300 bp in Kerma (19°30'), statistics regarding the percentage of cattle and ovicaprines diverge radically (Figure 4.3; Chaix, 1994). The same observation is valid for Wadi Howar (17°30'), where present-day Sahelian formations were in place as early as the end of the second millennium BC. Prosperity was gradually reduced to the vicinity of Wadi Howar's permanent pools, where there were still giraffes at around 3800 and cattle towards the end of the millennium, confirming the progressive character of desertification, as well as suggesting that the vegetation cover remained dense in the wadi around 3000 bp, that water tables played an essential role, and that man was indeed able to adapt to relatively arid conditions (Van Neer and Uerpmann, 1989; Kröpelin, 1993). Keding (1997) confirms these findings for the site of Djabarona, located in the same area.

From 3000 onwards, precipitation became weaker and more violent, lakes and marshes surfaces began to evolve into sebkhas, the vegetation rapidly deteriorated and dunes started to reform. Rainfall in Tichitt, however, was probably about 200% of its present-day level around 4000 bp. By 2600 it had dropped to 125% of that level (Munson, 1981). The largest animals migrated to the south. A few remissions occurred, in particular around 2800 bp.

Southeast of Lake Chad, a certain degradation of climate was also at work around 2900 bp (Breunig et al., 1996), followed by a remission around 2750 bp. The presence of Prosopis africana—traces of which are dated to the same period by Rolando and Dupuy (1998) near Kayes (Mali)—indicates the progression of savanna over forests.

Between 2500 and 1600 bp, with some differences from one region to another as well as a few brief reversions, the long and globally positive episode of the Holocene came to an end. Landscapes began to resemble what we observe today, even though anthropic pressure obviously did not yet fully come into play.

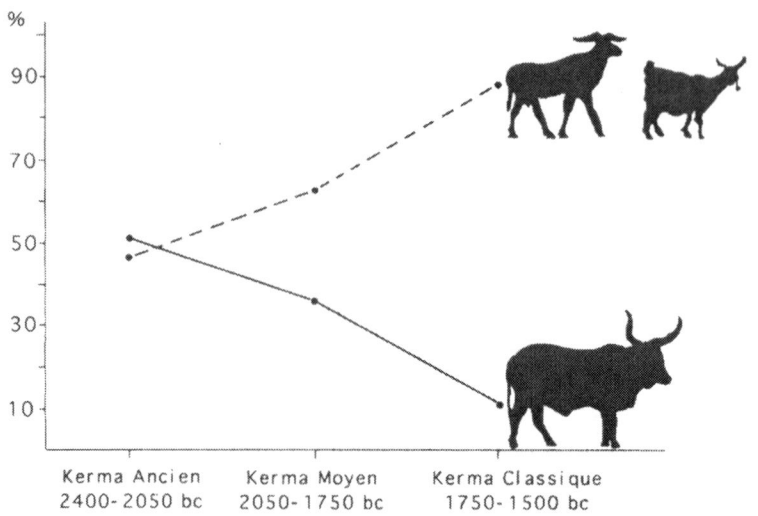

Figure 4.3. Diachronic evolution of economic stock species in Kerma (Sudan): bovines and ovicaprides (Chaix, 1994, p. 109).

"As expressed by their global character, affecting most parts of tropical Africa, these phenomena [regression of forests and erosion] are the result of an abrupt climatic change, in the sense of a climatic deterioration probably resulting from a more marked seasonality of rainfall, up into the equatorial zone" (Maley, 1996, p. 529; translation by author).

From 2500 bp on, aridity (associated with a marked north-south gradient) is quite clear in the southern Sahara as well as on the Nile, the latter reaching extremely low levels between 2500 and 2000 bp (Butzer, 1976). The same applies in West Africa, where maximum regression of the forest cover was reached during that period: savanna *graminaceae* then represented 30 to 40% of pollens, in a milieu where seasonal shifts had become more marked (Maley and Brenac, 1998). Deterioration of forests is just as visible in the Sahel, as observed in the case of the inner delta of the Niger, of the middle valley of the Senegal River (McIntosh, 1992; Bocoum *et al.*, 1995), or of southeastern Burkina Faso, where significant changes are recorded (Ballouche *et al.*, 1993).

In Senegal, a marked rupture is seen around 2000 bp. While vegetation typical of the Sudan was present from 4000 to 2000 bp, reinforced by the Guinea taxa, present-day semi-arid environments made their first appearance around 2000 bp. A biological threshold was reached, entailing the destruction of forests between 14° and 16°N, and the disappearance of species characteristic of the Sudan (Lezine, 1989a,b; Monteillet, 1988; Bocoum *et al.*, 1995). The same phenomenon occurred in the region of Lake Chad (Servant, 1973; Maley, 1981).

Further south, Lake Bosumtwi (Ghana) regressed between 2500 and 1800 bp (Maley, 1997). The same happened around 2000 bp in Lake Malha, in Sudan (Mees *et al.*, 1991), and the lakes of Ethiopia and Uganda (see Mohammed and Bonnefille, this volume).

A last phase, more humid than the present-day, occurred during the second half of the first millennium AD, playing a crucial role in the evolution of human settlement. Risks of arid crises had become less acute than during the preceding millennium, and the Sahel was a very favorable environment for agriculture and herding, with a noticeable increase in precipitation. The following developments are detected:

(1) from the seventh to the eleventh centuries, the inhabitants moved back to higher embankments in the Senegal Valley (Bocoum *et al.*, 1995);

(2) agriculture in the alluvial plain of the inner Niger delta was most pronounced around AD 1000. Human settlements diminished after that period (McIntosh, 1992);

(3) precipitation increased in the region of Timbuktu (Catella, 1988; McIntosh and McIntosh, 1986);

(4) evolution in the basin of the Chad during that period shows that the Bahr el Ghazal was functional from the third to the twelfth centuries, which means that rainfall was greater than in the present-day (Maley, 1981);

(5) Conrad (1969) recognized a humid phase in the middle of the first millennium AD in the Touat; and

(6) higher embankments are visible between AD 600 and 1000 in the Nile Valley (Butzer, 1980). The observation is confirmed in eastern Africa, where an optimum of precipitation in the upper basin of the Nile is observed during the second half of the first millennium AD (peaking around 1000), while lakes generally present high levels of water throughout the first millennium (Bonnefille, 1993).

This last humid period came to an end between the tenth and the thirteenth centuries, depending on the region.

EVOLUTION OF HUMAN SETTLEMENT

Between 7000 and 4000 bp, the Sahara was largely occupied by hunters and herders. Agriculture may have already been present in certain limited areas. The Sahel, by contrast, to our present knowledge (which, one has to agree, is quite fragmented) has very few Neolithic sites for that period. The beginning of the fifth millennium BC marks an irreversible rupture between the northern and southern Sahara. The number of radiocarbon age determinations is extremely low. However, the number of dates during the preceding millennium is significantly higher (Figures 4.4 and 4.5).

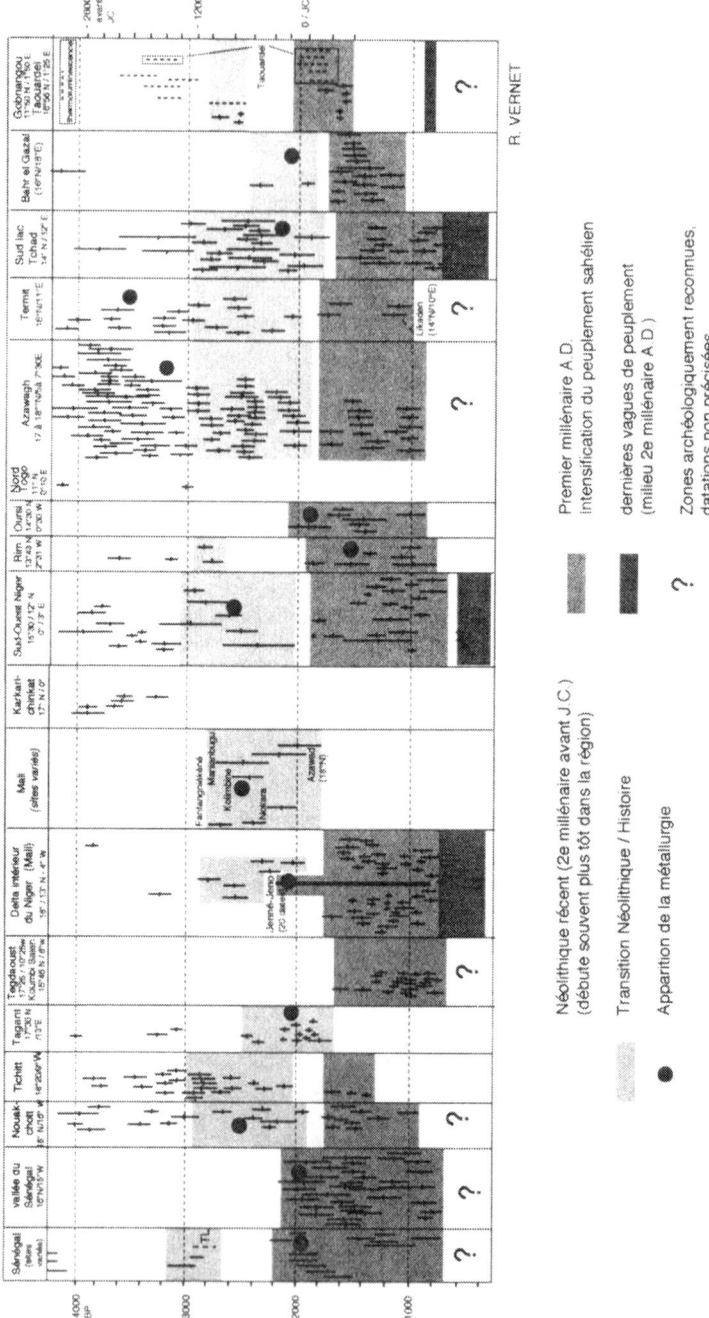

Figure 4.4. Human occupation in the south Sahara and north Sahel at the end of Neolithic and at the beginning of history, according to radiocarbon determinations (494 dates).

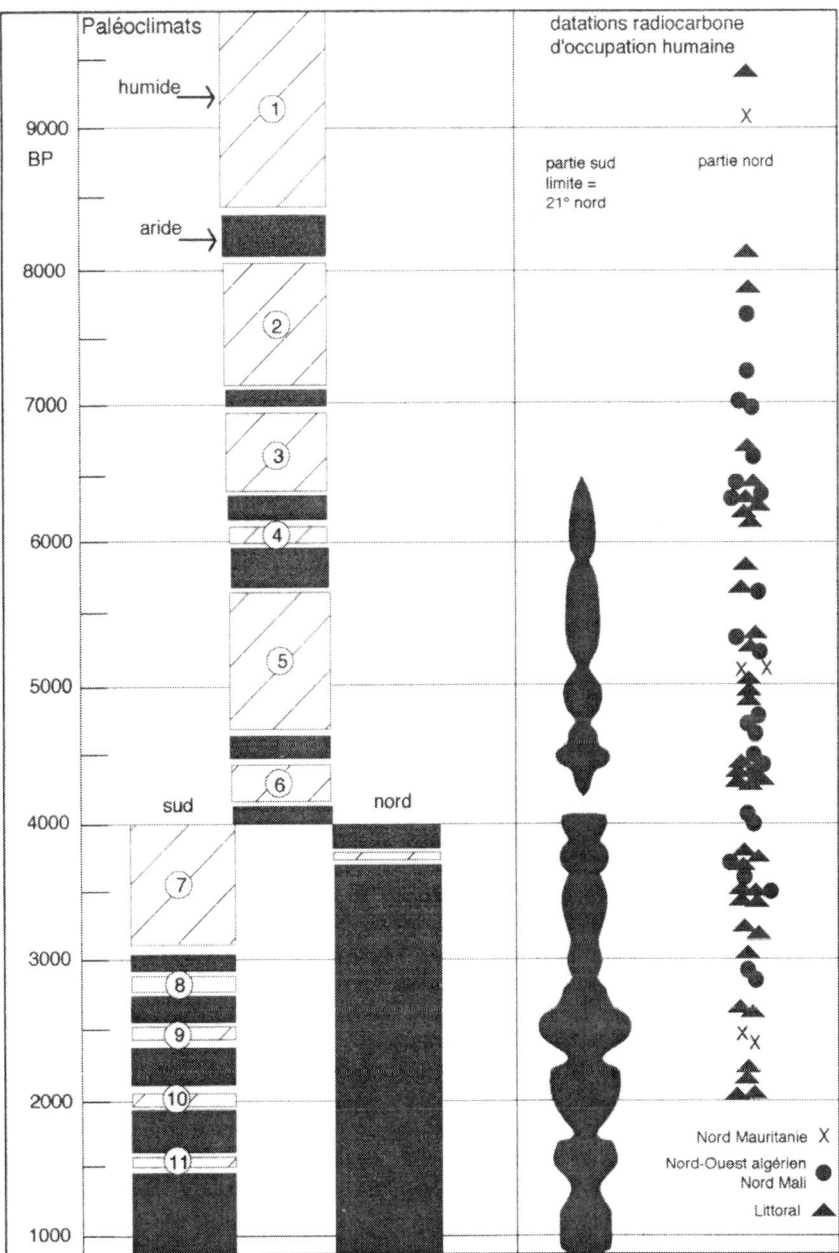

Figure 4.5. Palaeoenvironments and human occupation during the Holocene in the southwest Sahara.

At the end of the Neolithic, following the arid crisis of 4000 bp, we observe a progression of settlements following the isoyets, with an optimum of human occupation in the southern Sahara and the northern Sahel. Numerous groups settled south of the Sahara. They were apparently prosperous, practicing hunting, herding and agriculture. Fishing was still an activity in certain privileged zones (Tichitt's *bâten*, Azawad, Tilemsi, Azawagh, Lower Chad). The culture of Tichitt's stone villages, in Mauritania, along with the emergence, between 3500 and 2000 bp, in Niger and elsewhere, of iron metallurgy, are strong indicators of the evolution to come. Nomadism was common. However, sedentary life increases with demographic pressure (which reduces the size of settlement units), the gradual development of agriculture (millet, sorghum, rice, etc.), and the technical advances associated with metallurgy. Occupation of space probably remained discontinuous, for population density was low.

Though data regarding the Sahel is scarce, there are indications of the presence of herders and farmers along rivers. Here again, space does not seem to have been fully occupied. It is likely that the milieu remained too humid and too densely covered by forests for man to settle in it.

An arid episode around 3000 bp seriously affected human settlement in the southern Sahara. Populations sought refuge one or two degrees to the south, 18° of latitude north (even less in the east) becoming the new northern limit for sedentary settlement. The great Sahelian valleys were then occupied. Prior to that period, high flood levels had kept man at a prudent distance, as observed in the case of Djenne, near Niamey (Vernet, 1996).

Further north, nomadic herders were confined to areas where there were permanent sources of water (mountains, plateaus, valleys and depressions). There, human groups made use of the rainy season, which was getting shorter and shorter, settling in plains that were rich when it rained, but unattractive during the rest of the year for lack of surface (or near surface) water.

The migration movement is reinforced throughout the millennium, becoming irreversible after 2500 bp. The southern Sahara, which began to suffer from the lasting drought, was abandoned to the Berber nomads, who had by then acquired the dromedary yet were still able, around 2500 bp, to breed cattle (carved in stone alongside carts and horses) in the area of Bir Moghreïn, in the extreme north of Mauritania (25°N). In the southern Sahara, around valley oases in mountains and high plateaus, African populations of the Neolithic probably managed to maintain themselves. In the south, Sahelian populations settled in to their present-day geographical coordinates.

Around 2000 bp, at 18°N, copper metallurgy (Akjoujt, in Mauritania) or iron metallurgy (Tagant, Senegal Valley, Ighazer wan Agades, Termit) remained sufficiently important for us to believe that significant amounts of wood were still available. However, the ecological and human configuration of the southern Sahara and Sahel was not set until the first millennium BC, when the Sahara advanced up to 18°N.

During the second half of the first millennium AD, around the fifth century AD, an ultimate humid episode, lasting several hundred years, allowed Sahelian populations to move back toward the north, where they came in contact with the Berbers, and then with the Arabs. Core centers, however, remained located further south: Senegal River, the inner delta of the Niger, Gao, the southern border of Lake Chad. Different groups settled in the Tagant (18°) and the Adrar (21°); Songhaïs in the Ifoghas (20°); Haoussas in the Aïr (18°); Kanouris, Kanemis in the Kawar, the Djalo (18-20°) and as far as the Fezzan. Toubou groups—whose history remains obscure—moved back to the Tibesti, a region which they have never abandoned since. They, too, reached the Fezzan (Cuoq, 1984; Gado, 1980; Hamani, 1989).

During the same period, Berbers reached the Sahel, near the Atlantic, in the Hodh (Devisse, 1982), around the buckle of the Niger River, or the Nigerian Sahel. By the seventh century, Arab expeditions reached the Sahel. From the eighth century on, cities (city-states or capitals) were founded, in many cases by populations coming from the north as well as from the south. Tegdaoust (17°25'), Koumbi Saleh (15°46'), and Gao (16°) are among the most important of these. Others emerged in the Adrar of the Ifoghas, the Aïr or the Kawar, becoming important stopovers. Further to the east, fortresses and caravan relay points emerged far to the west of the Sudanese Nile.

At the beginning of the last millennium, during the eleventh century, a deterioration of climate probably took place. Between the twelfth and the fourteenth centuries, African groups began once more to move toward the south. Between 1100 and 1400, most sites of the region of Djenne were abandoned (McIntosh, 1993). In Niger, patterns of human settlement began to change during the fourteenth century, with the arrival of new ethno-cultural groups: Zarmas, arriving from the Azawagh (18°N), by now too arid for human settlement; Songhaïs, fleeing from the Adrar of the Ifoghas under the pressure of droughts; Tuaregs and other black ethnic groups. Mossi groups may also have arrived, a little earlier, from the Azawagh (Gado, 1980).

Mauritania was definitely abandoned to the Arabs and the Berbers at this time, Mali and Niger to the Tuaregs, and Chad to the Toubous. Traces of the patterns of human settlement during the first millennium are nonetheless visible. Human traces (the Kanouris of the Fachi or the Kawar), but also archaeological or ethnological traces (oral traditions associated with certain groups, for instance the Soninkes, the Toucouleurs, the Songhaïs, or the Haoussas). As shown through the work of Roset (1989) and Paris (1990) in the case of Iwelen, Haoussas are present in the Aïr up to the twelfth century, despite the fact that Berbers have by then been present for a long time.

Around 1000 bp (later as one moves toward the south), irreversible conditions of aridity set the frontier between the Sahara and sub-Sahara Africa. The limit is represented in the form of a line of contact between a nomadic and oasis dwelling world in the northern Sahara, and a sedentary world, mainly agricultural, located south the of the eighteenth parallel.

CONCLUSION

The onset of desert conditions in the Sahara began earlier in the north than in the south, where more favorable levels of precipitation led to the maintenance for a long time of northern Sahelian conditions. In order to understand the evolution of

Neolithic and post-Neolithic cultures, however, we need to gain a more precise knowledge of the geography and chronology of events.

In a context where water becomes a scarce resource, human groups have long attempted to postpone the inevitable, inventing new stock breeding techniques, developing new ways of using water (wells, hydro-agricultural innovations), metallurgy, or agriculture in the southern Sahara's mountain valleys and around Sahelian rivers where rice, for instance, was cultivated in the valley of the Niger prior to 2000 bp (McIntosh, 1993).

People have adapted, as long as it was possible and desirable. Yet, population movements toward the south are evident from 4000 bp on, intensifying a millennium later as levels of aridity become more acute. Many human groups, having reached the limits of their capacities to adapt, or wishing no longer to adapt to the new conditions, began, in some cases quite early, to migrate toward the south.

Relations between these different episodes are not well established. Many different groups may have joined in the process. What is certain, however, is that each arid phase has drawn herders and farmers of that time to seek refuge in the northern Sahel—between 20° and 18°N.

REFERENCES

Aumassip. G. (1986). *Le Bas Sahara dans la préhistoire*, CNRS, Paris.

Ballouche A., Küppers K., Neumann K., and Wotzka H.P. (1993). Apports de l'occupation humaine et de l'histoire de la végétation au cours de l'Holocène dans la région de la chaîne de Gobnangou, S.E. Burkina Faso. *Bericht des Sonderforschunsbereichs* 268, Bd 1, Frankfurt, pp. 13-31.

Bocoum H., McIntosh, S.K., and McIntosh, R.J. (1995). La réduction directe et ses rapports avec l'environnement dans la moyenne vallée du fleuve Sénégal, des origines au XVIe siècle. *Actes du coll. La Forge Catalane*, Barcelona, pp. 491-501.

Bonnefille R. (1993). Afrique, paléoclimats et déforestation. *Sécheresse* 4: 221-231.

Breunig, P., Neumann, K., and Van Neer, W. (1996). New research on the Holocene settlement and environment of the Chad Basin in Nigeria. *African Archaeological Review* 13(2): 111-145.

Butzer, K.W. (1976). *Early Hydraulic Civilization in Egypt: a Study in Cultural Ecology*, University of Chicago Press, Chicago.

Butzer, K.W. (1980). Pleistocene history of the Nile Valley in Egypt and lower Nubia. In Williams, M.A.J., and Faure, H (eds.), *The Sahara and the Nile*, Balkema, Rotterdam, pp. 253-280.

Callot, Y. (1987). *Géomorphologie des Paléoenvironnements de l'Atlas Saharien au Grand Erg Occidental. Dynamique Éolienne et Paléolacs Holocènes*, Thèse, Paris VI

Catella A.M. (1988). *Approche de l'Évolution du Milieu dans la Région de Tombouctou Depuis 10 Siècles*, Thèse, Aix-Marseille III

Chaix, L. (1994). Nouvelles données de l'archéozoologie au nord du Soudan. *Hommages à Jean Leclant. Vol. 2: Nubie, Soudan, Ethiopie*, IFAO 106/2, pp. 105-110

Conrad, G. (1969). *L'Évolution Continentale Post-Hercynienne du Sahara Algérien*, CNRS, Paris.

Cuoq J. (1984). *Histoire de l'Islamisation de l'Afrique de l'Ouest*, Geuthner, Paris.

Devisse, J. (1982). L'apport de l'archéologie à l'histoire de l'Afrique occidentale, entre le Ve et le XIIe siècle. *Comptes Rendus à l'Académie des Inscriptions et Belles Lettres*: 157-176

Gado, B. (1980). *Le Zarmaterey. Contribution à l'Étude des Populations d'entre Niger et Dalloll*, Mawri, Etudes Nigériennes, no. 45, Niamey.

Hamani, D. (1989). *Le Sultanat Touareg de l'Ayar*, Etudes Nigériennes, no. 55, IRSH, Niamey

Hassan, F. A. (1986). Holocene lakes and prehistoric settlements of the Western Faiyum. *Journal of Archaeological Science* 13: 483-501.

Hassan, F. A. (1988). The Predynastic of Egypt. *Journal of World Prehistory* 2: 123-185.

Hassan, F. A. (1997). Holocene palaeoclimates of Africa. *African Archaeological Review* **14**: 213-230.

Keding, B. (1997). *Djabarona 84/43*, Heinrich Barth Institut, Africa Praehistorica, 9, Köln

Kröpelin, S. (1993). Zur rekonstruktion der spätquartären umwelt am unteren Wadi Howar (Südöstliche Sahara/NW Sudan). *Berliner Geographische Abhandlungen*, 54, F.U., Berlin

Lezine, A.M. (1989a). Late quaternary vegetation and climate of the Sahel. *Quaternary Research* **32**: 317-334.

Lezine, A.M. (1989b). Le Sahel: 20.000 ans d'histoire de la végétation. *Bulletin de la Société Géologique de France* **1**(8): 35-42

Maley, J. (1981). *Etudes Palynologiques dans le Bassin du Tchad et Paléoclimatologie de l'Afrique Nord-tropicale de 30000 Ans à l'Époque Actuelle*. Travails et documents ORSTOM, no. 129, Paris.

Maley, J. (1996). Le cadre paléoenvironnemental des refuges forestiers africains: quelques données et hypothèses. In van der Maesen, L.J.G. (ed.), *The Biodiversity of African Plants*, Kluwer Academic, Netherlands, pp. 519-535.

Maley, J. (1997). Middle to late Holocene changes in tropical Africa and other continents: paleomonsoon and sea surface temperature variations. *NATO Advanced Study Institute* **149**: 611-640

Maley J., and Brenac, P. (1998). Vegetation dynamics, palaeoenvironments and climatic changes in the forests of western Cameroon during the last 28,000 years bp. *Review of Palaeobotany and Palynology* **99**: 157-187

McIntosh, R.J. (1992). Historical view of the semiarid tropics. *Carter Lecture Series*. Center for African Studies, University of Florida.

McIntosh, R.J. (1993). The pulse model: genesis and accommodation of specialization in the Middle Niger *Journal of African History* **34**: 181-220

McIntosh S.K., and McIntosh, R.J. (1986). Archeological reconnaissance in the region of Timbuktu, Mali. *National Geographic Research* **2**(3): 302-319

Mees F., Verschuren, D., Nijs, R., and Dumont, H.J. (1991). Holocene evolution of the crater lake at the Malha, northwest Sudan. *Journal of Palaeolimnology* **5**: 227-253.

Monteillet, J. (1988). *Environnements Sédimentaires et Paléoécologie du Delta du Sénégal au Quaternaire*, Thèse de doctorat d'Etat, Université de Perpignan.

Munson, P.J. (1981). A late Holocene (c.4500-2300 bp) climatic chronology for the southwestern Sahara. *Palaeoecology of Africa*. **13**: 53-60.

Neumann, K. (1989). Holocene vegetation of the Eastern Sahara: charcoal from prehistoric sites. *African Archaeological Review* **7**: 97-116

Neumann, K. (1992). Vegetationgeschichte der Ostsahara im Holozän. Holzkohlen aus prähistorischen fundstellen. In Kuper, R. (ed.), *Forschungen zur Umweltgeschichte der Ostsahara*, Heinrich-Barth Institut, Cologne, pp. 13-81.

Ould Khattar, M. (1995). *La Fin des Temps Préhistoriques dans le Sud-Est Mauritanien*, These, Université de Paris I.

Paris, F. (1990). Les sépultures monumentales d'Iwelen. *Journal des Africanistes* **60**(1): 47-74

Rolando, C., and Dupuy, C. (1998). Un four de métallurgie du fer en stratigraphie à Koussané (Mali): fouille, anthracologie et datations sur charbon. Pré-Actes du *3è Congrès International ¹⁴C et Archéologie*, Lyon, avril 1998, pp.168-169

Roset, J.P. (1989). Iwelen, site archéologique de l'époque des chars dans l'Aïr septentrional. *Libya Antiqua*, General History of Africa, Studies and Documents II, UNESCO, Paris, pp. 121-155

Servant, M. (1973). *Séquences Continentales et Variations Climatiques: Évolution du Bassin du Tchad au Cénozoïque Supérieur*, Thèse, ORSTOM, Paris

Trousset, P. (1985). Limes et "frontière climatique". *110e Congrès des Sociétés Savantes*, Montpellier. Colloque sur l'histoire et l'archéologie de l'Afrique du Nord, CTHS, 196, pp. 55-84

Van Neer, W., and Uerpmann, H.P. (1989). Palaeoecological significance of the Holocene faunal remains of the B.O.S. Missions *Africa*. *Praehistorica* **2**: 308-341.

Vernet, R. (1996). *Le Sud-Ouest du Niger, de la Préhistoire au Début de l'Histoire*, Etudes Nigériennes no. 56, IRSH-Sépia, Niamey-Paris.

5

LATE PLEISTOCENE AND HOLOCENE CLIMATIC CHANGES IN THE CENTRAL SAHARA. THE CASE STUDY OF THE SOUTHWESTERN FEZZAN, LIBYA

M. Cremaschi

GENERAL INFORMATION

Five field seasons of geoarchaeological survey in the Tadrart Acacus mountains, Messak Settafet Plateau, surrounding dune seas (Erg Uan Kasa, Edeyen of Murzuq) and adjoining Wadi Tanezzuft have yielded much palaeoclimatic information for the late Pleistocene and Holocene in a region hitherto poorly known from this point of view (Cremaschi and di Lernia 1998; Figure 5.1).

The area surveyed is roughly included in a frame of co-ordinates latitude North 24°30'-26°; longitude East 10°-12• . It has, currently, an hyperarid climate, with mean annual precipitation of 10 mm and mean annual temperature of 30°C (Walther and Lieth, 1960). It is composed of a wide range of physiographic features: the Tadrart Acacus consisting of a deeply dissected mountain range, the Messak Settafet of a relict plateau, the ergs Uan Kasa and Murzuq of sand seas and the Wadi Tanezzuft, a large wadi today temporarily active.

EVIDENCE OF CLIMATIC CHANGE INSIDE MOUNTAIN AREAS (TADRART ACACUS AND MESSAK SETTAFET)

Both in the Messak and in the Acacus areas, geomorphological, sedimentological and pedological evidence exists indicating a shift from dry to wet environmental conditions between the late Pleistocene and middle Holocene (Cremaschi, 1994, 1998). However, the most detailed information was obtained from the fill of the caves and shelters, which may act as reference sequences to integrate the proxies from other geomorphologic units of the landscape.

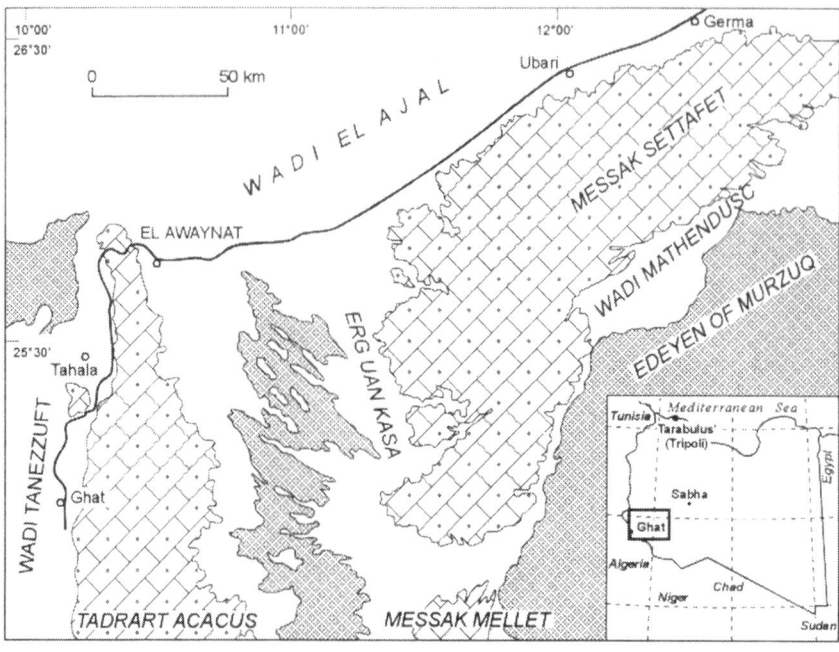

Figure 5.1. Location of the studied area. Mountain ranges and ergs (dotted areas).

A Dry Late Pleistocene

The oldest stratigraphic unit recovered to date in shelters and caves of the area (Figure 5.2) consists of reddish aeolian sand which has been observed at the base of the sequences of the Uan Afuda and Uan Tabu sites, and which can be correlated with remnants of fossil dunes locally existent outside the caves (Cremaschi, 1998). Both in the Uan Afuda and the Uan Tabu caves, this unit includes middle Palaeolithic artifacts which are attributed to the Aterian culture (Garcea, 1993; Cremaschi and di Lernia, 1995) and is dated using TL and OSL age determinations (Martini *et al.*, 1998) to the late Pleistocene: basal sand of Uan Tabu, 61,000 ± 10,000 years bp (OSL); basal sand of Uan Afuda (AF 1), 69,000 ± 7000 bp (OSL), and 70,500 ± 9500 - 73,000 ± 10,000 bp (TL); (AF 2), 90,000 ± 10,000 bp (OSL).

Late Pleistocene dunes have been observed also in the Messak Sattafet. At In Habeter (site MT22; Figure 5.3) they lie in a key stratigraphic position as they cover an Aterian living floor on top of fluviatile gravel.

These sand deposits have to be interpreted as evidence of a general desert expansion during late Pleistocene (stages 4-2); however, as indicated by collapsed blocks and artifacts inside the sequences of Uan Afuda and Uan Tabu, the climate may have had some moist oscillations during an arid interval.

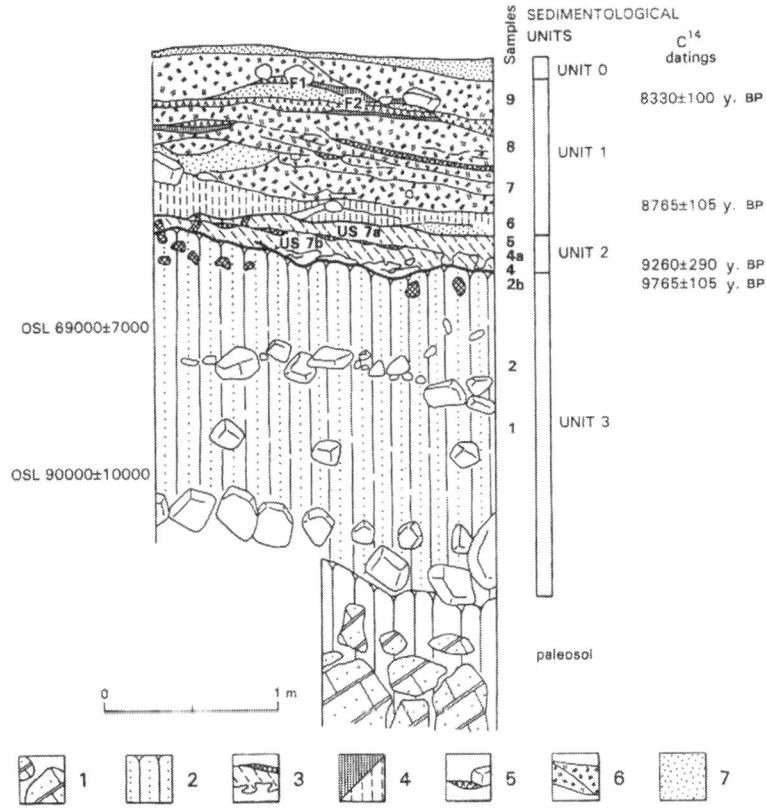

Figure 5.2. The stratigraphic sequence of Uan Afuda: 1. weathered bedrock and lower palaeosol; 2. aeolian sand and top soil; 3. Unit 2: colluvial deposits and gypsum lenses; 4. organic rich deposits; 5. hearth; 6. deposits with slightly decomposed plant remains; 7. present aeolian sand.

Terminal Pleistocene and Early Holocene Wet Conditions

Wet conditions in this period are recorded by the travertine deposits (Figure 5.4) which were observed in rock shelters of the Acacus in the form of flow stones, stalactites and stalagmites (Carrara *et al.*, 1998). They were deposited along the bedding planes of the sandstone bedrock as a consequence of the recharge of the hydrographic net inside the mountain and they implicate a high precipitation rate. A number of U/Th dates indicate that they were deposited from 15,600 to 9700 years bp.

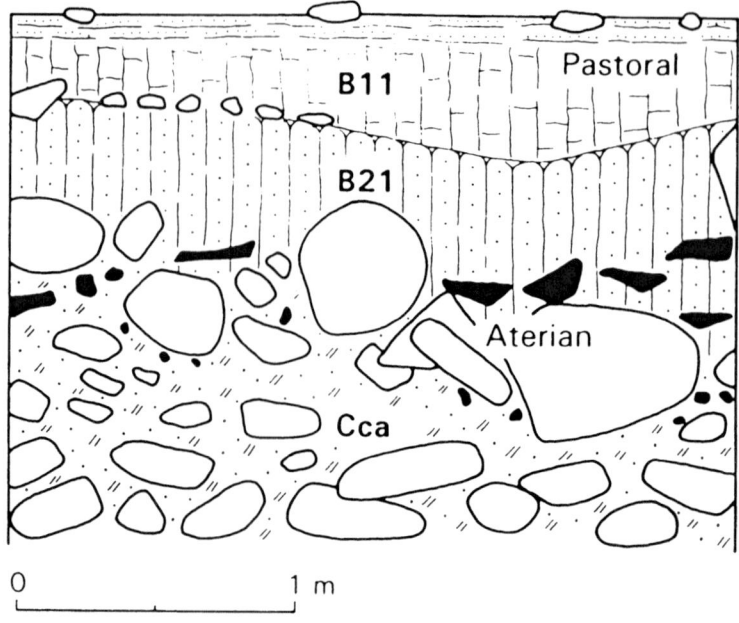

Figure 5.3. The stratigraphic sequence of In Habeter in the Messak Settafet. The fossil dune sands (B21) are superposed on alluvial gravel (Cca) and on an Aterian living floor (artifacts in black), and beneath colluvial deposits (B11) with pastoral artifacts.

Furthermore, palaeopedological evidence points to 'pluvial' conditions in this period: late Pleistocene dunes and wadi deposits have been affected by weathering which led to the formation of soil, the main pedological features of which are slight rubification and clay translocation, requiring seasonality in precipitation (Fedoroff and Courty, 1989; Cremaschi and Trombino, 1998). According to the stratigraphic sequence of Uan Afuda, in which this type of soil was observed at the top of Unit 3, the development of the soil should have been accomplished before 9765 ± 105 yr bp, the date for the basal anthropogenic deposit of Unit 2.

Two Wet-Dry Cycles During the Early Middle Holocene

Anthropogenic deposits in the cavities and rock shelters are mainly constituted of plant fragments, coprolites, finely subdivided organic matter in different stages of

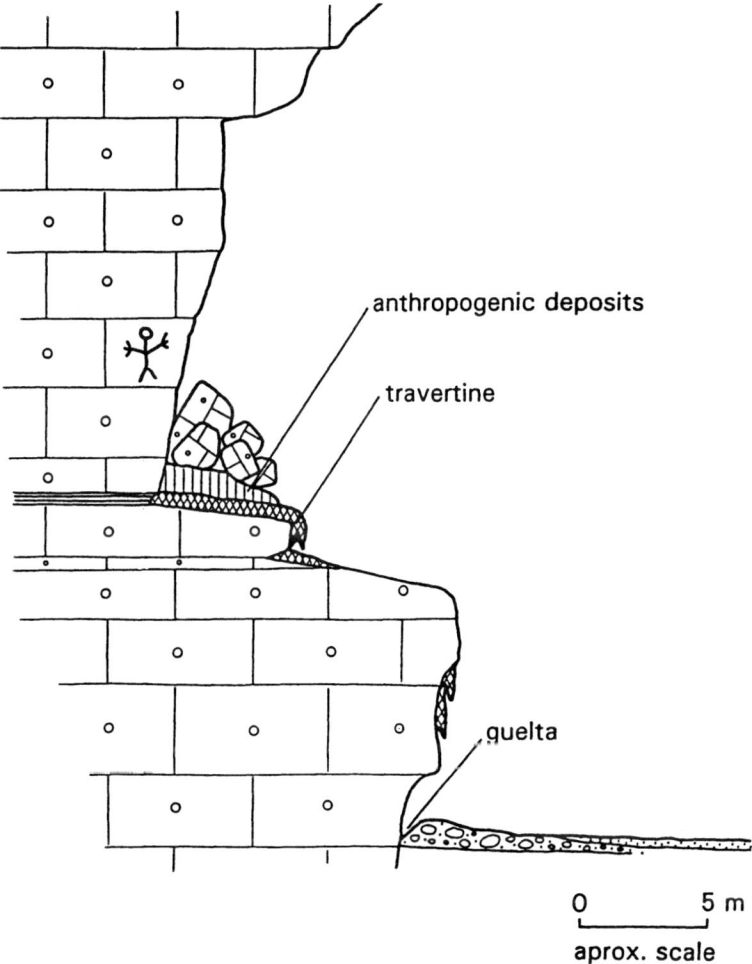

anthropogenic deposits

travertine

quelta

0   5 m

aprox. scale

Figure 5.4. The travertine and the anthropogenic deposits at site 96/50, Wadi Tanshalt.

humification, ash, charcoal, and to a lesser extent chemical precipitation and wind blown sand (Cremaschi and di Lernia, 1995, 1998). The main processes responsible for their formation are the artificial accumulation of fodder, mainly grass and plant fragments, and its transformation through fire, trampling and organic material addition, due to the use of the cave as a penning site and for human dwelling. Also, the climate played a primary role during the sedimentation as it controlled the degree of humification of the organic matter, the precipitation of solutes from percolating water and the input of aeolian sand by wind.

Although the anthropogenic deposits may differ slightly in architectural features related to the local morphology of the cave or shelter, they have rather similar lithostratigraphic characteristics and a similar trend in sedimentation has been observed in most cases. The sequences observed at Uan Tabu, Uan Afuda, Uan Muhuggiag and Uan Telocat, in the Acacus mountains (Pasa and Pasa Durante, 1962; Barich, 1987; Garcea, 1993; Cremaschi and di Lernia, 1995; Cremaschi, 1998) in the Mathendusc cave, and in the Messak Settafet (Trevisan Grandi *et al.*, 1995; Cremaschi, 1994, 1996) can be regarded as representative of more than one hundred cavities and shelters located and sounded during the survey (Cremaschi and di Lernia, 1998).

On the basis of radiocarbon age determination, sedimentological and micromorphological characteristics, two cycles of anthropogenic sediment accumulation can be distinguished (Cremaschi *et al.*, 1996). Each cycle starts from conditions of relative wetness and evolves toward aridity.

The first cycle has been recorded in the caves of Uan Afuda and Uan Tabu (Figure 5.5). It begins at 9765 ± 105 yr bp (Uan Afuda, Unit 2) with colluvial sediments (Uan Afuda Unit 2, Uan Tabu Unit 3), while upper sediments containing a large quantity of undecomposed plant fragments indicate drier conditions (Uan Afuda Unit 1; Uan Tabu Units 2 and 1). The dung at the top of the sequence of Uan Afuda, dated to 8000 ± 110 bp, may represent the terminal deposit of this cycle in an arid environment (Trevisan Grandi *et al.*, 1992; Cremaschi and di Lernia, 1995; Castelletti and Cottini, 1996).

The second cycle begins at about 7200 yr bp. Dates for its beginning are: Uan Muhuggiag ext. 7200 ± 210 yr bp, Uan Muhuggiag shelter (trench B) 6900 ± 220 yr bp, Uan Telocat 6745 ± 175 yr bp and Mathendusc cave 6825 ± 90 yr bp (Cremaschi and di Lernia, 1998), and goes up to 3770 ± 200 yr bp (dung at the top of the Uan Muhuggiag sequence.

The Uan Muhuggiag deposit may be regarded as a reference sequence (Figure 5.6) for this cycle, and three different units can be distinguished in it, according to their sedimentary characteristics, which indicate a progressive shift from wet to dry conditions.

The base unit consists of dark gray sand, rich in humified organic matter with gypsum and carbonate concretions but few preserved plant remains, which was deposited in a wet environment.

The second unit has less humified organic matter, the fragments of undecomposed grass and plant are much better preserved, indicating drier conditions. In Uan Muhuggiag, as in many cases, an erosive unconformity delimits its top. Radiocarbon dates of this eventspan from 6480 ± 50 to 5585 ± 195 yr bp.

The third unit consists of consolidated laminated dung, the organic matter of which did not suffer any bacterial degradation and was deposited in a very dry environment. Thin aeolian sand layers are intercalated within the dung indicating both discontinuous frequentation and aridity.

Pollen analysis from many sites of the Acacus and of the Messak further supports climatic interpretation of the sediments, as it reveals *Typha* and Cyperaceae in lower units, while *Pocaceae*, including *Acacia*, Capparaceae and

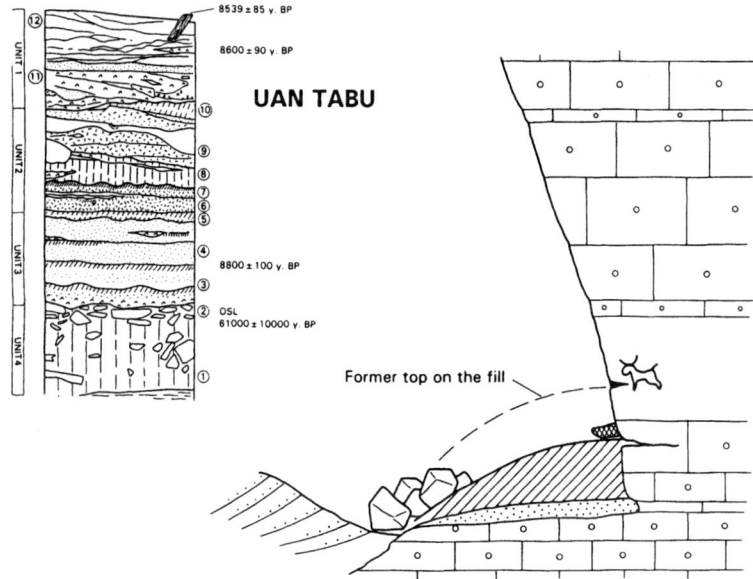

Figure 5.5. The first cycle of the rock shelter's fill: the Uan Tabu sequence. Unit 4 is composed of reddish sand with Aterian artifacts; Unit 3 is composed of colluvial sand and layers of plant remains; Unit 2 is mainly composed of layer of undecomposed plant remains; Unit 1 is mainly composed of thick ash layers.

Artemisia, indicative of dry steppe, are dominant in the upper ones (Trevisan Grandi *et al.*, 1992, 1998; Mercuri *et al.*, 1988).

Inside Unit 2 of the Uan Muhuggiag sequence (pollen zone UM2a in Mercuri *et al.*, 1998) a peak of *Panicum* dated < 6690 ± 130 and > 6035 ± 100 yr bp is indicative of a dry spell. Geological evidence of this episode is rare, but it may be correlated with the roof collapse and aeolian sand ingression represented in layer III of Ti-n-Torha North (Barich, 1987). Furthermore this period is concomitant with a main break from 6410 to 6082 yr bp in the radiocarbon dating curve of the studied area (see Figure 5.11). Due to the large number of dates involved (more than one hundred, Cremaschi and di Lernia, 1998) this gap cannot be considered a random bias, but could be related to a reduction in human occupation and to a short dry period.

The onset of severe dry conditions at about 5000 years bp is indicated by other proxies:the collapse of the roofs of a large number of shelters in the Acacus, which is indicative of enhanced thermal contrast on cliffs (Cremaschi, 1998) due to the destruction of vegetal cover and to the lack of protective films of water and the end of the deposition of the desert varnish in the Messak (Cremaschi, 1996), which developed mainly during the sixth millennium bp in a semi-arid climate.

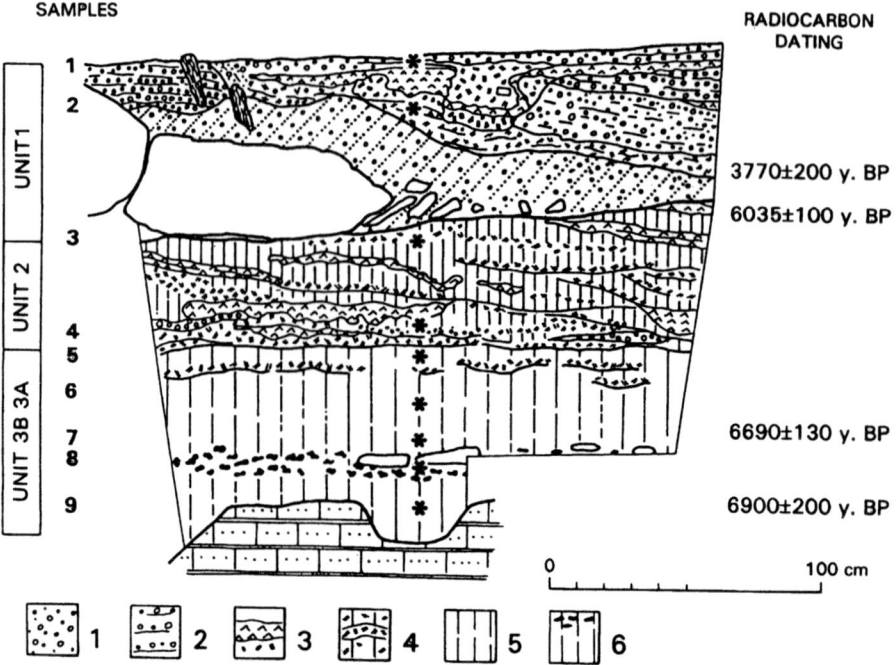

SAMPLES

RADIOCARBON
DATING

3770±200 y. BP

6035±100 y. BP

6690±130 y. BP

6900±200 y. BP

0   100 cm

Figure 5.6. The second cycle of the rock shelter's fill: the stratigraphic sequence of Uan Muhuggiag: 1. loose aeolian sand; 2. slightly cemented sand and coprolites: dung; 3. hearths; 4. hearths in slightly humified organic sand and lenses of fresh plant remains; 5. humified organic sand; 6. strongly humified organic sand and gypsum concretions.

EVIDENCE OF CLIMATIC CHANGES FROM DRIED LAKE DEPOSITS INSIDE THE ERGS

Landforms and sediments related to Holocene dried lakes and ponds have been identified both in the Erg Uan Kasa and in the northeastern part of the Edeyen of Murzuq (Cremaschi, 1994, 1998). The deposits are strongly affected by wind erosion and they represent only a small part of the extent of lakes or swamps which, during the early and middle Holocene, may have covered about one third of the whole area. The lakes were formed mostly as a consequence of the rise of the water table inside the ergs and can be compared to those indicated by Pachur and Hoelzmann (1991) as piezometre lakes. Lake deposits consist of bioturbated black silty sand, including vegetal fibers, indicating shallow water, very dark bioturbated organic sand silt and hydromorphic silty sand of shore facies; furthermore, laminated calcareous silt, including mollusc shells, are indicative of full lacustrine sedimentation. The lacustrine sequences are often sealed by a thick gypsum crust indicating the drying of the water body and a shift to sebkha conditions. From an environmental point of view, the mollusc fauna is rather homogeneous (Girod, 1998). It is mainly composed of a few species (*Biomphalaria pfeifferi, Lymnaea natalensis, Bulinus truncatus, Afrogyrus oasiensis, Segmentorbis angustus*) indicating permanent shallow and closed lakes with fresh or hypohaline water.

Holocene archaeological sites are systematically connected to the lake basins (Cremaschi and di Lernia, 1998). Epipalaeolithic sites were found systematically distributed along the shores, but buried by early-middle Holocene organic deposits. Hundreds of pastoral sites occur on the dune ridges along the higher reaches of the lacustrine shores.

Two Cycles of Lacustrine Sedimentation in the Murzuq Edeyen

Lacustrine sequences occur inside the large interdune corridors in the northwest part of the Murzuq sand sea. They are typically terraced, and at least three orders of terraces can be distinguished. Archaeological evidence indicates that the upper two orders, whose deposits include Acheulean artifacts, date back to the middle Pleistocene. Holocene deposits occur in deflation hollows in the lower parts of the lowest terrace and in the most depressed part of the corridors.

The stratigraphic section indicated in Figure 5.7 refers to an interdune corridor, at around the point of the co-ordinates 25°53'38"N, 12°48'23"E, and it can be regarded as typical of the area. The lacustrine sedimentation may be subdivided into two cycles. The first consists of a basal peat covering an Epipalaeolithic site (Cremaschi and di Lernia, 1998), radiocarbon dated to 8445 ± 160 yr bp. The peat is covered by laminated lacustrine calcareous silt which indicates a rise in water level. It is followed by an organic layer indicative of a later drop in the level of the lake. A mollusc shell included in it dates to 7390 ± 105 yr bp. An erosion phase concludes the first cycle. The base sediments of the subsequent cycle were found inside a blow-out cavity and consist of deposits of shallow water radiocarbon dated to 7325 ± 130 yr bp.

Littoral organic mud—dated to 6665 ± 100 yr bp—indicates that the highest level of the lake was reached during the second cycle. At this time large pastoral sites were connected to the lake shore.

The period of the desiccation of the lakes has not been directly dated up to now, but occurrence of late pastoral sites inside the previously deposited littoral deposits of the lake suggests that the lake level was dropping at 5500 yr bp and the most recent date for the pastoral site in the area is (site MT113) 5220 ± 100 yr bp.

Swamps and Lakes of the Erg Uan Kasa

At the eastern margin of the erg Uan Kasa some lacustrine basins were formed at the mouth of the wadis descending from the Acacus mountains as a result of the damming effect caused by the large dune of the erg. These basins develop on the front of the dunes and extend along the wadis cut into the bedrock. The fill of

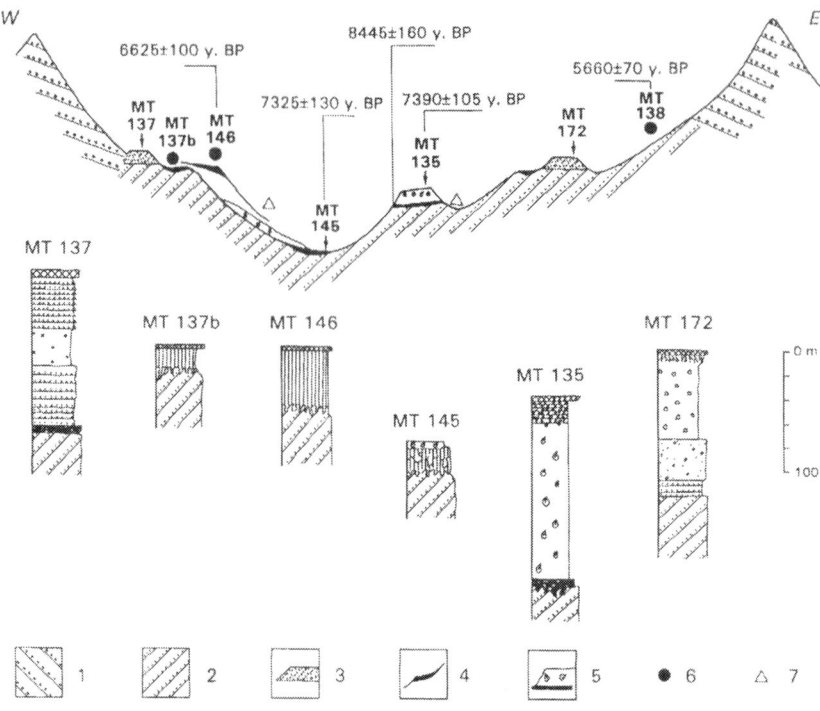

Figure 5.7. Schematic stratigraphic sequence of a lacustrine basin in the Murzuq Edeyen: 1. dune ridge; 2. bleached sand; 3. upper terrace; 4. organic mud interpreted as shore deposits; 5. calcareous silt including mollusc shell and base peat; 6. Pastoral sites; 7. Epipalaeolithic sites.

these basins is mainly composed of terrigenous deposits: at the base it consists of massive and planar laminated mottled green sand, with strong bioturbation in the form of root tubules, covered by a gypsum and sodium carbonate crust. In some cases, at the top of the green sand, black massive silty sand occurs rich in organic matter covered, in turn, by a thick gypsum crust. The organic layer which indicates the phase of drying of the lakes is radiocarbon dated to 7515 ± 210 yr bp, therefore correlating with the first cycle of the Murzuq lake sedimentation (Figure 5.7).

Inside the dune corridors, lake basins occur in closed depressions, delimited by bedrock irregularities, by middle Pleistocene alluvial terraces and by the longitudinal dunes which border the corridors. They range in length from some hundred meters to a few kilometers and their bedrock is bleached, due to persistent standing of a water table.

The stratigraphic sequence observed at the point of co-ordinates 24°09' 09"N, 10°54'06"E can be regarded as typical for the Erg Uan Kasa (Figure 5.8). At the base of the sequence there is a thin layer of green hydromorphic sand covered by organic deposits. White calcareous silt is best represented in the middle of the lake, while organic sandy silts occur at the basin margins, and are transgressive at the top of the sequence. Shore ridges delimit the margin of the lake. The lacustrine sequence is sealed by a crust of gypsum and sodium carbonate (about 10 cm thick) which covers the lacustrine deposits. Radiocarbon dating correlates Uan Kasa deposits to the second lacustrine cycle as observed in the Murzuq Edeyen. The organic basal layer yielded an age of 7250 ± 220 yr bp; a thin organic layer at the top of the calcareous mud gave 6520 ± 200 yr bp; and the organic humic acid of the shore 6005 ± 90 yr bp which should indicate the highest level of the lake or the beginning of its retreat. A date of 2375 ± 110 yr bp for *Acacia* roots penetrating the gypsum crust at the top of the sequence is the *ante quem* date for the complete drying out of the area.

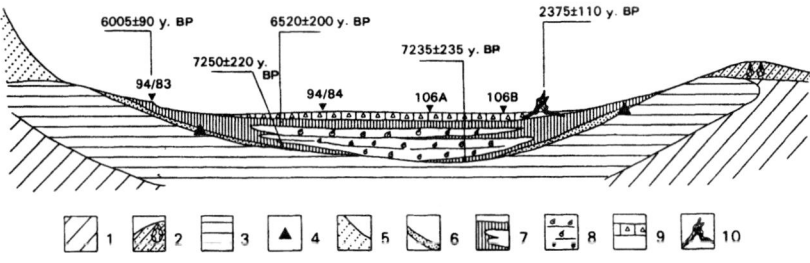

Figure 5.8. Schematic stratigraphic sequence of a lacustrine basin in the Erg Uan Kasa: 1. bedrock; 2. fossil dune including Acheulean artifacts; 3. bleached bedrock; 4. Epipalaeolithic artifacts; 5. sand ridge; 6. green sand; 7. organic mud; 8. calcareous silt, including mollusc shell; 9. gypsum crust; 10. Acacia tree.

Fluvial Activity in the Wadi Tanezzuft Valley

The Wadi Tanezzuft runs parallel to the western fringe of the Acacus. It sporadically receives water during major precipitation events. Along its course are located the present oases of Ghat and Tahala. Inside its valley, below a discontinuous sand veneer, extensive fluviatile deposits exist which are to be related to strong water discharge of the wadi during the Holocene. Gravel bars have been observed, but silt and sand overbank deposits dominate. They occur in wide lenticular sheets and are separated by poorly developed alluvial soil. Some hundreds of pastoral and late-pastoral sites have been found on the surface or buried in these deposits and research on them is still in progress (Cremaschi and di Lernia, 1998). Radiocarbon dating indicates that fluvial sedimentation was fully active from 4200 ± 180 yr bp to 2775 ± 125 yr bp, but the first evidence of desert conditions, indicated by phytogenic dunes lying on the alluvial plain, is much later and has been dated to 615 ± 45 yr bp (Figure 5.9).

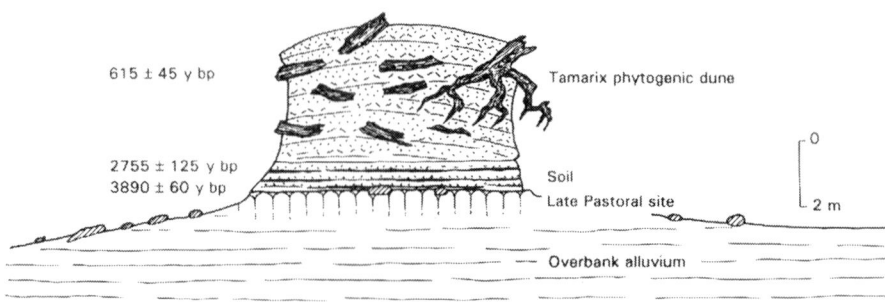

Figure 5.9. Stratigraphy of a phytogenic dune at the site TAH4 along Wadi Tanezzuft at Tahala. The phytogenic dune including *Tamarix* wood covers a soil, at the top of the Holocene alluvium. Inside the soil a pastoral site is buried.

The Wadi Tanezzuft was still active for millennia after the surrounding areas were dry. For this reason, the alluvial plain of the wadi during the fourth and third millennia bp was settled by the latest pastoral communities which experienced some forms of sedentism, and during classical and historical periods a main caravan route was established along it.

DISCUSSION AND CONCLUSIONS

The stratigraphic record for the late Pleistocene is quite fragmentary. However, remnants of fossil dunes indicate desert conditions inside the Tadrart Acacus Mountain contemporary to Mousterian and Aterian communities.

Evidence for a wet final late Pleistocene and early Holocene is indicated by travertine sedimentation, by enhanced fluvial activity and by rubification and clay translocation in soils developed at this time (Figure 5.10). Since no travertine was deposited after 9000 bp, this period should be regarded as the beginning of decreasing wetness in the mountain range.

In the erg areas, the rise of the water table and the consequent formation of lakes is delayed in comparison with the onset of the wet period in the mountain range, as probably the recharge of the erg acquifers lasted longer. Lakes and ponds were certainly present from 8500 bp up to about 5500 bp. A first cycle of high water level lakes can be dated to the second half of the ninth

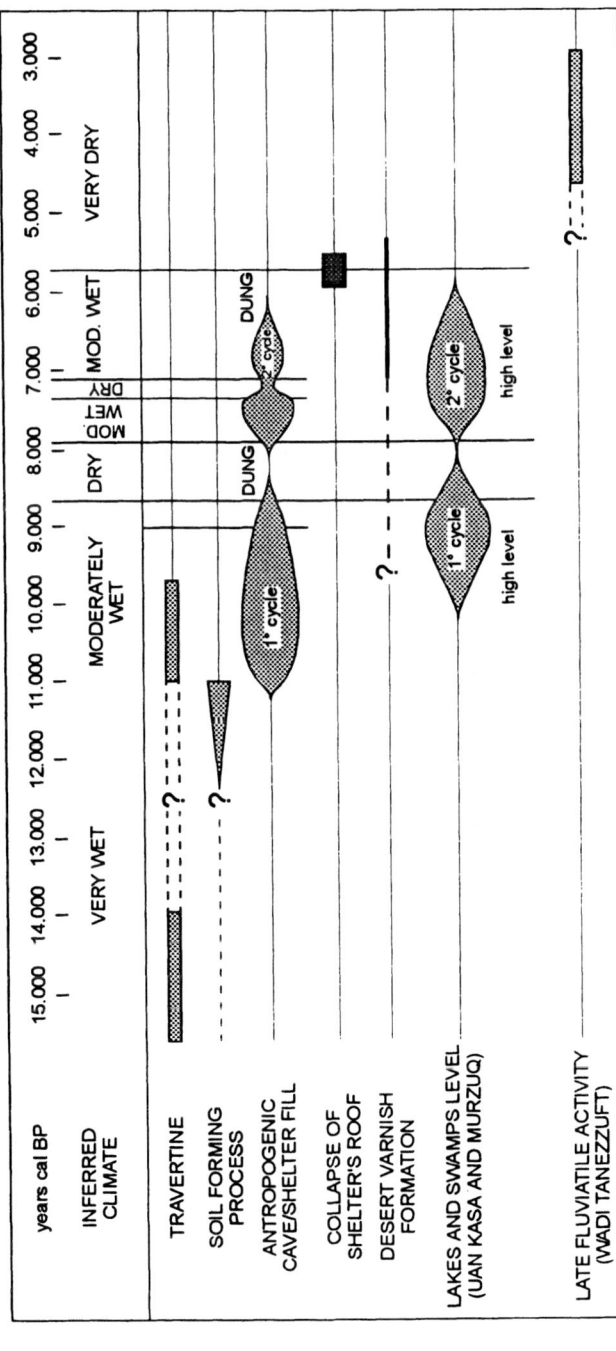

Figure 5.10. The climate changes as reconstructed by geological proxy.

millennium bp. It is partially contemporaneous with the older phase of deposition of the anthropogenic fill in the caves of the Tadrart Acacus, which is also indicative of wet conditions.

Both lakes and shelter fills record dry conditions during the first half of the eighth millennium bp (after 8000 bp before 7500 bp). If considered in calibrated age, (after 8990-8610 cal BP, before 8370-8140 cal BP), this period overlaps to some extent with the dry event dated 8.4-8.0 cal BP (Figure 5.11), recorded by windblown chemical indicators together with a significant decrease in methane in the Greenland ice-core proxy. It may be interpreted as the local effect of the 'middle Holocene' dry period (see Muzzolini, 1993), and correlated with dry conditions in broad monsoon regions (Alley *et al.*, 1997; Lamb *et al.*, 1995).

The second half of the eighth and the seventh millennia bp is a period of high water level in the lakes and of a wet climate in the mountain range, as indicated by the humified deposits inside caves and shelters, further supported by the pollen record which indicated wet savanna vegetation. A possible dry spell may have interrupted this period at 6410-6082 yr bp (7250-6950 years cal BP), indicating a tendency towards increasing aridity. The onset of dry conditions seems to have been a dramatic event which happened around 5000 bp (5750 years cal BP). To the contrary, hydrological activity and sedimentation of overbank deposits continued for at least two millennia in the alluvial valley of Wadi Tanezzuft.

The result obtained confirms a scenario of wet savanna landscape in the mountain and of a continuous belt of pond and lakes in the ergs during the terminal Pleistocene and early Holocene, further supporting the idea of dissolution of the desert during the Holocene 'Interglacial'. Following this period, the environment experienced a progressive decline from wet conditions toward the extreme dry conditions at 5000 bp. However, this trend was not consistent, as it was interrupted by at least two dry spells which accelerated the onset of aridity.

In comparison with surrounding areas of the Saharan belt (Gasse *et al.*, 1987; Kröpelin, 1987; Petit-Maire and Risier, 1981; Lezine *et al.*, 1990; Pachur and Wunneman, 1997; Hassan, 1997) the trend of climatic change appears quite similar but the timing is different to some extent (Cremaschi, 1998). The peak of precipitation preceded the rise of the Malian lake by about a thousand years and the onset of wet conditions in western Nubia and in the Egyptian desert. It is contemporaneous with the rise of the Ethiopian lakes, particularly the lake Afar and those of western Nubia (Meibob Hills). The onset of dry conditions took place more than a thousand years earlier than in the southern Malian belt and only about five hundred years before the reduction of the Ethiopian lakes.

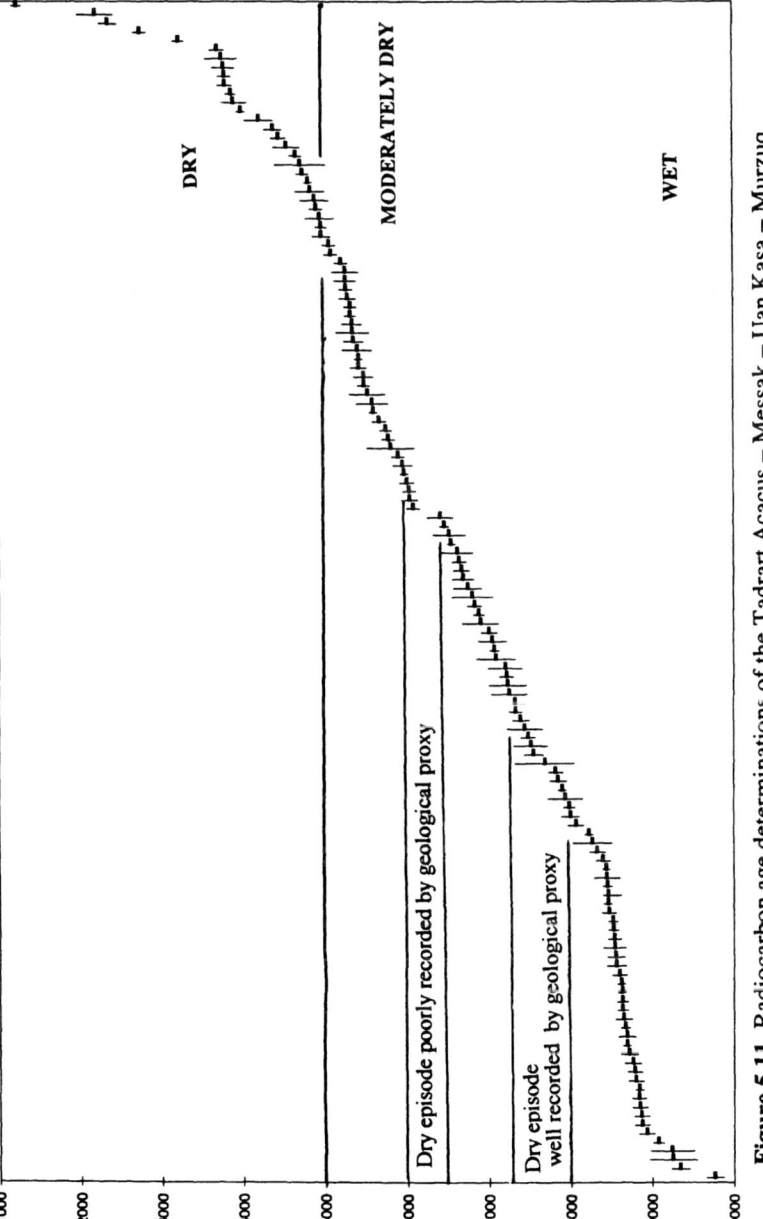

Figure 5.11. Radiocarbon age determinations of the Tadrart Acacus – Messak – Uan Kasa – Murzuq.

REFERENCES

Alley, R.B., Mayewski, P.A., Sowerst, A, Stuiver, M., Taylor, K.C., and Clarl, P.U. (1997). Holocene climatic instability. A prominent widespread event 8200 yr ago. *Geology* 25(6): 483-486.

Barich, B.E. (ed.) (1987). *Archaeology and Environment in the Libyan Sahara. The Excavations in the Tadrart Acacus, 1978 - 1983*, British Archaeological Reports International Series 368, Oxford.

Carrara, C., Cremaschi, M., and Quinif, Y. (1998). The travertine in the Tadrart Acacus mountains. In Cremaschi, M., and di Lernia, S. (eds.), *Wadi Teshuinat. Palaeoenvironment and Prehistory in South-western Fezzan (Libyan Sahara)*, Quaderni di Geodinamica Alpina e Quaternaria 7, C.N.R, Milan, pp. 59-68.

Castelletti, L., and Cottini, M. (1996). From occasional to systematic gathering: evidence from the early Holocene wooden flora at Uan Afuda cave (Tadrart Acacus, Libya). In di Lernia, S., and Manzi, G. (eds.), *Before Food Production in North Africa. Proceedings of the Homonymous Workshop held in Forli, September 1996, within the XIII World Congress of the International Union of the Prehistoric and Protohistoric Sciences*, Abstracts 2, Workshop 14-15-16, ABACO Edizioni, Rome, pp. 149-150.

Cremaschi, M. (1994). Le Paleo-Environment du Tertiaire tardif a l'Holocene. In Van Albada, A., and Van Albada, A.M. (eds.), *Art Rupestre du Sahara*, Les Dossiers d'Archeologie 197, pp. 4-14.

Cremaschi, M. (1996). The rock varnish in the Messak Settafet (Fezzan, Libyan Sahara), Age, archaeological context and paleo-environmental implication. *Geoarchaeology* 11(5): 393-421.

Cremaschi, M. (1998). Late Quaternary geological evidence for environmental changes in western Fezzan. In Cremaschi, M., and di Lernia, S. (eds.), *Wadi Teshuinat. Palaeoenvironment and Prehistory in South-western Fezzan (Libyan Sahara)*, Quaderni di Geodinamica Alpina e Quaternaria 7, C.N.R. Milan, pp. 13-48.

Cremaschi, M., and di Lernia, S. (1995). The transition between late Pleistocene and early Holocene in the Uan Afuda Cave (Tadrart Acacus, Libyan Sahara). Environmental changes and human occupation. *Quaternaire* 6(3-4): 173-189.

Cremaschi, M., and di Lernia, S. (1998). The Geo-Archaeological survey in central Tadrart Acacus and surroundings (Libyan Sahara). In Cremaschi, M., and di Lernia, S. (eds.), *Wadi Teshuinat. Palaeoenvironment and Prehistory in south-western Fezzan (Libyan Sahara)*, Quaderni di Geodinamica Alpina e Quaternaria 7, C.N.R. Milan, pp. 245-298.

Cremaschi, M., and Trombino, L. (1998). The palaeoclimatic significance of paleosols in southern Fezzan (Libyan Sahara): morphological and micromorphological aspects. *Catena* 355: (in press).

Cremaschi, M., di Lernia, S., and Trombino, L. (1996). From taming to pastoralism in a drying environment. Site formation processes in the shelters of the Tadrart Acacus massif (Libya, Central Sahara). In Castelletti, L., and Cremaschi, M. (eds.), *Micromorphology of Deposits of Anthropogenic Origin*, XIII International Congress of the UISPP, Vol. 3, Paleoecology, Colloquium VI, Forlì 1995, ABACO Edizioni, Forlì, pp. 87-106.

Fedoroff, N., and Courty, M.A. (1989). Indicateur pedologiques d'aridification. Exemples du Sahara. *Bulletin de la Société Géologique de France 8* 5(1): 43 -53.

Garcea, E.A.A. (1993). *Cultural Dynamics in the Saharo-Sudanese Prehistory*, Gruppo Editoriale Internazionale, Rome.

Gasse, F., Fontes, J.C., Plaziat, J.C., Carbonel, P., Kaczmarska, I., De Dekker, P., Soulie Marsche, I., Callot, Y., and Dupeuble, P.A. (1987). Biological remains, geochemistry and stable isotopes for reconstruction of environmental and hydrological changes in the Holocene lakes from north Sahara. *Paloegeography, Paleoclimatology, Palaeocology* 60: 1- 46.

Girod, A. (1998). The mollusc fauna in the Holocene late deposits of the Erg Uan Kasa and the Murzuq Edeyen. In Cremaschi, M., and di Lernia, S. (eds.), *Wadi Teshuinat. Palaeoenvironment and Prehistory in South-western Fezzan (Libyan Sahara)*, Quaderni di Geodinamica Alpina e Quaternaria 7, C.N.R. Milan, pp. 73-88.

Hassan, F. (1997). Holocene palaeoclimates of Africa. *African Archaeological Review* 14(4): 213 - 230.

Kröpelin, S. (1987). Palaeoclimatic evidence for early to mid-Holocene playas in the Gilf Kebir (SW Egypt). *Palaeoecology of Africa* 18: 189-208.

Lamb, H.F., Gasse, F., Benkaddour, A., El Hamouti, N., Van de Kaars, S., Perkins, W.T., Pearce, N.J., and Roberts C.N. (1995). Relation between century-scale Holocene arid intervals in tropical and temperate zones. *Nature* **373**: 134 - 137.

Lezine, A.M., Casanova, J., and Hillaire-Marcel, C. (1990). Across an early Holocene humid phase in Western Sahara: pollen and isotope stratigraphy. *Geology* **18**: 264-267.

Martini, M., Sibilia, E., Zelaschi, C., Troja, S.O., Forzese, R., Gueli, A.M., Cro, A., and Foti, F. (1998). TL and OSL dating of fossil dune sand in the Uan Afuda and Uan Tabu rockshelters, Tadrart Acacus (Libyan Sahara). In Cremaschi, M., and di Lernia, S. (eds.), *Wadi Teshuinat. Palaeoenvironment and Prehistory in South-western Fezzan (Libyan Sahara)*, Quaderni di Geodinamica Alpina e Quaternaria 7, C.N.R. Milan, pp. 67-72.

Mercuri, A.M., Trevisan Grandi, G., Mariotti Lippi, M., and Cremaschi, M. (1998). New pollen data from the Uan Muhuggiag rockshelter (Libyan Sahara). In Cremaschi, M., and di Lernia, S. (eds.), *Wadi Teshuinat. Palaeoenvironment and Prehistory in South-western Fezzan (Libyan Sahara)*, Quaderni di Geodinamica Alpina e Quaternaria 7, C.N.R. Milan, pp. 107-122.

Muzzolini, A. (1993). The emergence of a food-producing economy in the Sahara. In Shaw, T. A., Sinclair, P., Andah, B., and Okpoko, A. (eds.), *The Archaeology of Africa: Food, Metals and Towns*, Routledge, London and New York, pp. 227-239.

Pachur, H.J., and Hoelzmann, B. (1991). Paleoclimatic implications of late Quaternary lacustrine sediments in western Nubia, Sudan. *Quaternary Research* **36**: 257-276.

Pachur, H.J., and Wunneman, B. (1997). Reconstruction of the palaeoclimate along 30° in the eastern Sahara during the Pleistocene/ Holocene transition. *Palaeoecology of Africa* **24**: 1-33.

Pasa, A., and Pasa Durante, M.V. (1962). Analisi paleoclimatiche nel deposito di Uan Muhuggiag, nel massiccio del Tadrart Acacus (Fezzan Meridionale). *Memorie del Museo Civico di Storia Naturale* **10**: 251 - 255.

Petit-Maire, N., and Risier, J. (1981). Holocene lake deposits and paleoenvironment in central Sahara, North-Eastern Mali. *Palaeogeography, Paleoeclimatology, Palaeoecology* **35**: 45 - 61.

Trevisan Grandi, G., and Mercuri, A. (1992). Ricerche geobotanico-paletnobotaniche nel Sahara libico: siti neolitici ed epipaleolitici nel Tadrart Acacus (Fezzan Sud- Occidentale). *Atti della Società Naturalisti e Matematici di Modena* **123**: 53-71.

Trevisan Grandi, G., Mercuri, A.M., and Cremaschi, M. (1995). Quaternary studies in Libyan Sahara massifs: first palynological data on Messak Settafet MTS (VII - V millennia bp). *Atti Società Naturalisti e Matematici di Modena* **126**: 29-39.

Trevisan Grandi, G., Mariotti Lippi, M., and Mercuri, A.M. (1998). Pollen in dung layers from rockshelters and caves of Wadi Teshuinat (Libyan Sahara) In Cremaschi, M., and di Lernia, S. (eds.), *Wadi Teshuinat. Palaeoenvironment and Prehistory in South-western Fezzan (Libyan Sahara)*, Quaderni di Geodinamica Alpina e Quaternaria, 7, C.N.R. Milan, pp. 95-106.

Walter, H., and Lieth, H. (1960). *Klimadiagramm. Weltatlas*, Jena.

6

LATE HOLOCENE CLIMATIC FLUCTUATIONS AND HISTORICAL RECORDS OF FAMINE IN ETHIOPIA

M.U. Mohammed and R. Bonnefille

INTRODUCTION

In Ethiopia, millennial scale late Quaternary climatic changes have been reconstructed from lake level, stratigraphic and limnological data. They show the presence of early Holocene high stands interrupted by abrupt arid intervals (Gillepsie *et al.*, 1983; Gasse and Street, 1978; Lamb *et al.*, 2000). The timing of high lake levels corresponds to phases of organic-rich soil formation in northern Ethiopia (Berakhi *et al.*, 1998; Mohammed and Bonnefille, 1998). A rapid fall in lake levels was recorded after 4 kyr bp following increased aridity (Gillespie *et al.*, 1983). A phase of soil degradation was also noted during the same period (Belay, 1997).

However, fine resolution palaeoenvironmental data showing low amplitude short term climate fluctuations are rare. Such data are useful because they can be linked to historical and/or archaeological events. Here we present such records from a late Holocene pollen stratigraphy of two sites: Dega Sala and Orgoba 4, both found in the South Ethiopian mountains (Arsi and Bale respectively; Figure 6.1). The aim is first to compare the previously published Dega Sala record (Bonnefille and Mohammed, 1994), from the Arsi Mountains, with that of new data from the Orgoba 4 site on the Bale Mountains in order to evaluate the climatic significance of the results. Secondly, since the data has sufficient resolution and radiocarbon dates, it is possible to make links with events of historical famine in Ethiopia.

METHODS

Sediment cores were taken from the swamps using a Russian corer. This has a semi-cylindrical tube at its end, which is kept open when the sediment is recovered, and closed by turning it 180° before the corer is pulled out from a certain depth. This prevents contamination of the core by overlying sediment. Sampling was done at close intervals ranging from 5 to 2 cm. Pollen was analyzed following previously described methods (Bonnefille *et al.*, 1993).

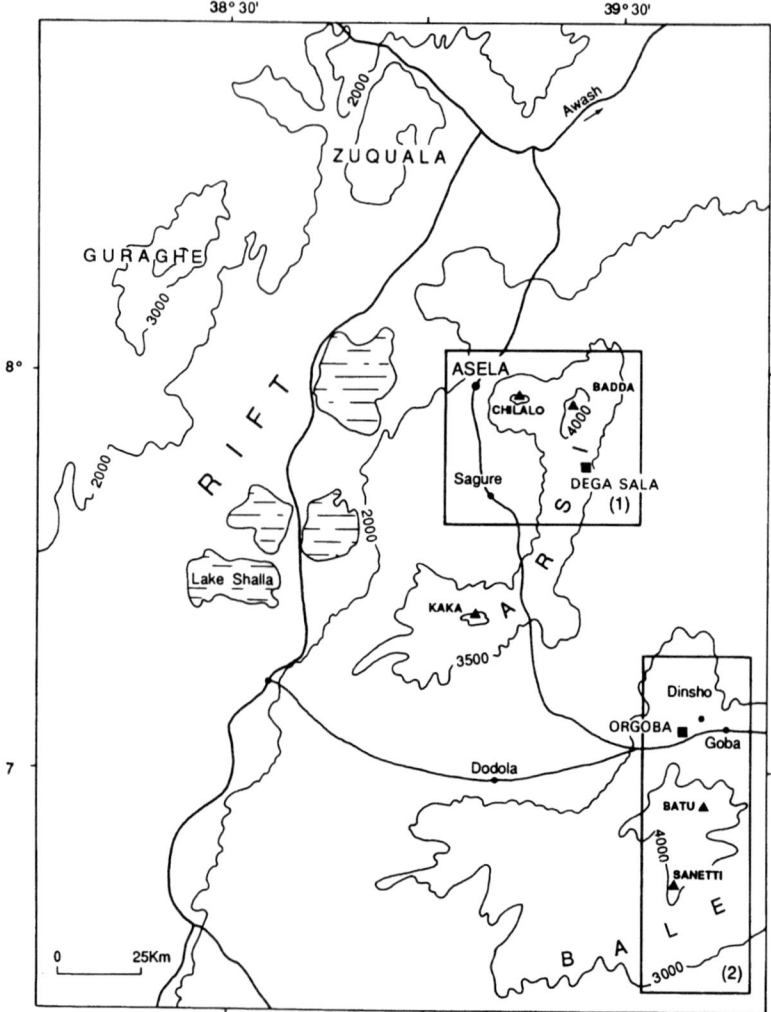

Figure 6.1. Location map of the core sites: 1. the Dega Sala site on the Arsi Mountains; 2. the Orgoba 4 site on the Bale Mountains in Ethiopia.

Several radiocarbon dates were obtained from the two cores using conventional methods and two were done by an AMS method (Table 6.1). The results are presented on a pollen diagram as percentages of the different taxa identified while counting pollen under a light microscope, at 1000 magnification.

Table 6.1. Uncalibrated ^{14}C dates in yr bp of the studied cores (D1 and O4)

Cores	Depth (cm)	^{14}C dates (yr bp)	Lab numbers
Dega Sala (D1)	1-10	Modern	LGQ 343
	40-50	Modern	LGQ 736
	58-60	560 ± 120 AMS	GIF
	81-90	590 ± 220	LGQ 342
	120-130	2050 ± 200	LGQ 737
	170-180	1560 ± 284	LGQ 284
	190-200	2920 ± 240	LGQ 738
Orgoba (O4)	0-10	Modern	LGQ 347
	40-50	Modern	LGQ 587
	50-60	370 ± 150	LGQ 627
	68-78	1040 ± 200	LGQ 348
	100-110	2210 ± 190	LGQ 740
	140-147	1550 ± 180	LGQ 280
	155-160	2470 ± 170	LGQ 567
	180-190	3060 ± 210	LGQ 281

THE REGIONS STUDIED

The Arsi and Bale mountains are located east of the Main Ethiopian Rift Valley where the Ziway-Shalla basin lakes are located (Figure 6.1). Both mountains rise above 4000 m. In forested areas the tree line could reach as high as 3200 m, above which is the Ericaceous belt going up to about 4000 m. Afro-alpine vegetation is found above this limit, mainly on the Bale Mountains. On the wet side of the mountains, such as on the southern slope of the Bale, a mixed humid forest is characteristic of the forest vegetation. On the drier slopes, coniferous dominated forest exists. It is characterized by *Podocarpus falcatus graalior* in the lower altitudes (2000-2500 m) while *Juniperus*, *Hagenia* and *Hypericum* cover the areas around the forest limit.

THE CORE SITES

The Dega Sala core is taken from a small swamp on the Galama ridge of the Arsi mountains at an altitude of 3600 m (Figure 6.2). The site is located in the Ericaceous belt. The swamp vegetation is dominated by Cyperaceae while *Alchemilla*, Poaceae, Caryophyllaceae, Apiaceae and *Helichrysum* grow around the valley.

The Orgoba 4 site is located in a valley at the foot of a basaltic hill at 3380 m on the northern side of the Bale Mountains (Figure 6.1). It is found in the Ericaceous belt just above the forest limit. Even some trees of *Juniperus* have been seen growing near the site. The swamp vegetation is predominantly composed of Cyperaceae. Poaceae, *Alchemilla* and also Primulaceae grow around the site.

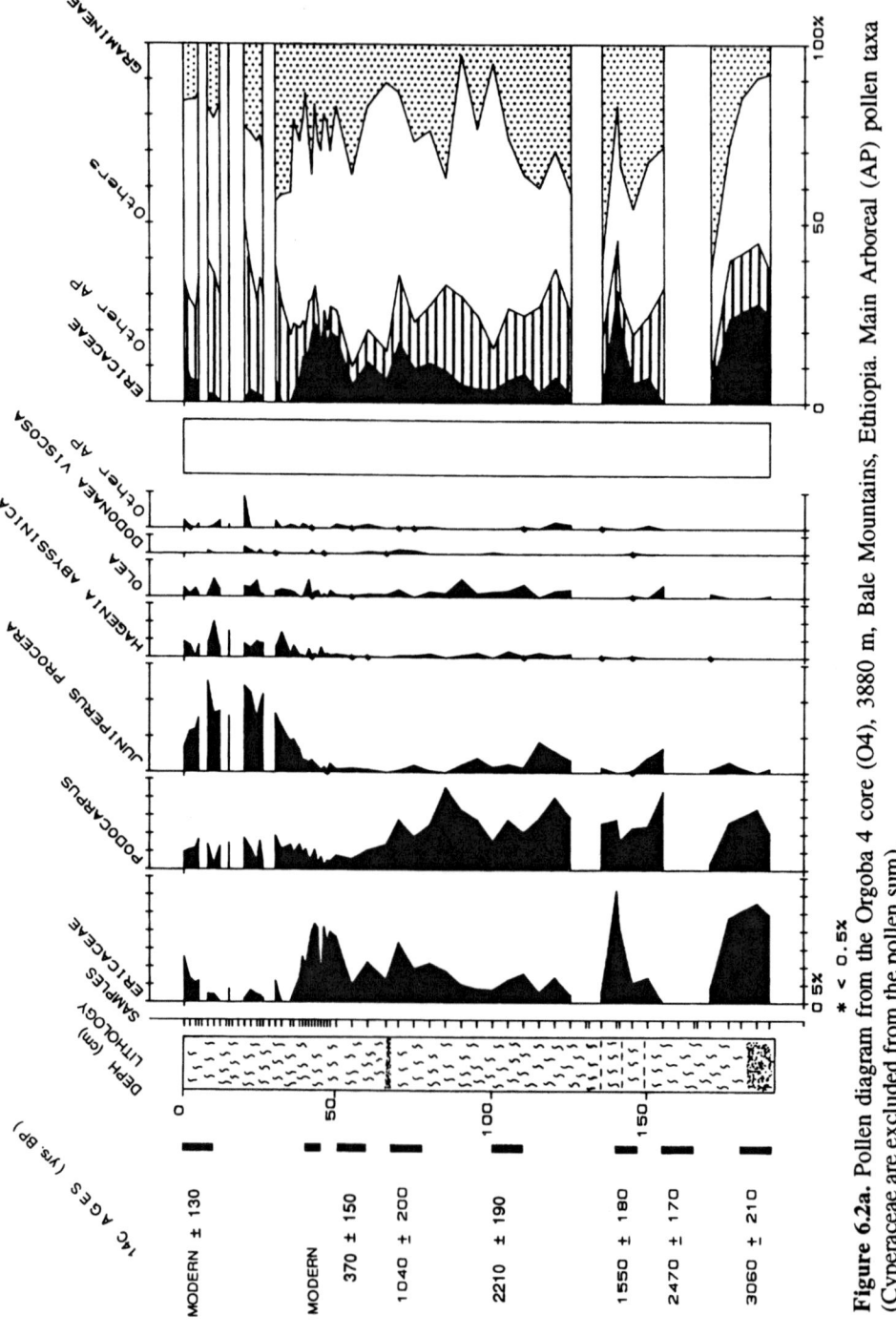

Figure 6.2a. Pollen diagram from the Orgoba 4 core (O4), 3880 m, Bale Mountains, Ethiopia. Main Arboreal (AP) pollen taxa (Cyperaceae are excluded from the pollen sum).

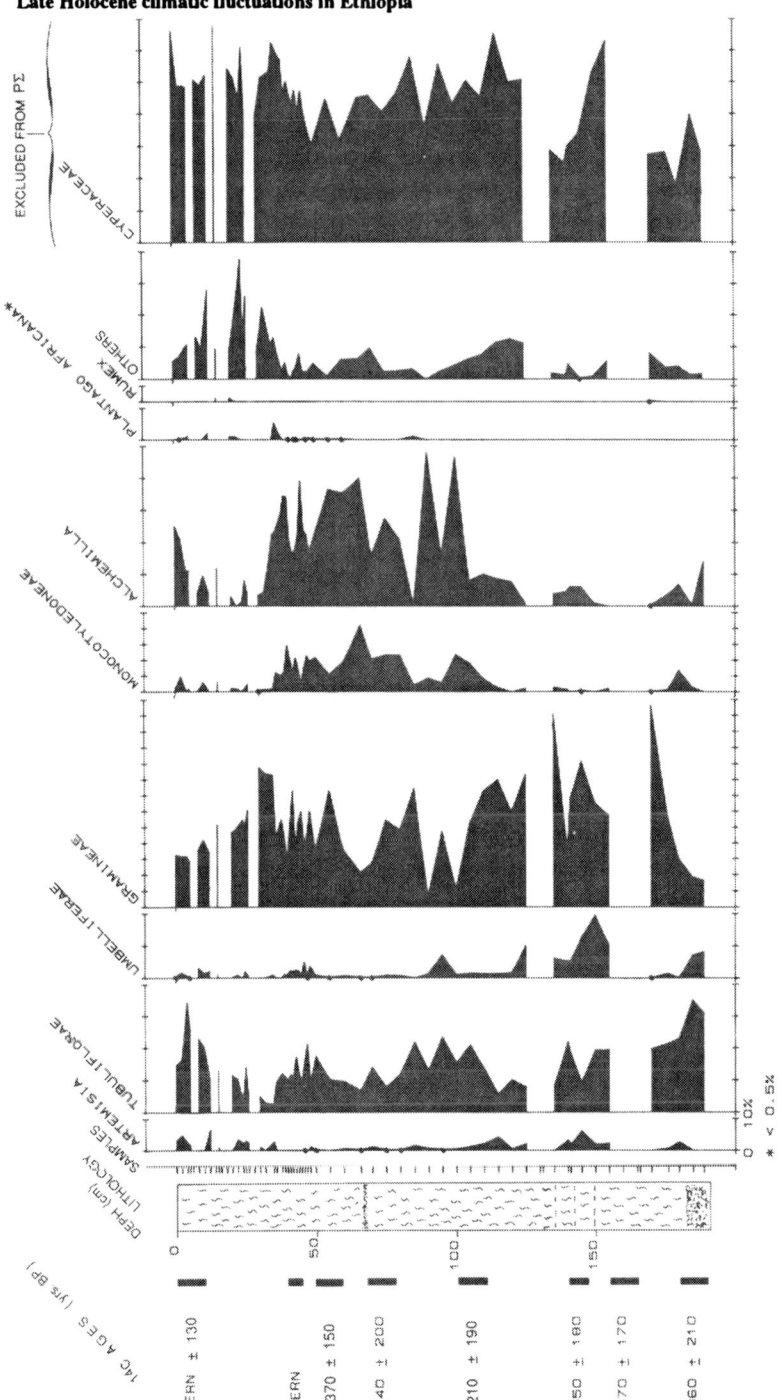

Figure 6.2b. Pollen diagram from the Orgoba 4 core (O4), 3880 m, Bale Mountains, Ethiopia. Non-Arboreal (NAP) pollen taxa (Cyperaceae are excluded from the pollen sum).

SEDIMENT AND CHRONOLOGY

Both cores contain organic rich sediment towards the top and become more and more mineral rich towards the bottom. At the very bottom, coarse sand and gravel prevented further coring. A thin layer of volcanic ash material appears in both cores, at around 68 cm in core O4 and 82 cm in core D1. Average sedimentation rate for core D1 is estimated to be 0.8 mm/yr. For core O4 the sedimentation rate varies from 1.5 mm/yr to about 0.15mm/yr from top to bottom. The time resolution is 7 years in the upper 50 cm and reaches 66 yrs towards the bottom.

We consider that dates of 1560 ± 284 yr bp and 1550 ± 180 yr bp, respectively obtained at 170 cm in D1 and at 140 cm in O4, are erroneous and may correspond to contamination by younger material, probably penetrating root systems.

POLLEN RECORDS

The pattern of change observed in the pollen diagrams from both cores is similar (Figures 6.2 and 6.3). However, the record is more continuous in D1 as compared to that in O4. This could be due to a difference in the sensitivity of the sites to very short (decadal?) changes in either the hydrology of the swamp or climate variability. This is particularly evident in pollen Zone 1 of the O4 pollen diagram. In this work we will discuss only the major trends of changes seen in the pollen diagrams.

In both cores there is a general pattern of increase in arboreal pollen and a decrease in Ericaceae from the bottom of the cores to the top part. This trend is interrupted in both cores in pollen Zone 2, where a marked decrease in arboreal pollen and an increase in Ericaceae was observed. Table 6.2 shows a summary of the characteristic features of the pollen zones in both cores.

Table 6.2. Synchronous pollen events in cores Orgoba 4, Bale Mountains and Dega Sala 1, Arsi Mountains

Pollen Zones		^{14}C dates	Pollen Events
O4	D1	(yr bp)	
1	A-1	~ 100- (Modern)	Low Ericaceae, high forest limit, increasing *Alchemilla* towards the top of the cores
2	A-2	500-(200-100)	High Ericaceae, low forest limit, low *Alchemilla* mainly in the upper part of the zones
3	B-1	1500-500	Low Ericaceae, increasing forest limit, high aquatics
5 and 4	B-4	3000-2000	High Ericaceae low forest limit, low
	B-3	(1500)	aquatics particularly *Alchemilla*.
	B-2		
		<3000	No pollen record in coarse sand and gravel

DISCUSSION

The above results show similar patterns of fluctuation in the forest limit taxa (*Juniperus* and *Hagenia*) and Ericaceae from the Ericaceous belt in both cores taken from two distant mountains. Moreover, the sites are well above the area currently under intensive human impact. These allow us to confidently interpret the results in terms of climate change. Knowing that the forest limit is governed by temperature (Holtmeier, 1992), the change in the forest limit taxa with respect to the higher Ericaceous belt should also indicate temperature variations during the last 3000 years. Moreover the development of a near swamp (*Alchemilla*) and aquatic (Cyperaceae) type of vegetation with respect to Ericaceae should indicate periods of rising water table and swamp development. From these it can be said that the general climate gradually got warmer and wetter in the mountains, except for an interruption in this trend in the period of pollen Zone 2 when cold conditions prevailed. This period was dated at 0.5 kyr bp to the last century. More particularly, well marked increases in tree and aquatic pollen were observed after about 2000 bp when organic rich sediment was deposited. A rise in temperature and a higher water table in the swamps are inferred. The climatic phases identified in the two cores are summarized in Table 6.3.

Table 6.3. Climatic phases inferred from the synchronous pollen zones of cores Orgoba 4, Bale Mountains and Dega Sala 1, Arsi Mountains

Phases	^{14}C dates (yr BP)	Climate
I	~ 100-(Modern)	Warm (wetter during the first half and droughts of the 70s and 80s during the second half)
II	500-(200-100)	Cold (dry intervals in low altitudes)
III	1500-500	Warm
IV	3000-2000 (1500)	Dry and cold (gradually changing to wet conditions)
V	<3000	?

Phase V of the South Ethiopian Mountains may correspond to the abrupt arid phase observed from a sudden drop in lake levels of the Ziway-Shalla basin in the adjacent rift valley. Phase IV coincides with neo-glacial activity on Mount Kenya, indicating a period of glacial extension between 3200-2500 yr bp (Rosqvist, 1990). High stand of low amplitude has been evidenced in multidisciplinary studies from cores of the Lake Turkana at around 1600 yr bp (Mohammed *et al.*, 1996) in good agreement with the lower date of climatic Phase III in this work.

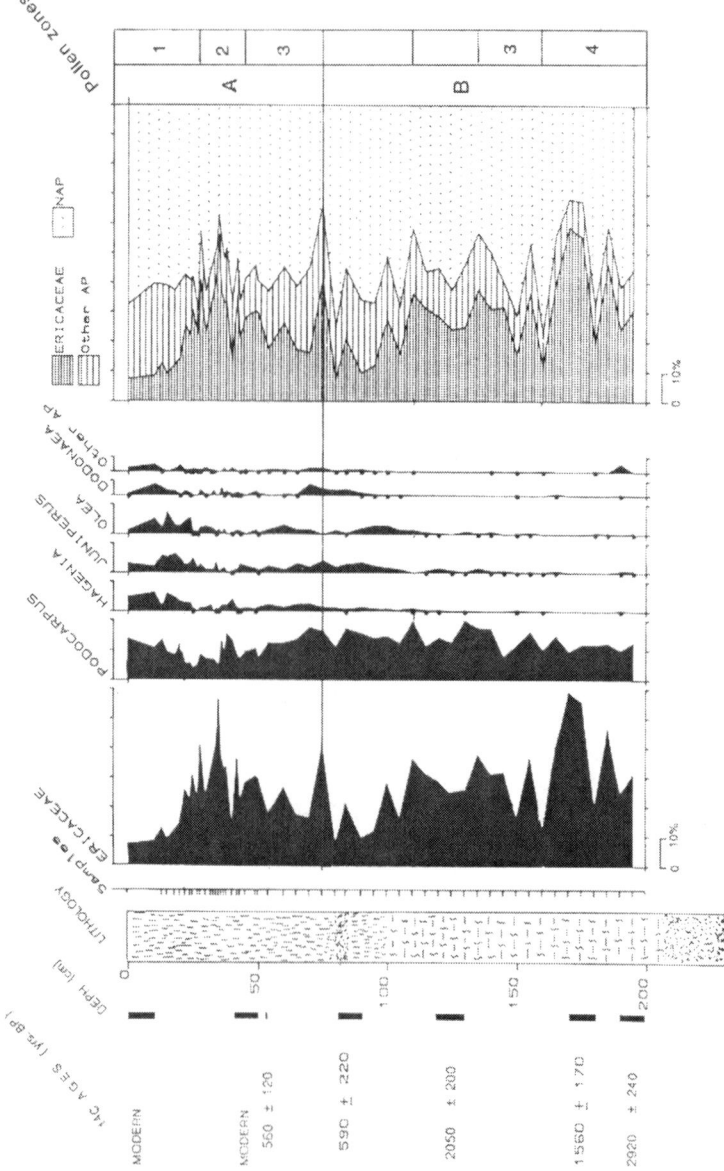

Figure 6.3a. Pollen diagram from the Dega Sala core (D1), 3600 m, Arsi Mountains, Ethiopia. Arboreal (AP) pollen taxa (Cyperaceae is excluded from the pollen sum).

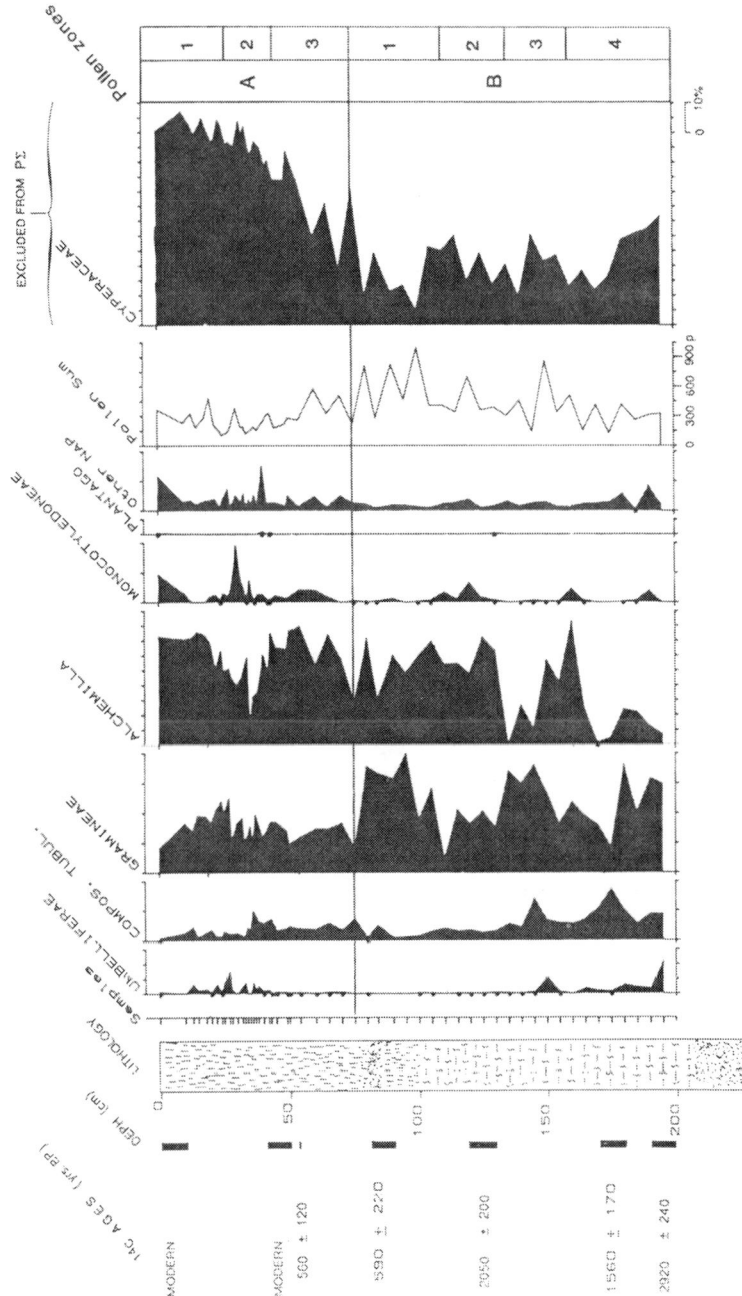

Figure 6.3b. Pollen diagram from the Dega Sala core (D1), 3600 m, Arsi Mountains, Ethiopia. Non-arboreal (NAP) pollen taxa (Cyperaceae is excluded from the pollen sum).

Evidence of Cold Climate During the Little Ice Age

The most important change in the pollen record is observed in pollen Zone 2 with a radiocarbon date after 0.4-0.5 kyr bp. Evidence of cold climate has also been suggested from other East African mountains from neo-glacial advances between AD 1500-1800 (Rosqvist, 1990). During this time the area covered by ice was estimated at 2 km^2 (Hastenrath, 1984). Historical records show that snow fell on the Ethiopian mountains above 3000 m during the seventeenth century AD where such events are never experienced today (Hovermann, 1954). The changing climate at that time has been explained by the fact that summer Westerlies weakened by about 30% or shifted southward by about 5°, while the subtropical monsoon rains decreased and also receded southwards. Low amplitude Nile floods were mentioned for the period 1470-1500 and 1640-1720 (Hassan, 1981), which could be attributed to a reduction of rainfall in the Ethiopian highlands (Bonnefille and Mohammed, 1994). All these data point to the persistence of cold climatic conditions at high altitudes and short dry intervals at low altitude sites. These dry intervals might have occurred within a generally wetter climate of the "Little Ice Age", as it has been recently shown (Verchuren *et al.*, 2000; Lamb *et al.*, 2001; Dapnachew *et al.*, in press).

Events of Famine in Ethiopia During Phase II

Historical data on palaeoclimatic conditions in Ethiopia can also be related to events of famine, epidemics and drought from Emperors' chronicles, Church acts and travellers' reports (Pankhurst, 1985). According to such data, frequent famines were recorded in the seventeenth century, in particular between AD 1611 and 1636, including a three-year drought between AD 1625 and 1627 and four other drier periods in AD 1252-1274, 1540-1567, 1772-1828, culminating in the 'Great Ethiopian Famine' of 1888-1892. These events also correspond to phases of soil instability in northern Ethiopia (Machando *et al.*, 1998). The cold climate inferred from the pollen stratigraphy in the Ethiopian highlands a few centuries ago is contemporaneous with a severe dry climate all over the country. Intensified droughts of the nineteenth century (Machando *et al.*, 1998) parallel those reported from the Sahelian regions (Hassan, 1997).

CONCLUSION

The pollen record presented here gives a high resolution proxy data for recent climate changes in Ethiopia. This study emphasizes fluctuations of strong amplitude that affected the vegetation of the highland countries far above the tree line on the mountains during the last three thousand years that include the historical period. Because these vegetation changes occurred in unpopulated areas, they are more likely attesting climatic changes. Indeed many of them are in good chronological correspondence with other palaeoenvironmental changes deduced from lake level changes, soil degradation, Nile floods, etc. They may have had significant impact on human population. Particularly significant are the increase and impact of droughts and famine during the seventeenth century, in the interval of the Little Ice Age. Our study indicates that climate is the long-term driving factor of environmental changes, even during the historical period.

ACKNOWLEDGMENTS

This work was financed by the Ministry of Foreign affairs (France) and the L.S.B Leakey foundation. Funding for laboratory sample analysis and radiocarbon dating was secured by CNRS/PNEDC (grant to R. Bonnefille). We acknowledge the help of G. Riollet and G. Buchet for pollen identification and thank the Department of Geology and Geophysics, and the Ethiopian Flora Projects in Addis-Ababa, for providing field logistics and facilities.

REFERENCES

Belay, T. (1997). Variabilities of soil catena on degraded hill slopes of Watiya catchement Wollo, Ethiopia. *SINET: Ethiopian Journal of Science* **20**(2): 151-175.

Berakhi, O., Brancaccio, L., Calderoni, G., Coltorti, M., Dramis, F., and Mohammed, M.U. (1998). The Mai Mekden sedimentary sequence. A reference point for the environmental evolution of the highlands of Northern Ethiopia. *Geomorphology* **23**(2-4): 127+138.

Bonnefille, R., and Mohammed, M.U. (1994). Pollen inferred climatic fluctuations in Ethiopia during the last 3000 yrs. *Paleogeography, Paleoclimatology, Paleoecology* **109**: 331-343.

Bonnefille, R., Buchet, G., Friis, I. B., Ensermu, K., and Mohammed, M.U. (1993). Modern pollen rain on an altitudinal range of forests and woodlands in South West Ethiopia. *Opera Botanica* **121**: 71-84.

Dapnachew, L., Gasse, F., Radakovitch, O., Vallet-Coulomb, C., Bonnefille, R., Verchuren, D., and Barker, P. (in press). Environmental changes in a tropical lake (Lake Abyata, Ethiopia) during recent centuries. *Paleogeography, Paleoclimatology, Paleoecology*.

Gasse, F., and Street, A. (1978). Late Quaternary lake level fluctuations and environments of the Northern Rift Valley region (Ethiopia and Djibouti). *Paleogeography, Paleoclimatology, Paleoecology* **24**: 279-325.

Gillespie, R., Street-Perrot, F.A., and Switsur, R. (1983). Post glacial arid episodes in Ethiopia have implications for climate prediction. *Nature* **306**: 680-683

Hassan, F.A. (1981). Historical Nile floods and their implications for climatic change. *Science* **212**: 1142-1145.

Hassan, F.A. (1997). Holocene paleoclimates of Africa. *African Archaeological Review* **14**(4): 213-230.

Hastenrath, S (1984). *The Glaciers of Equatorial East Africa*, Riedl, Dordrecht.

Holtmeier, F.K. (1992). How does the timber line respond to changing climate? In *Swiss Climate Abstracts*, Proclim, Bern.

Hoverman, J. (1954). Über die hohenlage der schneegrenze in Äethiopien und ihre schwankungen in historischer zeit. *Akademie der Wissenschaften in Göttingen* **11**(6): 111-137.

Lamb, H. F., Darbyshire, I., Mohammed, M. U., Leng, M., and Telford, R. (2001). 3000 years of climate change, forest clearance and regrowth in northern Ethiopia. Abstract. Paper given at *PAGES. PEP III. Past Climate Variability in Europe and Africa*, 27-31 August, Aix-en-Provence.

Lamb, L. A., Leng, M. J., Lamb, H. F., and Mohammed, M. U. (2000). A 9000-yr oxygen and carbon isotope record of hydrological change in a small lake. *The Holocene* **10**(2): 167-177.

Machando, M.J., Perez-Gonzalez, A., and Benito, G. (1998). Paleoenvironmental changes during the last 4000 years in the Tigray, Northern Ethiopia. *Quaternary Research* **49**: 312-321.

Mohammed, M.U., and Bonnefille, R. (1998). A late glacial to late Holocene pollen record from highland peat at Tamsaa, Bale Mountains, South Ethiopia. *Global and Planetary Change* **16-17**: 121-129.

Mohammed, M.U., Bonnefille, R., and Johnson, T. (1996). Pollen and isotopic records of late Holocene sediments from Lake Turkana, N. Kenya. *Paleogeography, Paleoclimatology, Paleoecology* **119**(3-4): 371-383.

Pankhurst, R. (1985). *The History of Famine and Epidemics in Ethiopia Prior to the Twentieth Century*, Relief and Rehabilitation Commission, Addis Ababa.

Rosqvist, G. (1990). Quaternary glaciations in Africa. *Quaternary Science Reviews* **9**: 281-297.

Verchuren, D., Larid, K. R., and Cumming, B. F. (2000). Rainfall and drought in equatorial East Africa during the past 1000 years. *Nature* **403**: 410-414.

7

ENVIRONMENTAL AND HUMAN RESPONSES TO CLIMATIC EVENTS IN WEST AND WEST CENTRAL AFRICA DURING THE LATE HOLOCENE

M. A. Sowunmi

INTRODUCTION

The comparatively few proxy data from the region considered here, the rainforest region of West and West Central Africa (Figure 7.1), indicate the return of wet climatic conditions and the re-establishment of the rainforest and its edaphic and topographic variants during the middle Holocene. This wet phase was followed by a drier one which became established in the more western, Guinean, sub-region at an earlier period than in the more eastern, Congolian, sector. Climatic changes, after the drier and subsequent wetter phases, were compounded by human activities, but conditions did not become as wet as in the middle Holocene period.

Archaeological and palaeoecological studies have shown that the late Holocene witnessed a series of complex environmental and cultural changes. Stahl (1993), in a review of the late Stone Age in West Africa, underscored the exasperating meager data. But the scanty evidence available indicates changes in subsistence economy including a more intensive use of resources. The responses were local and the timing of changes varied from area to area (Stahl, 1993, p. 273). In the more southern areas the pattern was of a "subsistence economy characterized by continuous reliance on hunting and gathering, perhaps supplemented by arboriculture (for example, exploitation of oleaginous species), horticulture and probably limited reliance on domestic animals (sheep/goat)... [in contrast] the subsistence base of sites in the Sahara/Sahel margins...", from what is known, was one "where investment in food production during the last millennia BC appears to have been greater" (Stahl, 1993, p. 272).

The focus of this paper is as follows: (1) an outline of the rather abrupt end to the middle Holocene Climatic Optimum, beginning from ca. 4500 bp, first in the Guinean sub-region of West and West Central Africa, and, after about 1500 years, in the Congolian sector; (2) effect of the climatic change on vegetation in

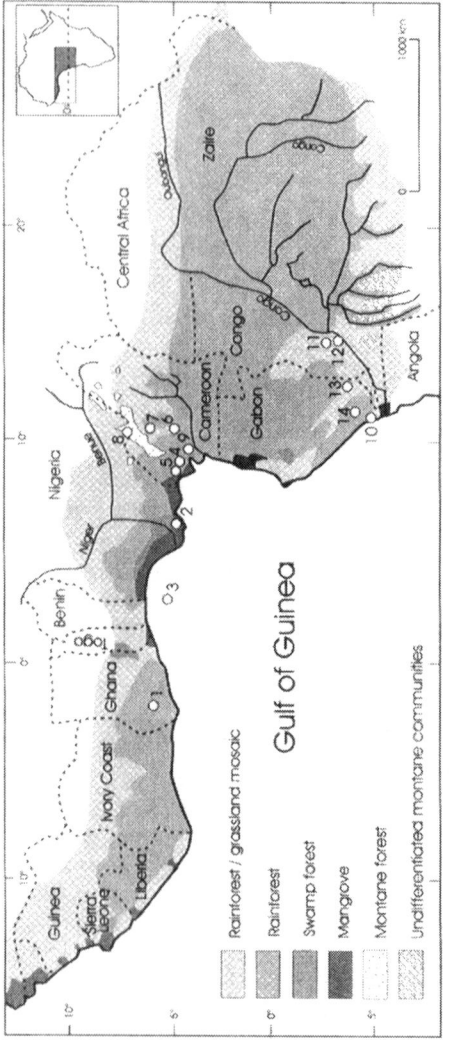

Figure 7.1. Map of West and West Central Africa showing the vegetation zones of the Guineo-Congolian region (modified after White, 1983) and pollen core sites: 1. Lake Bosumtwi, Ghana; 2. Niger Delta, Nigeria; 3. Deep-sea core off Niger Delta; 4. Mboandong, Cameroon; 5. Lake Barombi, Cameroon; 6. Bafounda Swamp, Bamileke Plateau, Cameroon; 7. Shum Laka Swamp, Cameroon; 8. Lake Njupi, Cameroon; 9. Lake Ossa, Cameroon; 10. Littoral Congo-Brazzaville; 11. Bilanko Depression, Bateké Plateau, Congo-Brazzaville; 12. Ngamalaka Pond, Bateké Plateau, Congo-Brazzaville; 13. Lake Sinnda, Congo-Brazzaville; 14. Lake Kitina, Congo-Brazzaville.

particular; and (3) inference on human activities which accentuated certain vegetational changes, based on palynological and geophysical evidence.

OUTLINE OF LATE HOLOCENE ENVIRONMENTAL CHANGES

Table 7.1 gives a summary of late Holocene environmental changes in West and West Central Africa, based on geomorphological, limnological, macrobotanical and palynological evidence. The proxy data show that the middle Holocene Climatic Optimum ended rather abruptly, first in the more western parts of the region, and subsequently in the eastern. Possible alternative interpretations to certain aspects of results from Cameroon are given below by the present author.

Shum Laka Region (Bamenda Highlands, Northwest Cameroon): Core From Altitude 1355 m

Pollen evidence (ca. 3500-3000 bp) indicates an increase/sudden appearance of Sudanic savanna trees, and a decrease in submontane forest. At ca. 3100 bp there was a slight increase in *Podocarpus* while another montane element (*Olea hochstetteri*) suddenly appeared and was well represented (ca. 12%). There was a sharp decrease in grasses and sedges (Kadomura and Kiyonaga, 1994). According to Kadomura and Kiyonaga, the expansion of montane-type woodland was a result of 'strong aridification' and a cooling of the climate. An alternative interpretation is that the climate might have been cooler but only mildly drier than that of the preceding phase. This suggestion is based on the following premises: (1) the savanna must have been densely wooded because of the low occurrence of grasses and an increase in arboreal species; (2) a decline in swamp sedges and grasses is suggestive of a higher level of water in the depression; and (3) an expansion of montane vegetation might indicate cooler conditions, and therefore reduced evaporation.

Lake Barombi Mbo (Cameroon)

Pollen evidence indicates a pronounced increase in grasses from ca. 3000 to 2000 bp with peak values at 2500 to 2000 bp and a marked decrease in closed forest trees, indicative of forest fragmentation. The first significant increase in oil palm, *Elaeis guineensis* occurs ca. 2800 bp (Maley, 1996). Maley did not attribute this oil palm increase to human action, although he did so for a subsequent increase ca. 2000 to 1600 bp. However, a human contribution to the first increase may not be ruled out.

Table 7.1. Summary of late Holocene environmental changes in West and West Central Africa

Age (year bp)/ Climatic inference	Country and locality of proxy data	Environmental indications
ca. 4500-3500 Dry phase (apparently no evidence of this dry phase from Cameroon and Congo-Brazzaville)	Sierra Leone—Koidu River Basin (Thomas, 1994)	Low fluvial activity, reduced discharges
	Côte d'Ivoire (Einsele *et al.* 1977)	Fall in sea level to 3.5 m below present level
	Ghana—Lake Bosumtwi (cf. Maley, 1996; Maley, personal communication, 1998; Talbot *et al.*, 1984); Birim River Valley (Thomas, 1994); Accra Beach (Talbot, 1981)	Slight increase in grasses with slight decrease in dense forest, phenomenal increase in oil palm, *Elaeis guineensis* (ca. 3800-3700); abrupt and sharp fall in lake level to 20-30 m below present level. Low fluvial activity, reduced discharges. Deposition of beach sand dunes.
	Nigeria—offshore, Eastern Niger Delta (Pastouret *et al.*, 1979); Eastern Niger Delta (Sowunmi, 1981a,b,c)	Abrupt cessation of freshwater discharge. Increase in grasses and open forest with decrease in dense forest and mangrove swamp forest.
	Cameroon—Mboandong (Richards, 1986); Lake Barombi (cf. Maley, 1996); Shum Laka Swamp (Kadomura and Kiyonaga, 1994); Lake Ossa (Reynaud-Farrera *et al.*, 1996)	Riverine and very diversified rainforest. Abundant dense forest, relative increase of *Podocarpus*. Submontane rainforest, low occurrence of grasses and *Podocarpus*. Abundant dense forest with slight degradation ca. 4770-4150; regular occurrence of *Podocarpus* and *Olea capensis*.
	Congo-Brazzaville— Coast (Caratini and Giresse, 1979; Dechamps *et al.*, 1988); Ngamalaka Pond (Elenga *et al.*, 1994); Lake Sinnda (Vincens *et al.*, 1994); Lake Kitina (Elenga *et al.*, 1996)	Abundant forest behind mangrove swamp forest; coastal, dense, primary forest, periodically flooded. Expansion of fresh water swamp forest, very low occurrence of grasses and other herbs. Abundant fresh water swamp and semi-deciduous, dense forests.
ca. 3500-3000 Wetter phase *(cont...)*	Ghana—Lake Bosumtwi (cf. Maley, 1996; Talbot *et al.*, 1984); Accra Beach (Talbot 1981)	Slight decrease in dense forest; increase in lake level to 25 m above present level. End of dune deposition.

	Nigeria—Eastern Niger Delta (Sowunmi, 1981a,b,c)	Decrease in grasses and dense forest, marked increase in mangrove swamp forest.
	Cameroon—Lake Barombi (cf. Maley, 1996); Shum Laka Swamp (Kadomura and Kiyonaga, 1994); Lake Ossa (Reynaud-Farrera et al., 1996)	Abundant dense forest. Increase/sudden appearance of Sudan savanna trees with sharp decrease in sedges and grasses, increase in montane forest. Abundant dense forest; regular occurrence of *Podocarpus* and *Olea capensis*.
	Congo-Brazzaville—Coast (Dechamps et al., 1988; Elenga et al., 1992); Lake Kitina (Elenga et al., 1996)	Coastal, dense, primary forest, periodically flooded; mangrove swamp forest, dominance of dense forest, periodically flooded. Abundant fresh water swamp and dense forests.
ca. 3000-2000 contrasting signals between western/eastern parts—severe dry phase in eastern part not recorded in the west.	Sierra Leone—Koidu River Basin (Thomas, 1994)	Increasing fluvial activity and discharges, sediment reworking and deposition, gravel aggradation.
	Ghana—Lake Bosumtwi (cf. Maley, 1996; Talbot et al., 1984)	Recovery of dense forest, very abrupt and first marked increase in oil palm (ca. 3200-2800) occurrence of *Podocarpus*, for the first time with another montane species, *Ilex mitis*.
	Cameroon—Mboandong (Richards, 1986); Lake Barombi (cf. Maley, 1996; Maley and Brenac, 1998); Shum Laka Swamp (Kadomura and Kiyonaga, 1994); Bafounda swamp (Farrera et al., 1996)	Opening of forest, followed by marked increases in grasses and some increase in oil palm (ca. 2000). Large increase in grasses and first significant rise in oil palm (ca. 2800-2400), marked decrease in dense forest, major regression in oil palm (ca. 2300-2100), abrupt disappearance of montane forest. Abrupt increase in grasses and sedge, sharp decrease in savanna trees, near disappearance of *Podocarpus* and expansion of *Raphia* swamp. Abrupt disappearance of *Podocarpus* and other species followed by increase in grasses, erosion down plateau slopes. Marked decrease in dense forest, *Podocarpus* and *O. capensis*, slight increase in grasses, beginning of increase in palm oil (ca. 2500-2000).

(cont...)

(table 1 cont.)

	Congo-Brazzaville— Coast (Dechamps *et al.*, 1988; Elenga *et al.*, 1992); Ngamalaka pond (Elenga *et al.*, 1994); Lake Kitina (Elenga *et al.*, 1996)	Rapid replacement of coastal, dense, periodically flooded forest by predominantly grassy savanna; coastal fresh water swamp forest reduced to woodland savanna with predominant grass cover. Abrupt and large increase in grasses, sedges and other herbs with sharp decrease in forest taxa, first occurrence of the oil palm (ca. 2850-2700). Significant increase in dense forest (ca. 2800-2600), very sharp decrease in dense forest and marked increases in grasses and other herbs (from ca. 2500), sharp increase in oil palm (ca. 2500-2000).
ca. 2000-1600 wet phase	Sierra Leone—Koidu River Basin (Thomas, 1994)	Increased fluvial activity and discharges, sediment reworking and deposition, gravel aggradation.
	Ghana—Lake Bosumtwi (cf. Maley, 1996)	No recorded change in dense forest.
	Nigeria—Eastern Niger Delta (Sowunmi, 1981a,b,c)	Relative decreases in oil palm, dense forest and grasses, increase in open forest (ca. 1900 ff.).
	Cameroon—Mboandong (Richards, 1986); Lake Barombi (cf. Maley and Brenac, 1998); Lake Ossa (Reynaud-Farrera *et al.*, 1996)	Marked increase in grasses followed by phenomenal rise in oil palm (ca. 1600). Continuation of expansion of woodland savanna at expense of forest. Sharp and abrupt decrease in oil palm (from ca. 1900), continued regression of forest (until ca. 400), slight decrease in grasses.

SOME OF THE EFFECTS OF CLIMATIC CHANGE ON VEGETATION

During the very dry phase in parts of Cameroon and Congo-Brazzaville, ca. 3000-2000 bp, forests were fragmented, apparently appreciably, while there were, concomitantly, marked increases in herbaceous elements, especially grasses. For example, based on evidence from Ngamakala pond, Bateké Plateau, southeastern Congo-Brazzaville (altitude 600-886 m), the fragmentation of the equatorial rainforest was more severe there than during the late Pleistocene arid phase (ca. 24000-13000 bp; Elenga *et al.*, 1994).

Several authors (see Kadomura and Kiyonaga, 1994) have suggested that the forest fragmentation ca. 3000-2000 bp in Cameroon and Congo-Brazzaville probably facilitated the movement of Bantu language peoples eastwards from their supposed homeland in the Grassfields region of West Cameroon. However,

the archaeological and palaeoecological data still remain too inconclusive and
scanty for definite conclusions to be drawn (Schwartz, 1992).

HUMAN ACTIVITIES AND CERTAIN VEGETATIONAL CHANGES

Attention will be focused here only on certain relevant vegetation changes which
occurred in forest communities and to which humans seemed to have contributed.
These changes are, firstly, the abrupt and marked increases in oil palm
populations, along with open forest vegetation, which occurred when forests
diminished during dry phases or before they became re-established at the
beginning of subsequent wet phases, and secondly, the development of the
Grassfields in southwestern Cameroon following a dry phase. In both cases,
processes involved in plant cultivation were the contributing factors.

As has been discussed at considerable length elsewhere (Sowunmi,
1999), the oil palm, a natural component of open forests, with a record of
occurrence dating back to the Eocene, (Zaklinkaya and Prokofyev, 1971, cited in
Maley and Brenac, 1998), is among one of the natural pioneers of forest
openings. However, the pattern of occurrence of its pollen from the Miocene to
the present, suggests that the abrupt and marked increase of the oil palm in the
late Holocene was due to both natural and anthropic factors. Furthermore, I have
proposed (Sowunmi, 1999), on the basis of available evidence, that the traditional
system of slash-and-burn and swidden plant cultivation in the forest zone was the
main anthropic factor which contributed significantly to late Holocene increase in
oil palm populations. The first of such marked increases occurred in the vicinity
of Lake Bosumtwi, Ghana, ca. 3800 3700 bp (Talbot et al., 1984), while the most
recent was recorded for the area around Lake Sinnda, southwestern Congo-
Brazzaville, ca. 800-1100 bp (Vincens et al., 1998). Apart from the comparatively
early date from Ghana, most of the other records place the first significant
increase in the oil palm between ca. 3200 and 2000 bp.

For various reasons, particularly the fact that oil palm is, today, mostly
only semi-domesticated (Hartley, 1977, p. 77), and the pollen of the wild and
cultivated forms are indistinguishable from the domesticate, it is virtually
impossible to determine when its domestication began. Furthermore, its fruits and
seeds are not food items in themselves, but supplementary; although there are
numerous uses of different parts of the tree (Table 7.2). Not only would this tree
most probably have been protected for its numerous uses, but the practice of
slash-and-burn, a necessary preparation for the cultivation of other plants,
coupled with short fallow periods, would have enhanced its proliferation and
spread. Of course, other animals, such as monkeys and birds, are also known to
disperse oil palm fruits.

A recent palynological and geophysical study of the sediments from the
Osaru Pond, West Central Nigeria (Oyelaran, 1997), has provided evidence
which seems to support the hypothesis referred to above. Although dates have not
yet been obtained for these sediments, there is a distinct synchronism between the
decline of lowland rainforest and the sharp increase in oil palm, charcoal, and the
magnetic susceptibility value of the soil. The increase in magnetic susceptibility

Table 7.2. Some of the uses of the oil palm tree

Category	Uses
Food	palm oil (from the epicarp)
	palm kernel oil (from the seed/kernel)
	palm kernel
	palm wine (fermented sap)
Fuel	Fibrous and oily mesocarp, matted and dried
	dried shell of kernel (endocarp)
Cosmetics/toiletries	body oil
	toilet soap (palm kernel oil as source of fatty acids)
Household items	broom (leaf midribs)
	soap (palm kernel as source of fatty acids)
House/shelter	verandah post (trunk)
construction	thatching for roof (leaves)
	wind-break
Divination	endocarp with kernel

value is "a phenomenon suggestive of soil magnetic enhancement through burning..." preparatory to cultivation (Oyelaran, 1997).

There is as yet no unequivocal and direct evidence for the beginning of agriculture in the forest zone of West and West Central Africa, but indirect or rather tenuous evidence seems to suggest that this practice was adopted after dry conditions had led to the fragmentation or opening up of forests, i.e. ca. 3500 bp in the Lake Bosumtwi area of Ghana (Talbot *et al.*, 1984) and ca. 3000 bp in southern Nigeria (Sowunmi, 1985), or ca. 3300-3000 in Kintampo, Ghana (Flight, 1976; Stahl, 1985).

The pollen analysis of a core from Shum Laka pond (Bamenda highlands) shows that, during a dry phase between ca. 3000-2000 bp, the change in vegetation consisted of an abrupt increase in grasses and sedges coupled with the disappearance/sharp decrease in savanna and forest trees, the near disappearance of *Podocarpus*, an increase in Myrtaceae (probably *Syzigium staudtii*) and an increase in *Raphia* palm (ca. 3000-1600 bp; Kadomura and Kiyonaga, 1994). The return of wetter conditions notwithstanding, grassland continued to expand at the expense of woody elements, resulting in the eventual development of the Grassfields landscape, which reached its maximum ca. 1700-1600 bp. Kadomura and Kiyonaga (1994) opined that the transformation in vegetation was probably due to slash-and-burn agriculture which might have started ca. 2600 bp. Kadomura and Kiyonaga further implied that the use of iron implements might have contributed to an intensification of deforestation.

CONCLUSION

In spite of the relative paucity of data, environmental and human responses to climatic events in the region considered can be deciphered. During the dry phases there was a relative reduction in the extent of the rainforest and its edaphic variety, coupled with an expansion of open forests or woodland/grassy savannas. Conversely, in the wet periods, there was a recovery of the forest vegetation at the expense of the savannas.

Reduction in the spatial extent of the forest created conditions which enhanced the remarkable spread of the oil palm, based on palynological evidence. Of particular significance is the pattern of occurrence of oil palm pollen, which strongly suggests that humans contributed to the expansion and spread of this economically important tree.

ACKNOWLEDGMENTS

Much of the material for this paper was developed while the author was on sabbatical leave from the University of Ibadan. The assistance of the University of Ibadan is gratefully acknowledged. Immense thanks are due to the following institutions and colleagues who provided both the academic stimulus and financial support without which the paper could not have been written— University of Uppsala, Sweden, University College London, and Johann Wolfgang Goethe-Universität, Frankfurt am Main, Germany, and Professor Paul Sinclair, Professor Peter Ucko and Dr. Katharina Neumann, respectively. My deep appreciation goes to Frau Barbara Voss for producing the figure.

REFERENCES

Caratini, C., and Giresse, T. M. (1979). Contribution palynologique à la connaissance des environnements continentaux et marins du Congo à la fin du quaternaire. *Comptes Rendus à l'Academie des Sciences de Paris* **288**: 379-382.

Dechamps, R., Guillet, B., and Schwartz, D. (1988). Découverte d'une flore forestière mid-Holocène (5800-3100 B.P.) conservée in situ sur le littoral ponténégrin (R.P. du Congo). *Comptes Rendus à l'Academie des Sciences de Paris* **306**: 615-618.

Einsele, G., Herms, D., and Schwartz, U. (1977). Variation du niveau de la mer sur la plate-forme continental et la côte Mauritanienne vers la fin de la glaciation de Wurm et a l' Holocene. *Bulletin de l'Association Sénégalese pour l'Etude du Quarternaire de l'Ouest Afrique* Dakar **51**: 35-48.

Elenga, H., Schwartz, D., and Vincens, A. (1992). Changement climatiques et action anthropique sur le littoral congolais au cours de l' Holocène. *Bulletin de la Société Géologique de France* **163**(1): 83-90.

Elenga, H., Schwartz, D., and Vincens, A. (1994). Pollen evidence of late Quaternary vegetation and inferred climate changes in Congo. *Palaeogeography, Palaeoclimatology, Palaeoecology* **109**: 345-356.

Elenga, H., Schwartz, D., Vincens, A., Bertaux, J., de Namur, C., Martin, L., Wirrmann, D., and Servant, M. (1996). Diagramme pollinique holocène du lac Kitina (Congo): mise en évidence de changements paléobotaniques et paléoclimatiques dans le massif forestier du Mayombe. *Comptes Rendus à l'Academie des Sciences de Paris* Séries IIa **323**: 403-410.

Flight, C. (1976). The Kintampo culture and its place in the economic prehistory of West Africa. In Harlan J.R., de Wet, J.M.J., and Stemler, A.B.L. (eds.), *Origins of African Plant Domestication*, Mouton, The Hague, pp. 211-221.

Hartley, C.W.S. (1977). *The Oil Palm*, Tropical agriculture series, 2nd edition, Longman, London, New York.

Kadomura, H., and Kiyonaga, J. (1994). Origin of Grassfields landscape in the west Cameroon highlands. In Kadomura, H. (ed.), *Savannization Process in Tropical Africa II*, University of Tokyo, Tokyo, pp. 47-85.

Maley, J. (1996). The African rain forest - main characteristics of changes in vegetation and climate from the Upper Cretaceous to the Quaternary. *Proceedings of the Royal Society of Edinburgh* **104B**: 31-73.

Maley, J., and Brenac (1998). Vegetation dynamics, palaeoenvironments and climatic changes in the forests of western Cameroon during the last 28,000 years B.P. *Review of Palaeobotany and Palynology* **99**: 157-187.

Oyelaran, A. (1997). Man-induced changes in the vegetation and the effects on the hydrological input of a catchment area - a study of Osaru pond in Iffe-Ijumu, Kogi State, Nigeria. [unpublished research paper]

Pastouret, L., Chamley, H., Delibrias, G., Duplessy, J. C., and Thiede, J. (1979). Late Quaternary climatic changes in Western Tropical Africa deduced from deep-sea sedimentation off Niger Delta. *Oceanologica Acta* **1**(2): 217-232.

Reynaud-Farrera, I., Maley, J., and Wirrmann, D. (1996). Végétation et climat dans les fôrets du sud-ouest Cameourn depuis 4770 ans BP: analyse pollinique des sédiments du lac Ossa. *Comptes Rendus à l'Academie des Sciences de Paris* Séries II **322**: 749-755.

Richards, K. (1986). Preliminary results of pollen analysis of a 6,000 year core from Mboandong, a crater lake in Cameroon. *Hull University Geography Department Miscellaneous Series* **32**: 14-28.

Schwartz, D. (1992). Assèchement climatique vers 3000 B.P. et expansion Bantu en Afrique centrale atlantique: queleues reflexions. *Bulletin de la Société Géologique de France* **163**(3): 353-361.

Sowunmi, M.A. (1981a). Nigerian vegetational history from the late Quaternary to the present day. *Palaeoecology of Africa* **13**: 217-234.

Sowunmi, M.A. (1981b). Aspects of late Quaternary vegetational changes in West Africa. *Journal of Biogeography* **8**: 457-474.

Sowunmi, M.A. (1981c). The late Quaternary environmental changes in Nigeria. *Pollen et Spores* **23**(1): 125-148.

Sowunmi, M.A. (1985). The beginnings of agriculture in West Africa: botanical evidence. *Current Anthropology* **26**(1): 127-129.

Sowunmi, M.A. (1999) The significance of the oil palm (*Elaeis guineensis*) in the late Holocene environments of west and west central Africa: a further consideration. *Vegetation History and Archaeobotany* **3**: 365-374.

Stahl, A.B. (1985). Reinvestigation of Kintampo 6 rock shelter, Ghana: implications for the nature of culture change. *African Archaeological Review* **3**: 117-150.

Stahl, A.B. (1993). Intensification in the west African Late Stone Age. In Shaw T., Sinclair P., Andah, B., and Okpoko, A. (eds.), *The Archaeology of Africa: Food, Metals and Towns*. Routledge, London, New York, pp. 261-273.

Talbot, M.R. (1981). Holocene changes in tropical wind intensity and rainfall: evidence from Southeast Ghana. *Quaternary Research* **16**: 201-220.

Talbot, M.R., Livingstone, D.A., Palmer, G., Maley, J., Melack, J.M., Delibrias, G., and Gulliksen, S. (1984). Preliminary results from sediment cores from Lake Bosumtwi, Ghana. *Palaeoecology of Africa* **16**: 173-192.

Thomas, M. F. (1994). *Geomorphology in the Tropics: A Study of Weathering and Denudation in Low Latitudes*, Wiley, Chichester.

Vincens, A., Guillaume, B., Elenga, H., Fournier, M., Martin, L., Namur, C. de, Schwartz, D., Servant, M., and Wirrmann, D. (1994). Changement majeur de la végétation du lac Sinnda (vallée du Niari, Sud-Congo) consécutif à l' assèchement climatique holocène supérieur: apport de la palynologie. *Comptes Rendus à l'Academie des Sciences de Paris* Séries II **318**: 1521-1526.

Vincens, A., Schwartz, D., Bertaux, J., Elenga, H., and Namur, C. de (1998) Late Holocene climatic changes in western equatorial Africa inferred from pollen from Lake Sinnda, south Congo. *Quaternary Research* **50**: 34-45.

SECTION II

PLANT CULTIVATION

He told me that many of the Mima were excellent hunters and knew the whereabouts of gazelle, the hares and wildcats. "But the game has gone since my father's time", he told me. "Now even the crops will not grow". I asked how it was possible for these people to survive in this drought. In answer he pointed out some tiny marks at intervals around the base of a bush. "See, the mouse of the desert!" he said. "He lives, even though there is no water in the land. He eats the little plants that grow in the wadis, that you hardly notice! My people are like the jerboa. As long as there is some moisture, they can survive". He told me how women would gather the wild grasses like *gau* and *haskanit* and grind them into flour on their stone hand-mills, and how the flour could be made into a polenta that was almost as that made from sorghum or millet.

Michael Asher, *A Desert Dies*, Viking, London, 1986, p. 45.

In this section, **Barakat** (Chapter 8) as well as **Breunig and Neumann** (Chapter 9) reveal, on the basis of archaeobotanical investigations, that there is no evidence for the cultivation of domesticated indigenous plants in Africa before 1200 cal BC. To the north of Burkina Faso, small-scale plant cultivation of millet was practiced by hunters and gatherers around 1000 cal BC. Sedentary, village-like, organized communities with domesticated animals and a strong economic dependence on cultivated plants did not appear until the beginning of our era, in an early Iron Age context.

In the Egyptian Sahara, as revealed by **Barakat**, recent archaeobotanical research provides evidence for intensive utilization of wild stands of indigenous sorghum and millet dating back to the eighth millennium bp under warm, moist conditions. However, according to **Barakat**, a return to relatively moist, and probably cool, conditions, perhaps associated with autumn and spring rain after 5.9 kyr bp, precluded the domestication of sorghum or millet in the Eastern Sahara. In the Nile Valley, wheat and barley from Southwest Asia were cultivated in preference to local grasses.

In the Nigerian Chad Basin at the transition to agriculture, as shown by **Breunig and Neumann**, the Holocene human occupation was initiated by pastoral communities colonizing the area after the retreat of Mega-Lake Chad some time between 1800 cal BC and 1500 cal BC. Under the stimulus of increasing aridity, for which there is good palynological evidence, the same communities began to expand their food producing economy by cultivating *Pennisetum americanum* around 1200 BC.

The archaeobotanical data indicate that agriculture started simultaneously in both areas around 1200 cal BC, and that it was most probably introduced from outside. *Pennisetum americanum* was the main crop, having its origin in the Central or Southcentral Sahara. The domestication area of *Sorghum bicolor* has to be sought east of Lake Chad, as domesticated *Sorghum bicolor* is absent from all early agricultural sites in the West African Sahel, including Northeast Nigeria.

Although indigenous plants were not cultivated before the fourth millennium bp, agriculture in Africa first appeared in the Nile Valley. Neolithic sites in Lower Egypt contain domesticated cereals (emmer wheat, barley) and also pulses imported from Southwest Asia. The introduction of cultigens is discussed here by **Rossignol-Strick**, who traces the first appearance of cereals from Southwest Asia to ca. 5200-4500 yr cal BC (7.15 to 6.5 kyr cal BP, 6.3 to 5.7 kyr bp) in the Fayum, and ca. 4800 yr cal BC (6.75 kyr cal BP, 5.9 kyr bp) in Merimde in the western Nile Delta. The cereal remains reveal the non-brittle, tough rachis of the ear, a recessive mutation considered the hallmark of domestication. In Egypt, tough-rachised plants are indicative of agriculture, because their wild progenitors, which have a brittle rachis ensuring wide dispersal of the seeds, are restricted to Southwest Asia, presently and probably also in the past. It is also in the latter region that the earliest domesticates are observed, dated 9700 yr bp (11,200 yr cal BP) in Tell Abu Hureyra on the Euphrates and 9800 bp in Aswad near Damascus.

Rossignol-Strick also examines the possible role of the rapid climate change from the arid and cold Younger Dryas (11-10 kyr bp) to the wet and warm early Holocene Optimum (9-6 kyr bp), in the emergence of cereal domestication in Southwest Asia. The monophyletic recessive mutation could have been naturally selected by the harsh climate conditions of the Younger Dryas, until forming natural populations large enough not to be missed by human gatherers at the dawn of the Holocene Optimum. Subsequently, the spread of cereal cultivation into the Nile Valley, ca. 3500 years later than in Southwest Asia, may be accounted for by its contingency upon the simultaneous introduction of domesticated sheep during the later part of the Climate Optimum, as those flocks required a continuous pasture land for their transit toward Africa.

Response to droughts in an ethnographic context, as a means of gaining an insight into the past, is provided in this section by **Butler** (Chapter 11), from her work in Ethiopia, where, within a monsoon climate generally considered favorable for agriculture, cycles of drought and unpredictable rains occur. These events have prompted the development of strategies to offset the effect of the crop losses which result. These measures appear to mitigate short-term and localized food shortages. However, should changing patterns of precipitation persist and become long term, further remedies may become essential. Ethnographic studies of traditional agriculture in the northern Ethiopian Highlands of Tigray, reported by Butler, reveal that grain crops cultivated today include indigenous domesticated African cereals together with an assemblage of crops domesticated in Southwest Asia. However, the dating and mechanisms of the beginnings of cultivation in Ethiopia are not fully understood.

8
REGIONAL PATHWAYS TO AGRICULTURE IN NORTHEAST AFRICA

H. N. Barakat

INTRODUCTION

The aim of this paper is to explore the transition to agriculture in Northeast Africa on the basis of archaeobotanical research conducted during the last two decades in Egypt and the Sudan. This is an attempt to show that the emergence of food production was not synchronous. In addition, the pathways and processes were not uniform in different regions, even if they were adjacent, such as in Southwest Asia and North Africa.

TERMINAL PLEISTOCENE

Our knowledge of plant use in Egypt begins 18,000 years ago in Wadi Kubbaniya, on the western bank of the Nile close to Aswan, the archaeological site from which the oldest plant macro-remains have been recovered (Figure 8.1). The radiocarbon dates for the site range from 19,000 to 17,000 years bp (Wendorf et al,. 1989; Haas, 1989). An extensive sampling strategy was carried out at the site and has yielded an impressive collection of floral and faunal remains. The floral remains were identified by G. Hillman and J. Hather (Hillman et al. 1989). Among the carbonized plant remains recovered from the site were the rhizomes of several sedges (*Cyperus rotundus, Scirpus maritimus* or *S. tuberosus*), and a Liliaceae. Dom palm (*Hyphaene thebaica*) fruit pericarp, and Anthemidae (*Compositae*) fruits featured among the identified remains. These plants must have been used as food and collected in the vicinity of the site. They were eaten fresh or were dried and then ground into flour, probably using the grinding stones abundant on the site (Hillman, 1989; Hillman et al., 1989). The faunal remains identified by Gautier and Van Neer (1989), which were dominated by fish bones, included birds, hare, gazelle, hartebeest and wild cattle. During the Terminal Pleistocene, the area of Wadi Kubbaniya consisted of a landscape of marshes with rushes, sedges and hydrophytes, while trees grew on the terraces (Wendorf et al., 1989; Hillman et al., 1989). The inhabitants of settlements along the Nile banks took up fishing, gathering of wild plants growing in the flood plain and, to a lesser extent, hunting. The subsistence economy of the site probably revolved

111

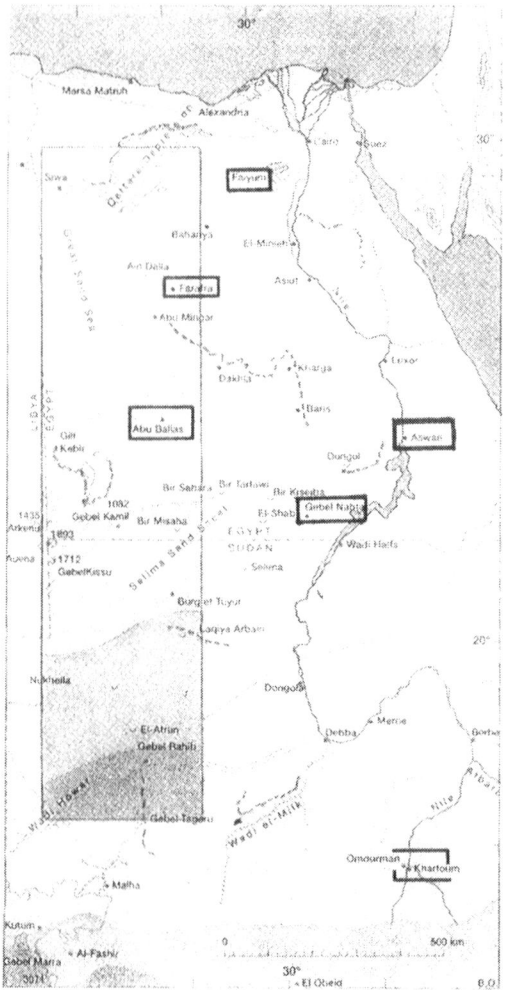

Figure 8.1. Location map showing sites mentioned in the text (after Neumann, 1989a, modified).

around the seasonal rhythm of the river, with a focus on fish and wetland root foods (Hillman, 1989; Wendorf *et al.*, 1989). A wide range of resources were available to the inhabitants. This pattern might have characterized other Epipalaeolithic settlements in the Nile Valley, although little is known about plants in other sites, probably due to poor preservation of wetland floral remains as well as inadequate sampling and recovery procedures.

By 12,000 bp, sites in the Nile Valley provide us with no plant remains but there are grinding stones, fish bones and hearths, as well as what might even be traces of controlled burning (Wetterstrom, 1998). The Nile Valley seems to have been unsuitable for settlement and agriculture between 15,000 and 12,000 years bp due to catastrophic floods (Butzer, 1980), which were also probably behind the movement of some groups to the Eastern Sahara.

THE HOLOCENE

During the Holocene, the Sahara desert of North Africa witnessed major climatic changes. Although it could generally be described as a wet/moist period (Hassan, 1986a, this volume), there is evidence that it consisted of several wet/moist episodes separated by dry interludes. These episodes varied in duration from 100 up to 1000 years (Hassan, 1986a, this volume). During the early Holocene, the wet episodes were wetter and longer than the dry intervals. A shift towards more dry conditions is evident by the beginning of the middle Holocene and this trend continued and was accentuated by 4500 bp, the end of the middle Holocene. During the late Holocene a much drier climate reigned over the whole region. The estimation of rainfall during wet episodes ranges from 200-300 mm in the south and 50-100 mm in the north, summer rain in the south and a combination of summer and winter rain further north (Hassan, 1986a; Neumann, 1989a; Ritchie and Haynes, 1987).

The identification of charcoal samples from archaeological sites along a north-south transect in Northeast Africa (Neumann, 1989a,b) enabled the reconstruction of the environment during the Holocene moist phases. It is only during these episodes that we have material that would provide charcoal samples. The data indicate a 400-500 km northward shift in the vegetation belts (Neumann, 1989a). A dry savanna ecosystem replaced the extreme desert in the hyperarid core of the present-day Eastern Sahara (see Figure 8.1). This environmental setting allowed settlement in the Sahara during the early and middle Holocene. The faunal remains from Neolithic Saharan sites show that the inhabitants practiced fishing in the ephemeral or permanent water bodies resulting from the monsoon summer rains, and the hunting of wild animals (Gautier, 1984a,b). Cattle keeping seems to have emerged in the Eastern Sahara around 9000 bp as a response to dry spells (Hassan, 2000). At the Nabta Playa and Bir Kiseiba, the bones of putatively domesticated cattle (Wendorf and Schild, 1994; Gautier, 1984a) were identified from early Neolithic sites. By 8000-7000 bp, cattle herding and pastoralism became a significant mode of subsistence, which entailed a mobile way of life in order to ensure pasture for their animals. Fireplaces, the so-called 'Steinplätze' dated from 9000-5000 bp, found all over the Sahara, have been interpreted as short term campsites for pastoralists (Gabriel, 1987). The identification of charcoal from Nabta and other sites in the Eastern Sahara provides evidence for the presence of oasis-like vegetation around temporary bodies of water (Figure 8.2; Barakat, 1995a). The role played by plants in the subsistence strategies of the Saharan inhabitants had until recently been almost unknown. However, recent archaeobotanical research in the Eastern Sahara

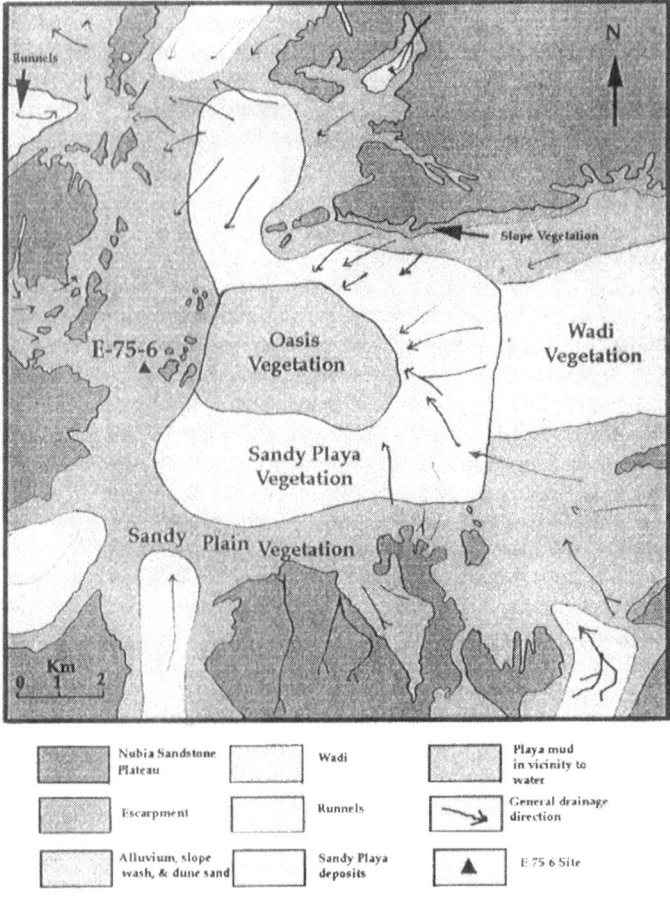

Figure 8.2. Palaeovegetation map of Nabta area ca. 8000 yr bp.

provides exciting new data from three Neolithic sites (Figure 8.1): Nabta Playa, the Hidden Valley in Farafra Oasis and Eastpans in the Abu Ballas ridge, dated to 8000, 7600 and 6200 bp respectively. The three sites are found in depressions and were covered by fine playa sediments which acted as a protective layer and led to the conservation of the plant material (Barakat and Fahmy, 1999). The botanical material was recovered through dry sieving of bulk samples of sediments (Barakat, 1995b; Barakat and Fahmy, 1999; Wasylikowa, 1997) and included seeds, grains, fruits, wood charcoal, rhizomes and other underground organs. The sediments were retrieved from hearths, in and around potholes, and in pits (Barakat and Fahmy, 1999; Wasylikowa, 1992, 1997; Wendorf et al., 1992), and the plant remains were thus probably brought to the site to be used as food and fuel.

In spite of difficulties in identification of the material (Wasylikowa, 1992, 1997), due to the lack of a comprehensive reference collection and the poor preservation of part of it, a large number of morphological types ranging between 25 (in the Hidden Valley and Eastpans) and almost 100 (in Nabta) were identified. The variation in the number of types (taxa) between the sites is probably a function of the number of samples examined. The high diversity in the material implies a wide resource palette, as well as an extensive knowledge of the uses of the different plants. The assemblages from the three sites were compared (Barakat and Fahmy, 1999). A striking resemblance, especially among the grass taxa, where identified, was noticed (Table 8.1). Sorghum was among the grasses identified from the three sites and because, at least at Nabta, there is further evidence that sorghum grains were harvested, used or stored separately from other plants, the question of whether sorghum was domesticated has been raised (Wendorf *et al.*, 1998). The wild progenitor of domesticated sorghum (*Sorghum bicolor* var. *arundinaceum* var. *verticillatum*) grows nowadays 600 km to the south of Nabta and the question has arisen as to whether it might have grown that far north during the early Holocene. Attempts to ascertain that the sorghum at Nabta was domesticated involved the use of gas chromatography/mass spectrometry in the analysis and identification of sorghum seeds (Biehl *et al.*, 1999). It is now confirmed that the grains were identified as *Sorghum bicolor* var. *arundinaceum* and are morphologically wild (Wasylikowa, 1997).

Table 8.1. Presence/absence of a variety of grass types at Nabta, Hidden Valley and Abu Ballas

Grass name	Nabta 8000 bp	Hidden Valley 6700 bp	Abu Ballas 6200 bp
Panicum turgidum	+	+	+
Setaria viridis	+	+	+
Echinochloa colona	+	+	+
Digitaria	+	+	+
Sorghum	+	+	+
Brachiaria	-	+	+
Urochloa	+	-	+
Cenchrus/Pennisetum	-	+	-

In addition to the archaeological evidence of the use of plants, including the grinding stones, hand stones and ground stones scattered among artifacts at Neolithic sites, there is now archaeobotanical evidence for the intensive gathering and use of wild plants by the inhabitants of these sites and probably at others yet to be discovered in the Eastern Sahara.

The harvesting of wild grass grains is a practice that was still very important among the Tuareg in the Central Sahara and Sub-Saharan Africa until the beginning of the twentieth century (Harlan, 1989a,b). It still persists today on a much reduced scale and provides an interesting ethnographic analogy to the subsistence strategy practiced during the Neolithic in the Sahara (Harlan, 1989a, b). The Tuareg collect several species of Graminae grains in large quantities. In the desert, they collect *Panicum, Aristida* and *Cenchrus*; in the savanna, there is Kreb: a complex of a dozen species including *Eragrostis* spp., *Brachiaria deflexa, Dactyloctenium aegyptium, Digitaria* spp., and *Panicum laetum*. While in the swampy areas the gathering of *Oryza, Paspalum, Bourgou, Echinochloa stagnina* is common, in Ethiopia the Abyssinian oat *Avena abyssinic* is endemic and much collected and used. The grains are made into porridge, ground into flour, used for couscous, sold commercially and stored in bell shaped pits during the dry season (Harlan, 1989a,b; Nicolaisen, 1963). This large scale harvest might have been the first step towards the domestication of these grasses, but the process does not begin until the grains harvested are deliberately planted and, according to Harlan, agriculture in the Sahara was unnecessary until quite recently. Deficits in the grain supply were complemented by importing or buying grains from the savanna (Kreb), and harvesting was carried out by slaves, women or children (Harlan, 1989a). Before European intervention, the wild cereals represented a major portion of the diet and were supplemented with milk and fat. We can now trace the antiquity of this practice back to 8000 years ago.

From the end of the middle Holocene, very few sites were recorded from the Eastern Sahara. The scarcity of sites dating to later than 6200 bp in the Eastern Sahara could be related to the trend towards drier conditions which started during the middle Holocene and was well established by the beginning of the late Holocene at 4500 bp (Ritchie *et al.*, 1985; Hassan, 1997).

THE SUDAN

According to the vegetation reconstruction by Neumann (1989a), and Ritchie and Haynes (1987), the present day semi-desert and dry savanna in Northern and Central Sudan were covered with deciduous savanna vegetation. It seems plausible that the inhabitants of prehistoric sites in the desert were hunter-gatherers, although our knowledge is still fragmentary since only insufficient evidence is available about the role of plants in the subsistence strategy. The scarcity of plant macro-remains in archaeological sites in the Sudan is mainly due to the poor preservation of organic material in the present day alternating wet/dry climate of the Central Sudan (Magid and Caneva, 1998).

The sites in the Sudanese Nile Valley belonging to the so-called Mesolithic, which extends from 8500 to 7700 bp, for example Aneibis, El Damer and Abu Darbein, yielded only a few plant macro-remains, mainly *Ziziphus* sp., and *Celtis* sp. stones, and *Setaria* sp. The study of imprints of plants in pottery provided evidence for the presence of several wild grasses: sorghum, panicum and setaria grains, and this suggests that the inhabitants were gathering wild plants in the vicinity of the sites (Magid, 1989, 1995). More plant impressions were examined from Sagai and Kabbashi al-Haitah, and dated between 7500 and 6100 bp. Similar results were obtained from earlier sites (Magid and Caneva, 1998).

Neolithic sites, such as Kadero, dated to 6000 bp, from which domestic livestock bones have been identified (Krzyzaniak, 1984, 1991; Gautier, 1984), offer little information about food plants. The plant macro-remains consist of a few *Celtis* stones and wild sorghum grains (Stemler, 1990; Magid, 1989). Attempts to identify plant impressions in pottery resulted in evidence for the use of wild grasses such as *Sorghum* sp., and *Setaria* sp., as well as another Panicoid grass, and stones of *Celtis* (Stemler, 1990; Magid, 1989). During the 1994 season, extensive recovery efforts yielded only charcoal (Barakat, 1995c). Through the analysis of the charcoal samples, I detected the establishment of a scrub and thorn savanna replacing the deciduous savanna in the vicinity of the site. This was interpreted as evidence for an intensive human impact on the vegetation. Human activities would have included the felling of trees, grazing and controlled burning of the vegetation (Barakat, 1995c), but there is no direct evidence for the domestication of plants on any of the Mesolithic or Neolithic sites. Neither do later sites from the Sudan offer any better information about the domestication of plants. The first solid evidence for the domestication is from Meroe, dating to the end of the first millennium BC (20 ± 160 BC), from which domesticated sorghum grains were identified (Stemler and Falk, 1981).

THE TRANSITION TO AGRICULTURE IN SOUTHWEST ASIA

During the period from 11,000 to 7500 bp, the Neolithic transformation was taking place in Southwest Asia, a relatively well known region, where the results of half a century of archaeobotanical research, carried out by eminent archaeobotanists, now allows us to construct a developmental model of the transition to agriculture in the region.

Recent studies by Hillman (Harris, 1998; Hillman, in press) show that cereal cultivation had begun during the late Epipalaeolithic. Due to the cool, dry climate of the Younger Dryas from 11,000 to 10,000 bp, there was a decrease in the food plants available, especially wild cereals. In the archaeobotanical assemblage from Abu Hureyra, Hillman found a notable increase in the number of weeds, an indication of cultivation, in addition to a few rye (*Secale cereale*) grains which belong to a domesticated form (Hillman, in press).

The emergence of agriculture, however, took place during the Pre-Pottery Neolithic A (PPNA), ca. 10,300-9500 bp, which corresponds to a warmer wetter climate. This transition was slow, as evidenced from the small number of grains and/or seeds showing domestic traits. The plants that were first domesticated are called 'founder crops' by Zohary (1996). These include barley (*Hordeum vulgare*), einkorn wheat (*Triticum monococcum*), emmer wheat (*Triticum dicoccum*), lentil (*Lens esculentus*), pea (*Pisum sativum*), chickpea (*Cicer arietinum*), bitter vetch (*Vicia sativa*) and flax (*Linum usitatissimum*). During the PPNA, both the collection of wild grasses and the small scale cultivation of wild cereals and pulses on small patches of ground were taking place simultaneously (Harris, 1998). With the employment of sickle harvesting in some sites, there is a gradual shift to agriculture and the appearance of domesticates. This shift might have happened first in the Southcentral Levant, although more evidence is needed before this could be firmly established (Harris, 1998).

From ca. 9500 bp onwards, there is clear evidence that agriculture, together with raising livestock, became the dominant subsistence strategy in Southwest Asia. The process was cumulative and developmental as more sites yielded remains of cultigens, and as more plant species were being added to the crop list. By the end of the PPNB, which lasted from 9500 to 7500 bp, agro-pastoralism was well established in the region and began to spread to other regions such as Central and South Asia, Europe (Harris, 1998) and eventually to Northeast Africa.

THE TRANSITION TO AGRICULTURE IN EGYPT

The beginning of agriculture in Egypt seems to have occurred ca. 6300 bp, or 5000 cal BC (Hassan, 1985; Wetterstrom, 1993). This coincided with the onset of aridity in the Sahara (Hassan, 1986a) and the emergence of agricultural communities in the Nile Valley. The first evidence for the transition comes from two sites: Merimde Beni Salama in the western Delta and Fayum, both dated to around 6300 bp. The archaeobotanical assemblage included emmer wheat, two-row and six-row barley, pea, lentil, flax, weeds and sedges (Caton-Thompson and Gardener, 1934; Zohary and Hopf, 1988). Later sites such as El Omari, on the eastern bank of the Nile south of Cairo, dated to 5220 bp, yielded vetch, flax, *Lathyrus*, pea, *Ficus sycomorus* and weeds (Täckholm, 1948; Barakat, 1990). Badari, the earliest known site in Upper Egypt, dates to 5200 bp. In addition to emmer, six-row barley and flax, weeds such as *Bromus* and *Vicia hirsuta* were apparently identified in the assemblage (Brunton, 1930). Nagada is another site in Upper Egypt from a slightly later date where cereals: emmer, six-row barley, pea and several wild plants were identified (Wetterstrom, 1993).

The earliest domesticates in Egypt were thus the founder crops of Southwest Asia. This implies that, at the transition to agriculture, a decision was taken not to domesticate or cultivate the indigenous sorghum and millets. This was probably due to their unsuitability for the Nile Valley. Nowadays sorghum is widely cultivated in Central Sudan rather than in Egypt because, as a summer crop, it requires higher temperatures than those present today, or by 5000 BC, in the lower Nile Valley. Moreover, in a habitat favorable for Near-Eastern cultivars, wheat and barley provided a much higher yield than sorghum. Both wheat and barley are also more palatable, and are used as human food in Egypt, whereas sorghum is mostly used nowadays as fodder. Sorghum was not cultivated until much later; the first firm evidence of domesticated sorghum (using DNA analysis) comes from the first century AD in Qasr Ibrim (Rowley-Conwy, 1991; Rowley-Conwy *et al.*, 1997).

The severe droughts from 7200 to 5900 bp (Hassan, 1986a) were apparently responsible for the dispersal of the desert inhabitants and their livestock to the Nile Valley. Once in the Nile Valley, the drifters from the Sahara encountered others from the Levant, who brought domesticated crops already under cultivation for more than 3000 years. From this package, a mixed population of indigenous dwellers and newcomers created a subsistence regime that used the incoming crops, the domesticated animals and employed their own agricultural system of cultivation in a riverine environment. By 6000 bp in Lower Egypt, and almost a thousand years later in Upper Egypt, Southwest Asian domesticates assumed a paramount role in the economic basis of the society. Crops were sown after the receding of the flood water, grew through the winter and were harvested in spring; this knowledge might have stemmed from the desert dwellers' experience of the filling up of lakes and the recession of water after rainfall in depressions in playa sites in the Sahara (Harlan and Pasquereau, 1969). They continued gathering wild plants alongside the cultivation of the crops brought in from Southeast Asia and did not domesticate the indigenous plants such as sorghum and millets. The transition to a full sedentary agrarian society was apparently faster in the Nile Valley than in marginal regions such as in the Fayum (Hassan, 1986; Wetterstrom, 1993).

CONCLUSIONS

Agriculture did not emerge in the Nile Valley during the early Holocene because:

(1) the riverine environment allowed a different way to ensure food security through fishing, collection of the root parts of water plants and sedges.
(2) unlike Southwest Asia, the wild progenitors of wheat and barley were absent in the Nile Valley.

Agriculture did not emerge in the Eastern Sahara because:

(1) high inter-annual variability in rainfall made farming unreliable and too risky.
(2) cattle keeping emerged in the Eastern Sahara but did not lead to specialized pastoralism until it had reached the Central Sahara and the Sudan where the dense savanna grasses were ideal for the grazing of cattle.
(3) the droughts between 7200 and 5900 bp led to the dispersal of the desert dwellers and the abortion of any agricultural initiatives.

The late emergence of food production in the Nile Valley, Delta and the Fayum after 6300 bp relied upon local fishing, cultivation of 'founder crops' and sheep/goat herding, introduced from Southwest Asia, and cattle herding from the Eastern Sahara.

REFERENCES

Barakat, H.N. (1990). Plant remains from El Omari. In Debono, F., and Mortensen, B. (eds.), *El Omari. A Neolithic settlement*, AV 85, pp. 109-114.

Barakat, H.N. (1995a). *Contribution Archéobotanique à l'Histoire de la Végétation dans le Sahara Oriental et dans le Soudan Central*, Doctoral thesis, University d'Aix-Marseille III. [unpublished]

Barakat, H.N. (1995b). Charcoals from the Neolithic Site at Nabta Playa (E-75-6), Egypt. *Acta Palaeobotanica* 35(1): 163-166.

Barakat, H.N. (1995c). Middle Holocene vegetation and Human Impact in Central Sudan: charcoal from the Neolithic site at Kadero. *Vegetation History and Archaeobotany* 4: 101-108.

Barakat, H.N., and Fahmy, A.G. (1999). Wild Grasses as 'Neolithic' Food resources in the Eastern Sahara: A Review of the Evidence from Egypt. In van der Veen, M. (ed.), *The Exploitation of Plant Resources in Ancient Africa*, Kluwer Academic/Plenum Publishers, New York, pp. 33-46.

Biehl, E., Wendorf, F., Handry, W., Desta, A., and Watrous, L. (1999). Use of gas chromatography and mass spectrometry in the identification of ancient Sorghum seeds. In van der Veen, M. (ed.), *The Exploitation of Plant Resources in Ancient Africa*, Kluwer Academic/Plenum Publishers, New York, pp. 47-53.

Brunton, G. (1930). *Qau and Badari*, III, Bernard Quatrich, London.

Butzer, K.W. (1980). Pleistocene History of the Nile Valley in Egypt and lower Nubia. In Martin A.J., Faure W., and H. Faure (eds.), *The Sahara and the Nile: Quaternary Environments and Prehistoric Occupation in Northern Africa*, Balkema, Rotterdam, pp. 253-280.

Caton-Thompson, G., and Gardener, E.W. (1934). *The Desert Fayum*, London, Royal Anthropological Institute of Great Britain and Ireland.

Gabriel, B. (1987). Palaeoecological Evidence from Neolithic Fireplaces in the Sahara. *African Archaeological Review* 5: 93-103.

Gautier, A. (1984a). Archaeozoology of the Bir Kiseiba region, Eastern Sahara. In Wendorf, F., and Schild, R. (ass.), Close, A.E. (ed.), *Cattle Keepers of the Eastern Sahara: the Neolithic of Bir Kisieba*, Southern Methodist University Press, Dallas, pp. 49-72.

Gautier, A. (1984b). The Fauna from the Neolithic site of Kadero (Central Sudan). In Krzyzaniak, L., and Kobusiewicz, M. (eds.), *Origin and Early Development of Food-producing Cultures in North-Eastern Africa*, Poznan Archaeological Museum, Poznan, pp. 317-319.

Gautier, A., and Van Neer, W. (1989). Animal Remains from the Late Paleolithic Sequence at Wadi Kubbaniya. In Wendorf, F., Schild, R., and Close, A. (eds.), *The Prehistory of Wadi Kubbaniya*, Southern Methodist University Press, Dallas, pp. 119-161.

Haas, H. (1989). The Radiocarbon Dates from Wadi Kubbaniya. In Wendorf, F., Schild, R., and Close, A. (eds.), *The Prehistory of Wadi Kubbaniya*, Southern Methodist University Press, Dallas, pp. 274-279.

Harlan J. (1989a). Wild grass seeds as food sources in the Sahara and sub-Sahara. *Sahara* 2: 69-74.

Harlan J. (1989b). Wild-grass seed harvesting in the Sahara and Sub-Sahara of Africa. In Harris, D.R., and Hillman, G. (eds.), *Foraging and Farming: the evolution of plant exploitation*, Unwin Hyman, London, pp. 79-98.

Harlan J., and Pasquereau, A. (1969). Décrue Agriculture in Mali. *Economic Botany* 23: 70-74.

Harris, D. (1998). The Origins of Agriculture in Southwest Asia. *The Review of Archaeology* 19(2): 5-11.

Hassan, F.A. (1985). A radiocarbon chronology of neolithic and predynastic sites in Upper Egypt and the Delta. *African Archaeological Review* 3: 95-116.

Hassan, F.A. (1986a). Desert environment and origins of agriculture in Egypt. *Norwegian Archaeological Review* 19(2): 63-76.

Hassan, F.A. (1986b). Holocene lakes and prehistoric settlements of the Western Faiyum. *Journal of Archaeological Science* 13: 483-501.

Hassan, F.A. (1997). Holocene Palaeoclimates of Africa. *African Archaeological Review* 14: 213-230.

Hassan, F.A. (2000). Climate and Cattle in North Africa. In Blench, R., and MacDonald, K. C. (eds.), *The Origins and Development of African Livestock: Archaeology, Genetics, Linguistics, and Ethnography*, University College London Press, London, pp. 61-86.

Hillman, G. (1989). Late Palaeolithic plant foods from Wadi Kubbaniya in Upper Egypt: Dietary diversity, infant weaning and seasonality in a riverine environment. In Harris, D.R., and Hillman, G. (eds.), *Foraging and Farming: the evolution of plant exploitation*, Unwin Hyman, London, pp. 207-239.

Hillman, G. (in press). Overview: The Plant-Based Components of Subsistence at Abu Hureyra 1 and 2. In Moore, A.M. T., Hillman, G.C., and Legge, A.J. (eds.), *Abu Hureyra and the Advent of Agriculture*. Oxford University Press, New York.

Hillman, G., Madeyska E., and Hather, J. (1989). Wild Plant Foods and Diet at Late Paleolithic Wadi Kubbaniya: The Evidence from Charred Remains. In Wendorf, F., Schild, R., and Close, A. (eds.), *The Prehistory of Wadi Kubbaniya*, Southern Methodist University Press, Dallas, pp. 162-242.

Krzyzaniak, L. (1984). The Neolithic Habitation at Kadero (Central Sudan). In Krzyzaniak, L., and Kobusiewicz, M. (eds.), *Origin and Early Development of Food-producing Cultures in North-Eastern Africa*, Poznan Archaeological Museum, Poland, pp. 309-316.

Krzyzaniak, L. (1991). Early farming in the middle Nile basin: recent discoveries in Kadero. *Antiquity* 65(248): 515-532.

Magid, A.A. (1989). *Plant domestication in the Middle Nile Basin*. Oxford, British Archaeological Reports.

Magid, A.A. (1995). Plant Remains and their Implications. In Haaland, R., and Magid, A.A. (eds.), *Aqualithic Sites along the Rivers Nile and Atbara, Sudan*, Alma Mater, Bergen, pp. 147-177.

Magid, A.A., and Caneva, I. (1998). Economic Strategy based on food-plants in the Early Holocene Central Sudan: a reconsideration. In di Lernia, S., and Manzi, G. (eds.), *Before Food Production in North Africa. Proceedings of the Homonymous Workshop held in Forlì, September 1996, within the XIII World Congress of the International Union of the Prehistoric and Protohistoric Sciences*, ABACO Edizioni, Rome, pp. 79-90.

Neumann, K. (1989a). Vegetationsgeschichte der Ostsahara im Holozän: Holzkohlen aus prähistorischen Fundstellen (mit einem Exkurs über die Holzkohlen von Fachi-Dongouboulo/Niger). *Africa Prähistorica* 2:341.

Neumann, K. (1989b). Holocene Vegetation of the Eastern Sahara. Charcoal from Prehistoric Sites. *The African Archaeological Review* 7: 97-116.

Nicolaisen, J. (1963). *Ecology and culture of the Pastoral Tuaregs*, Nationalmuseets Skrifter Ethnografisk Raekke, Copenhagen.

Ritchie, J.C., and Haynes, C.V. (1987). Holocene vegetation zonation in the Eastern Sahara. *Nature* 330: 645-647.

Ritchie, J.C., Eyles, C.H., and Haynes, C.V. (1985). Sediment and pollen evidence for a middle Holocene humid period in the Eastern Sahara. *Nature* 314: 352-355.

Rowley-Conwy, P. (1991). Sorghum from Qasr Ibrim, Egyptian Nubia, c. 8000 BC-AD 1811: A Preliminary Study. In Renfrew, J.M. (ed.), *New light on Early Farming: Recent Developments in Palaeoethnobotany,*. Edinburgh University Press, Edinburgh, pp. 211-234.

Rowley-Conwy, P., Deakin, W., and Shaw, C.H. (1997). Ancient DNA from archaeological Sorghum (Sorghum bicolor) from Qasr Ibrim, Nubia: Implications for Domestication and Evolution and a Review of the archaeological Evidence. *Sahara* 9: 23-34.

Stemler, A. (1990). A scanning Electron Microscopic analysis of plant impressions in pottery from Kadero, El Zakiab, Um Direiwa and Al Kadada. *Archéologie du Nil Moyen* 4: 87-105.

Stemler, A., and Falk R.H. (1981). SEM of archaeological Specimens. *Scanning Electron Microscopy* III: 191-196.

Täckholm V. (1948). Flore et Agriculture. In Debono, F. (ed.), *ElOmari (près Helouan)*, Ann. Seru. Ant. Egypt. Cairo.

Wasylikowa, K. (1992). Exploitation of wild plants by prehistoric peoples in the Sahara. *Würzburger Geographische Arbeiten* 84: 247-262.

Wasylikowa, K. (1997). Flora of the 8000 years old archaeological site E-75-6 at Nabta Playa, Western Desert, Southern Egypt. *Acta Palaeobotanica* 37(2): 99-205.

Wendorf, F., and Schild, R. (1994). Are the Early Holocene Cattle in the Eastern Sahara Domestic or Wild? *Evolutionary Anthropology* 3(4): 118-128.

Wendorf, F., Schild, R., and Close, A.E. (eds.) (1989). *The Prehistory of Wadi Kubbaniya*, Volumes II and III, Southern Methodist University Press, Dallas.

Wendorf, F., Schild, R., Wasylikowa, K., Dahlberg, J., Evans, J., and Biehl, E. (1998). The use of plants during the early Holocene in the Egyptian Sahara: Early Neolithic food economies. In di Lernian, S., and Manzi, G. (eds.), *Before Food Production in North Africa. Proceedings of the Homonymous Workshop held in Forli, September 1996, within the XIII World Congress of the International Union of the Prehistoric and Protohistoric Sciences*, ABACO Edizioni, Rome, pp. 71-78.

Wendorf, F., Close, A.E., Schild, R., Wasylikowa, K., Housley, R.A., Harlan, J.R., and Królik, H. (1992). Saharan exploitation of plants 8,000 years BP. *Nature* 359: 721-724.

Wetterstrom, W. (1993). Foraging and Farming in Egypt: The Transition from Hunting and Gathering to Horticulture in the Nile valley. In Shaw, T., Sinclair, P., Andah, B., and Okpoko, A. (eds.), *The Archaeology of Africa: Food, Metals and Towns*, Routledge, London, pp. 163-226.

Wetterstrom, W. (1998). The Origins of Agriculture in Africa: with particular reference to Sorghum and Pearl Millet. *The Review of Archaeology* 19(2): 30-46.

Zohary, D. (1996). The Mode of Domestication of the Founder Crops of Southwest Asian Agriculture. In Harris, D.R. (ed.), *The origins and Spread of Agriculture and Pastoralism in Eurasia*, London and Smithsonian Institution Press, Washington, pp. 142-158.

Zohary, D., and Hopf, M. (1988). *Domestication of Plants in the Old World: the Origin and Spread of cultivated plants in west Asia, Europe and the Nile valley*, Clarendon Press, Oxford.

9

FROM HUNTERS AND GATHERERS TO FOOD PRODUCERS: NEW ARCHAEOLOGICAL AND ARCHAEOBOTANICAL EVIDENCE FROM THE WEST AFRICAN SAHEL

P. Breunig and K. Neumann

INTRODUCTION

The relationship between language history, cultural development and environment in the West African savanna has been the focus of an interdisciplinary project at the University of Frankfurt (Germany) in co-operation with the Universities of Ouagadougou (Burkina Faso) and Maiduguri (Nigeria) since 1988 and Cotonou (Benin) since 1997. One major research question of the archaeological and archaeobotanical working groups is the transition from hunting and gathering to food production. Since the beginning of archaeological research in West Africa many local and interregional studies have dealt with this topic (e.g., Andah, 1987; Casey, 1998; Davies, 1968; Harris, 1976; Holl, 1998; McIntosh and McIntosh, 1988; Shaw, 1981; Smith, 1980; Sowunmi, 1985, 1999; Stahl, 1986, 1993). In most cases these papers were based on very limited direct evidence, as a multidisciplinary approach including archaeology, archaeozoology and archaeobotany had only been occasionally applied in Africa.

Initially, our own point of view was colored by the European and Near Eastern evidence, i.e. the idea that sedentism, livestock keeping, agriculture and pottery appear synchronously as a 'Neolithic package'. However, in the light of new research, the simultaneous emergence of Neolithic traits in Central Europe and the Mediterranean rather seems to be a unique case which cannot be transferred as a general model to other regions (see Harris, 1996) and apparently has no similarities with its West African counterpart. With an increasing amount of data it is becoming clear that even the West African version of the transition from hunting and gathering to food producing communities is quite heterogeneous. In this paper we would like to present two case studies from northern Burkina Faso (Prov. Oudalan) and from the Nigerian Chad Basin (Figure 9.1) which might illustrate the interregional variability in the West African Sahel zone.

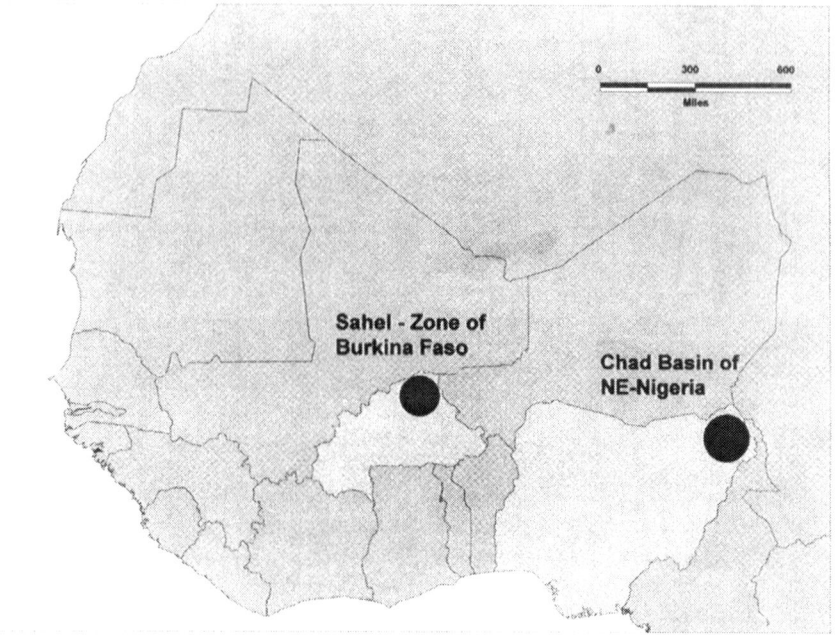

Figure 9.1. Study areas under consideration in this paper.

PALAEOCLIMATE AND PALAEOVEGETATION

The fourth millennium bp (2500-1200 cal BC) is usually regarded as a crucial period for the development of plant food production in the Sahara and the Sahel. Munson (1976), in establishing his Dhar Tichitt sequence, strongly claimed that the beginnings of cultivation in Northwest Africa were a response to climatic deterioration at the end of the fourth millennium bp. After the early and middle Holocene moister phase, a turn to arid conditions is visible in many pollen diagrams and palaeoecological data from the southern margin of the Sahara at least from 4800 bp onwards (see compilations by Lézine, 1989; Hassan, 1996, 1997). Vernet (this volume) states that an arid crisis around 4000 bp, followed by humid conditions in the fourth millennium, should have enabled large-scale prehistoric occupation in the southern Sahara. However, this millennium witnessed a general tendency towards lower rainfall (see Hassan, this volume), and severe dry spells occurred between 3800 to 3600 bp and around 3000 bp. Recent well-dated pollen profiles from the Sahel of Nigeria show an abrupt change to arid conditions around 3300 bp, indicated by the synchronous disappearance of southern (Sudanian and Sahelian) elements and the establishment of a modern Sahelian vegetation (Salzmann, 1996; Salzmann and Waller, 1998).

THE SAHEL ZONE OF BURKINA FASO

The study area is situated in northern Burkina Faso, close to the borders to Mali and Niger. The landscape is characterized by stabilized east-west trending, 1-10 km wide sand dunes of Pleistocene origin (Figure 9.2), which today are used for pasture and millet cultivation. From the archaeological point of view the dunes are specifically important because most prehistoric sites have been found here. Several reasons might be cited to explain this pattern, but the availability of water seems to be most significant. The flat crystalline plains with isolated granitic *inselbergs* and lateritic crusts extending between the dune ridges can hardly provide sufficient water supply in the dry season. In contrast, the vicinity of the dunes is different in this respect. Along the foot of some dunes there are springs which today permit horticulture. Some of the drainage channels crossing the plains had already been blocked by dunes in early Holocene times (Andres *et al.*, 1996) leading to the development of semi-permanent lakes. The largest among these depressions is Mare d'Oursi. Lakes like this, which dry up only in years with very low rainfall, are especially attractive for pastoralists, and clusters of archaeological sites in the surroundings testify that this is equally valid for the past.

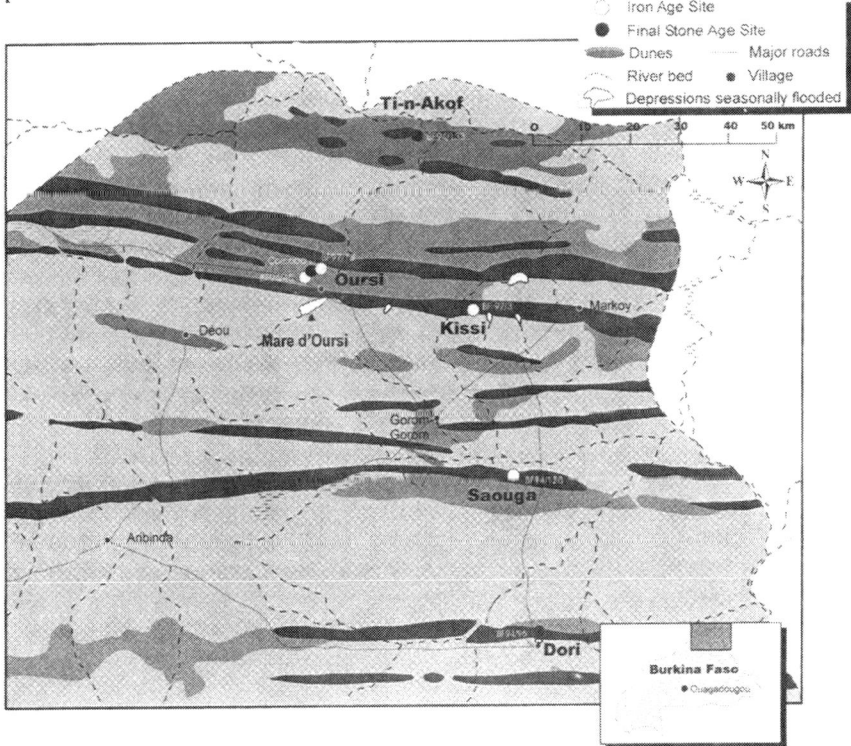

Figure 9.2. Map of northern Burkina Faso with major excavation sites (based on Vogelsang, 1995, p. 17).

Fieldwork carried out during the last few years (Neumann and Vogelsang, 1996; Vogelsang, 1995, 1996, 1997; Hallier, 1999; Thom, 1998; Vogelsang *et al.*, 1999) resulted in the discovery of several sites, the most important of which were found in the dune areas. With the exception of burials and scattered unstratified material, the occupation sequence summarized in Figure 9.3 is based on two types of sites: (1) dune sites mainly composed of quartz artifacts and some potsherds without any considerable accumulation of cultural deposits; and (2) settlement mounds with stratified cultural deposits up to 8 m in thickness. Type 1 belongs to the final Stone Age, and type 2 exclusively to the Iron Age. Both types are separated by the 'dark' first millennium BC, which is archaeologically completely unknown up till now.

FINAL STONE AGE HUNTERS AND GATHERERS

As Figure 9.3 and Table 9.1 show, there are two early dates (9230 ± 50 bp, UtC-5154; 6990 ± 50 bp, UtC-5162) from charcoal found in dune sites with microliths near Dori, which we do not consider to be correlated with archaeological material. We think that the confusing set of dates for these sites results from animal activities, and demonstrates the problems of dune sites. Apparently there are no sites with secure dates older than 2200 cal BC. Either the area was uninhabited during the early and middle Holocene, or the occupation during this period is not detectable so far by archaeological means.

Between 2200 cal BC and 1000 cal BC there is a cluster of dates from small final Stone Age dune sites (Vogelsang *et al.*, 1999). Besides the dated ones shown in Figure 9.3, many similar sites have been found but not excavated so far. We suppose they all belong to the same period, roughly between 2000 cal BC and 1000 cal BC, and that they are remnants of hunters and gatherers because they are of small spatial extent and have a rather low amount of cultural remains. In most cases one plastic bag is sufficient to store the whole surface finds, and stratified remains do not add up to very much. The cultural material (Figure 9.4) consists of a few potsherds, mainly comb-decorated, polished stone axes, some grinding equipment, and a microlithic industry dominated by segments and micro-points as well as bifacially retouched arrow-points of the Saharan type.

There are no visible settlement structures, except pits of unknown use which were found at two sites (Cocorba and Ti-n-Akof). They have a depth of up to one metre and cover an area of some square metres. The filling of the pits is slightly darker than the surrounding deposits. With regard to the preservation of organic materials, which are almost completely absent outside, the pits are a boon. Faunal remains are especially abundant at the Cocorba site. These faunal remains comprise fish, reedbuck, kob and large bovids, but no domesticated animals have yet been identified (Van Neer, personal communication).

Indirect evidence for agriculture in northern Burkina Faso is provided by the pollen diagram of Oursi (Ballouche and Neumann, 1995). For the period from 8000 to 3000 bp, the palynological data indicate a Sahelian grassland with few trees, and no human impact is detectable. But a significant change in the pollen diagram occurs between 3130 ± 80 bp (1495-1304 cal BC, UtC 2922) and 2850 ±

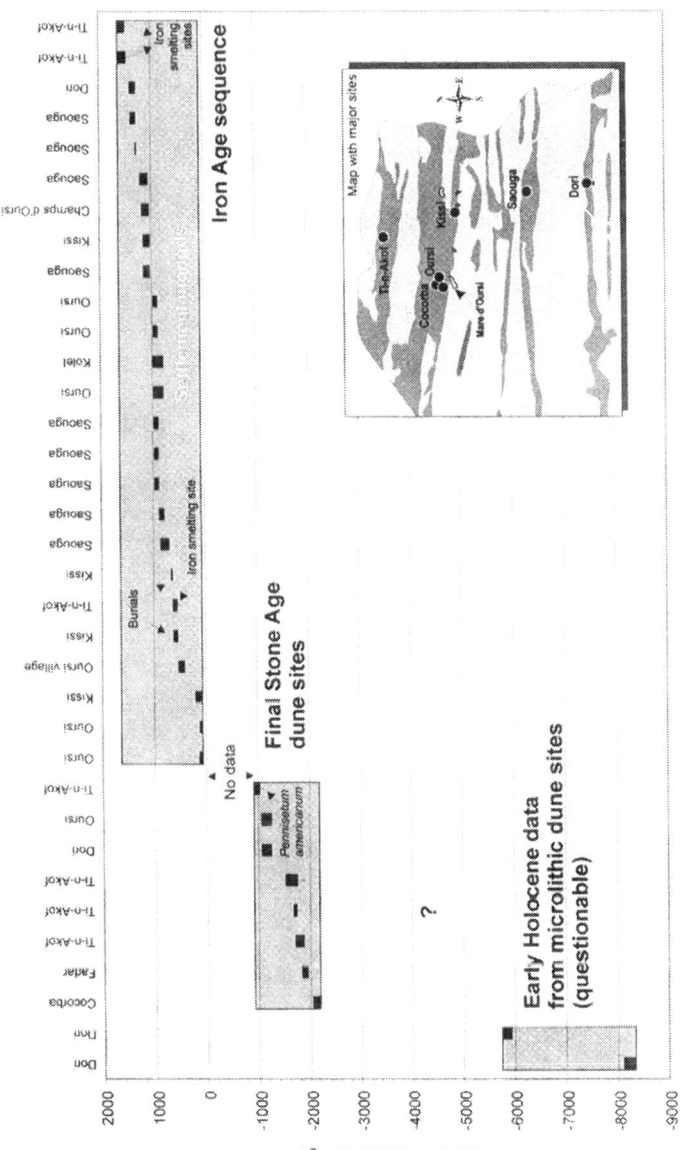

Figure 9.3. Radiocarbon chronology of the later prehistory of the Sahel Zone of Burkina Faso based on calibrated dates (I-Sigma: black rectangles; II-Sigma:vertical lines).

Table 9.1. Radiocarbon dates from archaeological sites in the Sahel zone of Burkina Faso

Site	Period	Lab code	^{14}C age bp	cal age (one-sigma std. dev)	Remarks
Dori (BF 94/96)		UtC-5154	9230 ± 50	8342-8096 BC	Charcoal from layer in dune site below microlithic industry and pottery
Dori (BF 94/40)	Final Stone Age	UtC-5162	6990 ± 50	5936-5752 BC	Charcoal from dune site with pottery and microlithic industry
Cocorba (BF 97/5)	Final Stone Age	UtC-7282	3722 ± 33	2201-2039 BC	Charcoal from pit in dune site with pottery and microlithic industry
Fadar (BF 96/40)	Final Stone Age	UtC-5743	3576 ± 40	1950-1828 BC	Charcoal from dune site with archaeological material similar to Ti-n-Akof
Ti-n-Akof (BF 94/133)	Final Stone Age	KN-4777	3479 ± 45	1877-1704 BC	Charcoal from dune site with pottery and stone arrowheads
Ti-n-Akof (BF 94/133)	Final Stone Age	UtC-6466	3413 ± 37	1743-1675 BC	See KN-4777
Ti-n-Akof (BF 94/133)	Final Stone Age	KN-4776	3380 ± 100	1759-1522 BC	See KN-4777
Dori (BF 94/96)	Final Stone Age	UtC-3949	2950 ± 40	1254-1060 BC	Charcoal from dune site with pottery and microlithic industry
Oursi (BF 94/45)	Final Stone Age	UtC-7218	2931 ± 32	1257-1053 BC	Charcoal from LSA layer at the base of an Iron Age mound
Ti-n-Akof (BF 94/133)	Final Stone Age	UtC-4906	2840 ± 49	1034-916 BC	Grain of *Pennisetum americanum*. Same site as KN-4777
Dori (BF 94/96)	Final Stone Age	UtC-5155	2316 ± 49	400-371 BC	Charcoal from dune site with pottery and microlithic industry
Oursi 1 (BF 97/26)	Iron Age	KN-4361	1950 ± 65	87 BC-AD 78	Charcoal from Iron Age mound close to Oursi (BF 94/45)
Dori (BF 94/40)	Final Stone Age	UtC-3950	1916 ± 55	AD 58-141	Charcoal from dune site with pottery and microlithic industry
Oursi (BF 94/45)	Iron Age	KN-4785	1907 ± 46	AD 69-141	Charcoal from Iron Age mound
Oursi (BF 94/45)	Iron Age	UtC-7353	1891 ± 33	AD 65-131	See KN-4785
Dori (BF 94/96)	Final Stone Age?	UtC-5157	1852 ± 49	AD 118-239	Charcoal from dune site with pottery and microlithic industry
Kissi (BF 96/22)	Iron Age	UtC-6467	1882 ± 34	AD 84-210	Seeds of *Vitex sp.* from mound close to graveyard (UtC-5670)
Oursi village (BF 97/13)	Iron Age	UtC-7354	1587 ± 35	AD 415-535	Charcoal from basal cultural layer of Iron Age mound
Kissi (BF 96/3)	Iron Age	UtC-5670	1495 ± 45	AD 543-630	Wood from Iron Age burial
Ti-n-Akof (BF 94/140)	Iron Age	KN-4803	1487 ± 48	AD 545-636	Charcoal from iron smelting site
Kissi (BF 96/3)	Iron Age	UtC-5671	1393 ± 33	AD 641-666	See UtC-5670

Site	Period	Lab code	Date BP	Calibrated	Comment
Dori (94/40)	Final Stone Age?	UtC-5161	1378 ± 33	AD 650-670	Charcoal from dune site with pottery and microlithic industry
Saouga (BF 95/7)	Iron Age	KI-4362	1230 ± 35	AD 692-861	Charcoal from Iron Age settlement (excavation on flat ground)
Saouga (BF 95/7)	Iron Age	UtC-6465	1197 ± 29	AD 7786-887	Seeds of *Butyrospermum paradoxum* from same excavation as above
Saouga (BF 94/120)	Iron Age	KN-4784	1139 ± 40	AD 883-974	Charcoal from mound of Iron Age settlement. Same site as dates above
Saouga (BF 95/7)	Iron Age	KN-4940	1135 ± 36	AD 885-974	See KI-4362
Saouga (BF 94/120)	Iron Age	KN-4775	1136 ± 44	AD 882-978	See KN-4784
Oursi 2 (BF 97/27)	Iron Age	KI-4360	1130 ± 30	AD 784-979	Charcoal from Iron Age mound close to Oursi (BF 94/45)
Kolel Nord (BF 97/23)	Iron Age	KI-4343	1110 ± 65	AD 783-994	Charcoal from Iron Age mound
Oursi 3 (BF 97/28)	Iron Age	KI-4358	1110 ± 30	AD 890-982	Charcoal from Iron Age mound close to Oursi (BF 94/45)
Oursi 4 (BF 97/29)	Iron Age	KI-4359	1110 ± 30	AD 890-982	Charcoal from Iron Age mound close to Oursi (BF 94/45)
Saouga (BF 94/120)	Iron Age	KN-4783	964 ± 35	AD 1022-1157	See KN-4784
Kissi (BF 97/31)	Iron Age	KI-4344	930 ± 35	AD 1025-1158	Charcoal from Iron Age mound in the area of the Kissi graves (BF 96/3)
Champs d'Oursi (97/25)	Iron Age	KI-4363	910 ± 30	AD 1035-1181	Charcoal from Iron Age mound
Saouga (BF 96/1)	Iron Age	UtC-5745	894 ± 34	AD 1053-1215	Charcoal from an isolated find of a clay figurine far off from BF 95/7
Saouga (BF 95/7)	Iron Age	KN-4939	720 ± 36	AD 1279-1297	Charcoal from Iron Age mound
Dori (BF 94/96)	Final Stone Age?	UtC-5156	698 ± 43	AD 1282-1304	Charcoal from dune site with pottery and microlithic industry
Saouga (BF 95/7)	Iron Age	KN-4941	657 ± 60	AD 1288-1396	Charcoal from Iron Age mound
Dori (BF 94/86)	Iron Age	KN-4786	634 ± 78	AD 1289-1406	Bone from surface of Iron Age settlement mound
Ti-n-Akof (BF 94/43)	Iron Age	UtC-5163	362 ± 37	AD 1471-1633	Charcoal from iron smelting site
Ti-n-Akof (BF 94/43)	Iron Age	UtC-5164	342 ± 27	AD 1486-1636	Charcoal from iron smelting site
BF 94/15	Final Stone Age?	UtC-6468	239 ± 34	modern	Ostrich eggshell from dune site near Markoy
Kissi (BF 96/7)	Final Stone Age?	UtC-5744	124 ± 33	modern	Charcoal from dune site with stone arrowheads
Dori (BF 94/96)	Final Stone Age?	UtC-5303	-460 ± 50	modern	Bone from dune site with microlithic industry

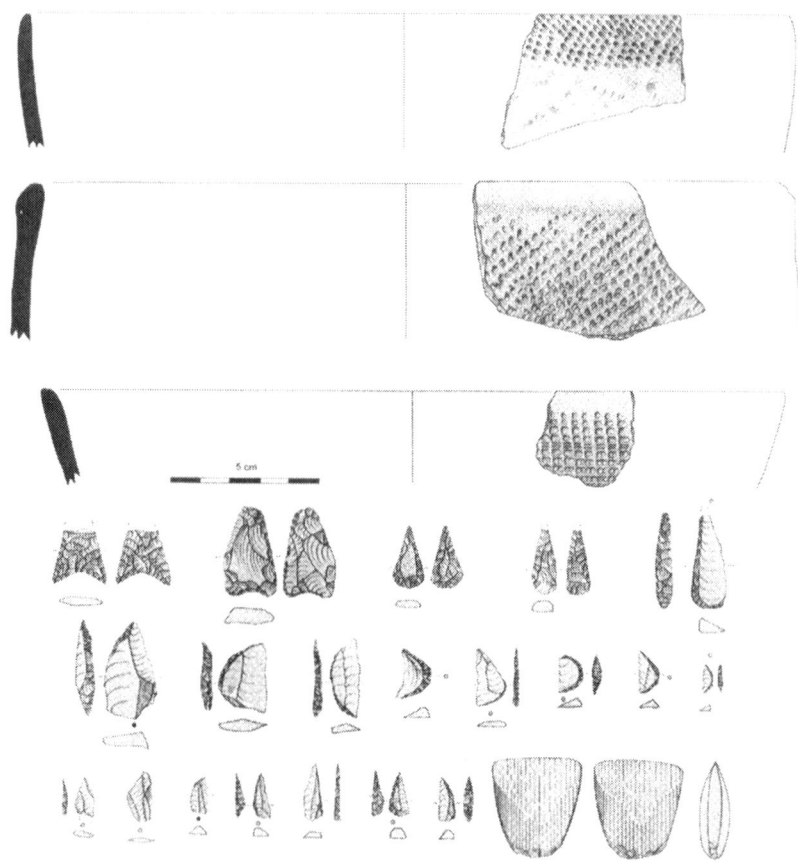

Figure 9.4. Cultural material from different final Stone Age dune sites in the Sahel Zone of Burkina Faso.

50 bp (1093-932 cal BC, UtC 2211). Increasing values of fallow plants and weeds in the upper part of the diagram suggest that the grasslands were opened by agricultural activities, such as clearance and shifting cultivation. The changes visible in the pollen diagram around 3000 bp (1200 cal BC) are quite abrupt, suggesting that agriculture started suddenly in this area. Direct archaeobotanical evidence for early agriculture in northern Burkina Faso is very sparse. On the dune sites, preservation of plant material is very bad, sediments are often disturbed, and despite an enormous effort with dry sieving and flotation almost no plant remains were recovered (Vogelsang, 1996; Neumann and Vogelsang, 1996; Neumann, 1999b). An exception is the dune site Ti-n-Akof, where grains of domesticated *Pennisetum americanum* have been found, directly dated by AMS

to 2840 ± 49 bp (UtC-4906, 1035-916 cal BC; Vogelsang, *et al.* 1999). Even though the *Pennisetum* grains are smaller than in the modern varieties, their shape clearly attributes them to a domesticated form (Kahlheber, personal communication), probably representing an early stage in the domestication process. The Ti-n-Akof *Pennisetum* confirms the hypothesis that agriculture was practiced on the dunes after 3000 bp, but the charcoal dates of this site indicate that it was occupied at least 500 years before (around 1700 cal BC), and for this period there is no evidence for cultivation so far. The charcoal of Ti-n-Akof mostly consists of gallery forest species whereas those species that would indicate anthropogenic activities are absent (Neumann *et al.*, 2000).

The contradiction between pollen evidence for agriculture from around 1200 cal BC onwards and the missing contemporary archaeological sites has not yet been solved; perhaps cultivation played only a minor role and apparently did not require a sedentary way of life in village-like settlements. There are no sites of sedentary agropastoral communities either for the second or the first millennia BC. We strongly claim this, because how should it be possible after nearly five years of surveys to come across microlithic assemblages all over the area and at the same time overlook the more substantial remains of sedentary farmers? Hence we conclude that, in the Sahel Zone of Burkina Faso, the final Stone Age economy was based on hunting, gathering and fishing throughout the second millennium BC, by the end of which domesticated plants were introduced.

The chances that the few available data are misleading cannot be excluded. One bone of cattle or ovicaprines will change the model of farming hunters and gatherers, and the sites dated between 2000 cal BC and 1000 cal BC would be attributed to a pastoral way of life like those from contemporary sites excavated by MacDonald in the neighboring Windé Koroji region of Mali (MacDonald, 1996).

After 1000 cal BC there is a complete lack of data for a thousand years until the beginning of our era, a fact which up to now is not well understood.

IRON AGE SEQUENCE

With the beginning of our era the situation changes completely. Large sites suddenly appear roughly 2000 years ago and are well recorded for the first millennium AD (Figure 9.3). They consist of groups of settlement mounds, separated by flat areas. Large amounts of cultural material and organic remains have accumulated both on the surface and in the mounds' deposits. It would be difficult not to interpret these agglomerations as villages or, at least, large hamlets.

From the beginning of this period, i.e. around 1 cal AD, the economy was strongly based on domesticated plants and animals, and intensive millet cultivation resulted in distinct vegetation changes (Neumann *et al.*, 1998). Grave-goods from the graveyard of Kissi point to far-reaching trade networks and prosperity, and weapons indicate conflicts in the sixth century AD (Thom, 1998).

In the meantime Iron Age villages have been found all over northern Burkina Faso, thus enabling studies on changes in settlement pattern and economy for more than a millennium as well as the influence of the emerging medieval empires in the neighborhood.

THE CHAD BASIN OF NORTHEAST NIGERIA

The study area in the Chad Basin of Northeast Nigeria also belongs to the Sahel Zone. Our work concentrated on the plains southwest and south of the lake (Figure 9.5). The only prominent landmark in the Chad Basin of Nigeria is the Bama Ridge, a sandy ridge of up to 12 m in height, supposed to be the shoreline of Mega Chad during the early and middle Holocene, which had flooded large areas of the Chad Basin during periods of higher rainfall. The existence of a lake with a water-table about 40 m higher than today has been questioned by Durand (1982), but recent geomorphological studies support the assumption that aquatic as well as tectonic processes contributed to the accumulation of the Bama Ridge sands (Thiemeyer, 1992, 1997).

East of the Bama Ridge stretch vast clay plains with lagoonal deposits which today are only seasonally flooded. Palaeoecological data (e.g., remains of fish and water birds from archaeological sites) demonstrate that these clay depressions had been permanent lakes between 2000 cal BC and 1000 cal BC, the period we are focusing on. In the clay plains, dispersed slightly elevated sand areas are found, possible settlement ground today as well as in the past.

Archaeological fieldwork carried out by Connah (1976, 1981, 1984) has provided a sequence of human occupation for the last 3000 years in the so-called *firki* area east of the Gajiganna region. Since 1990 studies in the Nigerian Chad Basin have been conducted by the Frankfurt project in cooperation with the University of Maiduguri and the National Commission for Museums and Monuments, Nigeria (Breunig *et al.*, 1992, 1993a,b, 1996; Breunig, 1995; Breunig and Neumann, 1996, 1999; Garba, 1997; Gronenborn, 1996a,b, 1997; Gronenborn *et al.*, 1995, 1996; Hambolu, 1996; Klee and Zach, 1999; Klee *et al.*, 2000; Wendt, 1997). During several extensive surveys, about 400 sites, most of them settlement mounds, have been identified (Figure 9.5; Breunig, in preparation).

EARLY AND MIDDLE HOLOCENE EVIDENCE

As most parts of the Chad Basin apparently were not accessible during the Mega Chad period, the bulk of the known archaeological sites are dated to the last 4000 years. Only two sites, Dufuna and Konduga, gave an insight into the Mega Chad period (Figure 9.6).

At Dufuna an 8000 year old dugout canoe was found in 1987 and finally excavated in 1998 (Breunig, 1996). The boat is completely preserved and has a length of 8.5 m. It is the oldest known boat in Africa, demonstrating the adaptation of early Holocene hunters and gatherers to an environment dominated by the waters of Mega Chad, at roughly the same time when logboats appeared in Europe (Lanting, 1997/98).

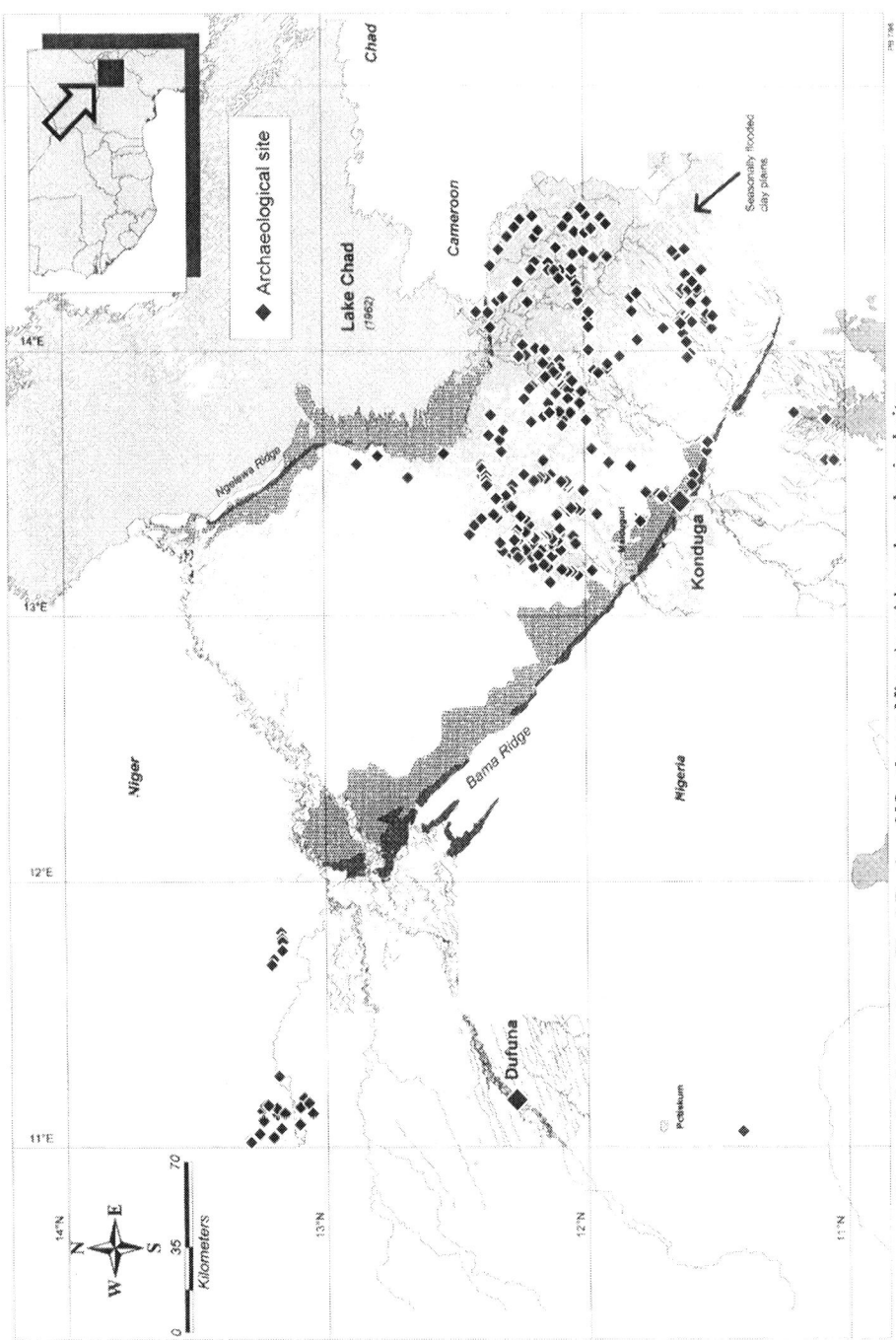

Figure 9.5. Map of the Chad Basin of Northeast Nigeria with archaeological sites.

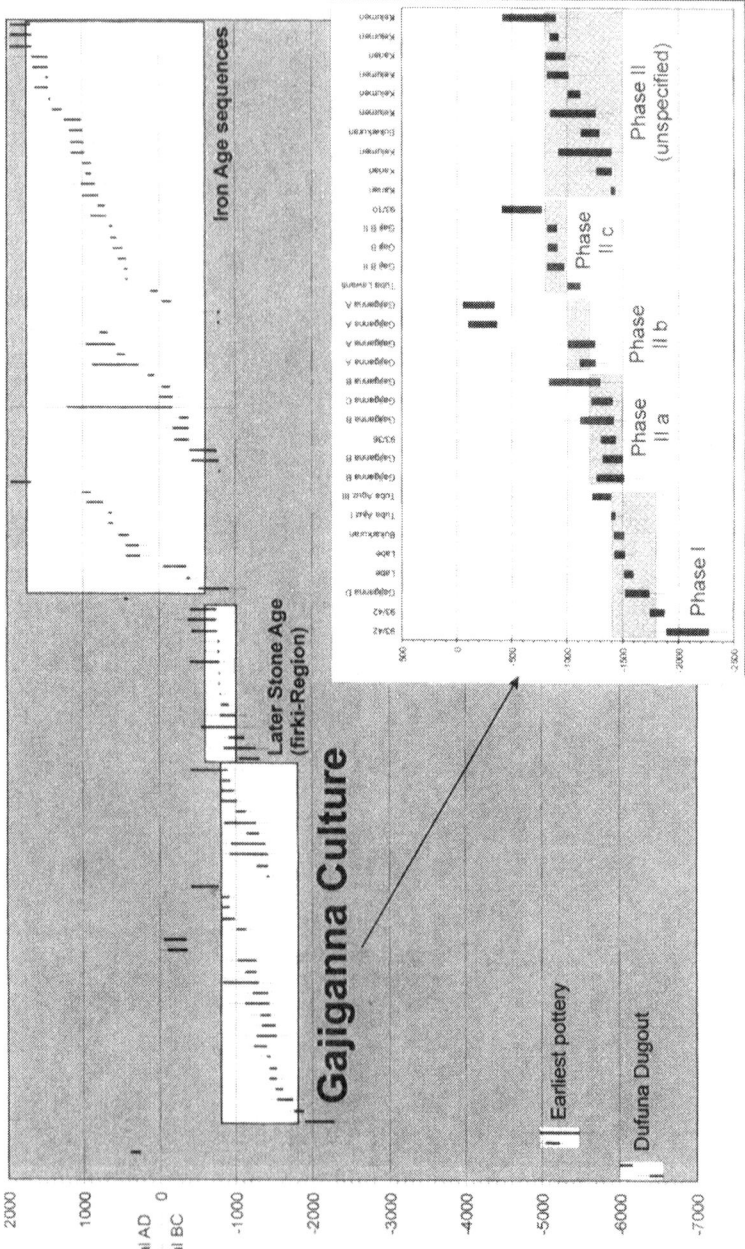

Figure 9.6. Radiocarbon chronology of the Holocene prehistory of Northeast Nigeria.

The pottery site of Konduga, located on top of the Bama Ridge, is roughly one millennium younger than the boat from Dufuna. In undisturbed sandy deposits pottery was recovered, associated with charcoal yielding dates between 5500 and 5000 cal BC (Breunig *et al.*, 1996, Thiemeyer, 1992; Ballouche and Neumann, 1995). The pottery is heavily tempered with quartz fragments and extensively decorated with a rocker stamp technique and shows striking similarities with Central Saharan material. Therefore it seems obvious that Central Saharan influence in the form of pottery reached the Chad Basin 3000 years before the emergence of food production.

GAJIGANNA CULTURE—THE FINAL STONE AGE COLONIZATION OF THE CHAD BASIN

After Konduga there is a lack of data for about 3000 years from 5000 to 2000 cal BC (Figure 9.6). With the disappearance of Mega Chad coinciding with increasing climatic aridity between 5000 and 2000 cal BC, a vast area which had been previously flooded was opened for human occupation. The colonization of the Southwest Chad Basin after 2000 cal BC provides a unique case study on prehistoric migration. The questions to be asked are: where were the colonists from and how did settlements develop on the previous lake floor? The colonization is represented by an archaeological complex named after the village of Gajiganna where the first discoveries were made (Breunig *et al.*, 1992). The presently available data point to an age between 1800 and 800 cal BC (Figure 9.6). The Gajiganna Complex represents the introduction of food production into this part of the Chad Basin. This complex is followed by an Iron Age sequence (Figure 9.6), which will not be considered here.

The Sites

So far around 120 Gajiganna sites, scattered all over the study area in several regional concentrations, have been located (Figure 9.7). After extensive surveys of thousands of kilometres covering the whole study area we assume that the site clusters shown in Figure 9.7 more or less provide a clue to the original settlement pattern at least on a larger regional scale. There is a dense concentration of sites in the northwest (Gajiganna group), where the core area of the complex might have been situated. From here to the east site density decreases. Another dense cluster of sites can be observed at the southeast fringe of the Gajiganna complex (Walasa group). On the Bama Ridge and in its vicinity to the north, the Bama-Konduga group is found. In comparison with the other regional groups, the Bama-Konduga sites are situated in a different environment. Instead of the clay plains as a source of water, there are rivers close by, in particular the Yedseram. Deltaic deposits of the Yedseram probably explain the lack of sites further north.

The sites are mainly settlement mounds. Some are quite flat and only appear as mounds because an underlying dune is coated by a thin cultural layer, others consist of cultural deposits with a maximum accumulation of 2-4 m. Apart

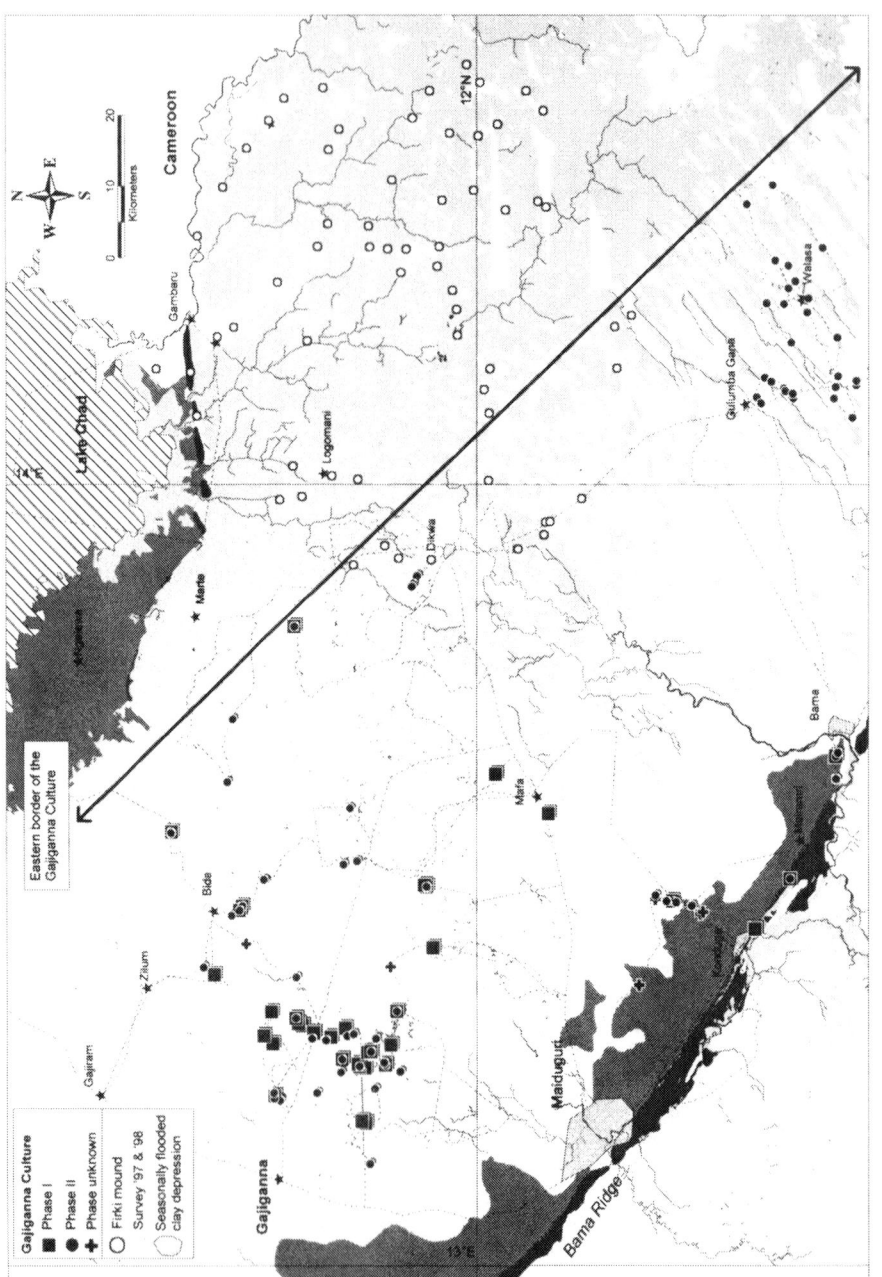

Figure 9.7. Distribution of firki mounds and of sites of the Gajiganna Culture, Northeast Nigeria (ca. 1800 cal BC – 800 cal BC) divided into pastoral (phase I) and agropastoral (phase II) stages.

from the sites in the vicinity of the Bama Ridge, most find-spots are situated on slightly elevated sand areas. These originated from aeolian sediments accumulated during the hyper-arid phase of the final Pleistocene (Thiemeyer and Buschbeck, 1993), and deltaic deposits reworked by the high waters of Mega Chad and subsequent erosion. The sand plains are surrounded by seasonally inundated clay depressions. According to geomorphological and archaeozoological evidence (Breunig et al., 1996; Thiemeyer, 1997), it seems reasonable to suppose that during the time of the Gajiganna complex the clay depressions were perennially filled with water. If the Gajiganna people came from arid lands —and there is also evidence to assume this (see below)—they enjoyed a well-watered environment in the Chad Basin.

About 20 sites have been excavated and provide sufficient data (including 33 radiocarbon dates—see Table 9.2) to establish a sequence with two main phases, I and II, mainly on the basis of pottery; the younger phase is further sub-divided into three stages—IIa, IIb and IIc (Figure 9.6).

Chronology

The sub-division of the phases of the Gajiganna complex is based on stylistic and technological analysis of about 5000 individual pots from sites representing the northwest group of the complex (north of Maiduguri, see Figure 9.7; Wendt, 1995, 1997). Work on material from other excavations is in progress.

Considering all available radiocarbon dates, we have to claim that phase I existed from 2200 to 1200 cal BC (Figure 9.6), a time-span which seems slightly too long. The oldest date especially might be wrong, because it is contradicted by a re-measurement. If the oldest and youngest dates are rejected, a duration of phase I from 1800 to 1400 cal BC is proposed which seems to be more reasonable. According to the radiocarbon dates, phase II is dated to 1500 to 800 cal BC. Phase IIa has dates clustering from 1500 to 1200 cal BC. Phase IIb is dated only by two measurements covering the period from 1200 to 1000 cal BC. The calibrated dates of phase IIc reach from 1100 to 400 cal BC. The long duration is caused by an isolated and quite young date. We suppose that the central group of dates, from 1000 to 800 cal BC, represents the actual duration of Phase IIc, but this will be checked by excavations of specific sites of this phase in the future, as the whole chronology might be slightly modified by further studies.

With regard to the technological aspects, the distinguishing features of the pottery of phase I are the predominance of inorganic temper and a fairly often slipped and polished surface. Polishing occurs sometimes in a very diagnostic way, which has been labeled as cross polishing and was probably made by smoothing incised cross pattern (Figure 9.8). The most significant pottery shape is a pot with flat shoulder. Decoration in general is concentrated on the upper part of the vessel, in most cases close to the rim. Zigzag bands made either of comb impression or triangular stamping are typical of phase I pottery.

There are some distinguishing differences between the pottery of the two phases. The most striking difference is the increase in organic temper coupled with the decrease of polishing within the course of phase II, resulting in a general

Table 9.2. Radiocarbon dates of the Gajiganna complex in the Chad Basin of Northeast Nigeria

Site	Phase	Lab code	^{14}C age bp	cal age (one-sigma std. dev.)	Dated material
NA 93/42	I	UtC-3515	3690 ± 120	2272-1891 BC	Charcoal
NA 93/42	I	UtC-5297	3489 ± 34	1876-1743 BC	Charcoal
Gajiganna D	I	UtC-2801	3360 ± 80	1739-1522 BC	Charcoal
Bukarkurari	I	UtC-6779	3205 ± 43	1515-1419 BC	Charcoal
Labe	I	UtC-6964	3275 ± 33	1597-1513 BC	Charcoal
Labe	I	UtC-6781	3223 ± 33	1521-1423 BC	Charcoal
Tuba Ajuz I	I	UtC-6778	3150 ± 38	1435-1396 BC	Charcoal
Tuba Ajuz III	I	UtC-6783	3059 ± 50	1396-1225 BC	Charcoal
Gajiganna B	IIa	UtC-2332	3140 ± 110	1515-1264 BC	Charcoal
Gajiganna B	IIa	UtC-2330	3150 ± 70	1505-1320 BC	Charcoal
NA 93/36	IIa	UtC-3514	3120 ± 70	1436-1304 BC	Charcoal
Gajiganna B	IIa	KI-3605	3040 ± 120	1419-1219 BC	Charcoal
Gajiganna C	IIa	UtC-2799	3070 ± 70	1410-1219 BC	Charcoal
Gajiganna B	IIa	KN-4675	2880 ± 165	1300-834 BC	Cattle bone
Gajiganna A	IIb	UtC-2795	2960 ± 50	1259-1111 BC	Charcoal
Gajiganna A	IIb	UtC-2329	2930 ± 60	1253-1009 BC	Charcoal
Gajiganna A	IIb	KIA-603	2180 ± 80	368-101 BC	Human bone
Gajiganna A	IIb	KN-4674	2141 ± 66	344-53 BC	Cattle bone
Tuba Lawanti	IIc	UtC-6780	2897 ± 40	1124-1002 BC	Charcoal
Gajiganna B II	IIc	UtC-2797	2750 ± 70	976-817 BC	Charcoal
Gajiganna B	IIc	UtC-2331	2740 ± 50	916-822 BC	Charcoal
Gajiganna B II	IIc	UtC-2796	2730 ± 50	911-817 BC	Charcoal
NA 93/10	IIc	UtC-3513	2470 ± 70	772-407 BC	Charcoal
Kariari	II	UtC-5158	3146 ± 38	1432-1394 BC	Charcoal
Kariari	II	UtC-5159	3076 ± 43	1401-1265 BC	Charcoal
Kelumeri (Borno 38)	II	N-794	2960 ± 160	1400-921 BC	Charcoal
Kelumeri (Borno 38)	II	KI-4206	2940 ± 125	1375-942 BC	Charcoal
Bukarkurari	II	UtC-6841	2991 ± 46	1293-848 BC	Charcoal
Kelumeri (Borno 38)	II	N-795	2880 ± 140	1262-848 BC	Charcoal
Kelumeri (Borno 38)	II	UtC-5746	2902 ± 35	1124-1007 BC	Charcoal
Kelumeri (Borno 38)	II	KI-4204	2780 ± 95	1016-820 BC	Charcoal
Kariari	II	KI-4073	2740 ± 90	987-807 BC	Charcoal
Kelumeri (Borno 38)	II	UtC-5747	2773 ± 38	928-844 BC	Charcoal
Kelumeri (Borno 38)	II	N-796	2590 ± 170	901-413 BC	Charcoal

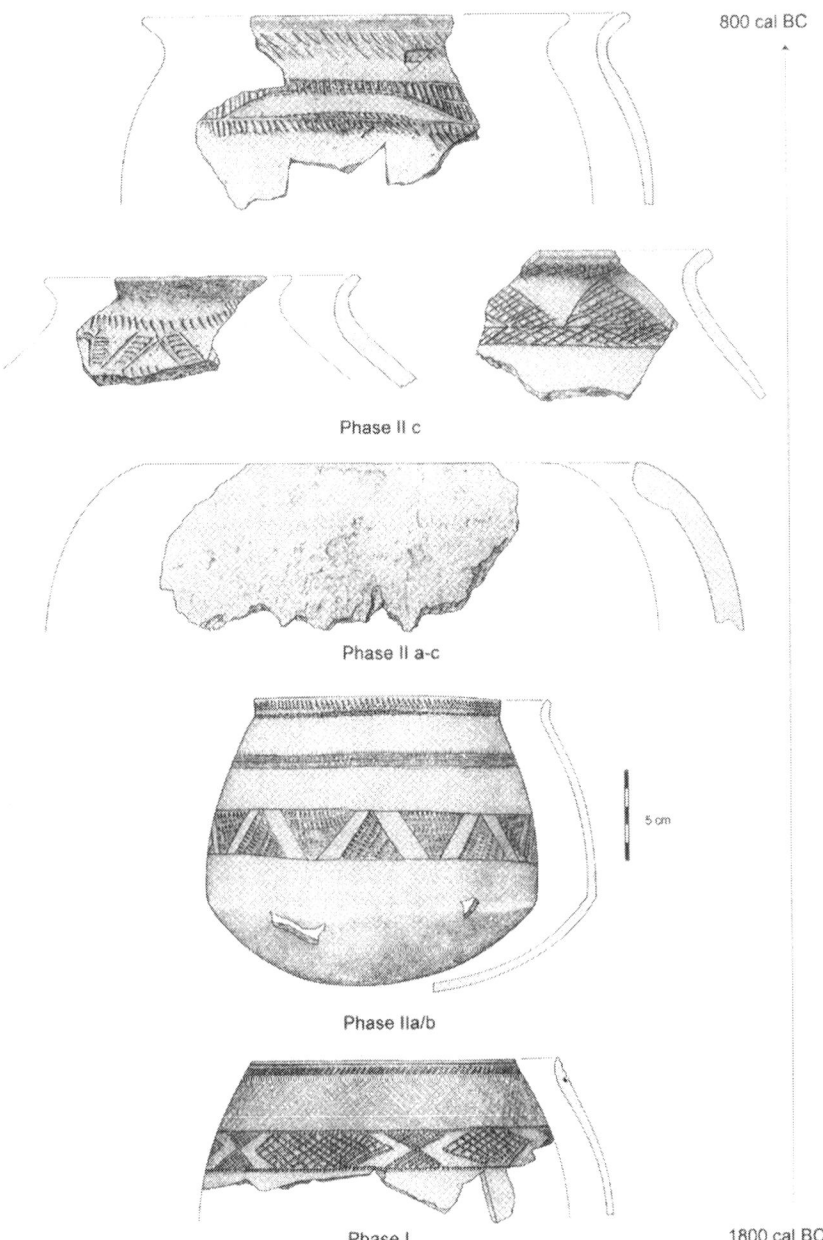

Figure 9.8. Pottery of phases I and II of the Gajiganna Culture.

decline of quality towards the end of the phase (Figure 9.8). Another technological aspect is the appearance of mat impressions which had been unknown before. Decoration is made in the tradition of phase I, but there is a larger variety of geometric pattern, either made by comb impression or incision and a tendency to include more parts of the vessel's body. Concerning shape, an almost exclusively undecorated pot with inverted rim is diagnostic for phase II.

The Economy and its Archaeological Context

Archaeobotanical and archaeozoological data furnish information on the economy and environment of the Gajiganna complex (Figure 9.9). The animal bones from most sites are well preserved, especially in the lower spits of the mounds, which had not been affected by soil formation processes. So far, the material has not been studied completely (by Wim Van Neer and Sven Lambrecht, Tervuren, Belgium), therefore changes in sample composition and frequency of livestock, game, fish, molluscs, birds and reptiles during the economic development of the Gajiganna Culture as well as environmental change cannot be evaluated yet. But the data are sufficient to establish some general outlines on the management of animal resources (Van Neer, this volume).

The first settlers arriving in the Chad Basin were pastoralists. The earliest sites of phase I already have domesticated animals. Cattle are most common, accounting for up to 60% of all mammal bones in the subsequent phases IIa and IIb (Breunig *et al.*, 1996). Besides pastoralism, hunting and especially fishing was practiced throughout the Gajiganna Culture. The lakes must have been an almost inexhaustible source of food. The pastoral stage lasted for about half a millennium.

As carbonized plant remains are very badly preserved in the Gajiganna sites, archaeobotanical work has concentrated on impressions in ceramics (Figure 9.10). The interpretation of plant impressions in ceramics is limited by the fact that they represent the special activity of tempering and cannot be directly translated into statements on diet and economy. However, some significant general outlines can be given (Klee and Zach, 1999).

In phase I, pottery is exclusively tempered with inorganic material, and plant impressions are preserved on the sherds only by accident. In phase II, the majority of sherds are tempered with plant material, rising from 40% in phase IIa to 90% in phase IIc. In the pottery from phases I and IIa, only wild grasses—rice (*Oryza* sp.) and Paniceae—are represented. Paniceae are a large group of the grass plant family (Gramineae/Poaceae) with 270 species in West Africa, of which 60 can be collected for food. The famous West African *fonio* or *kreb* contains several species all belonging to Paniceae. From ethnographic sources we know that in some areas *fonio* has been an important part of the diet up to modern times, and that it can easily replace cultivated cereals as a staple (Harlan, 1989). Hence we suggest that in phases I and IIa wild grasses were the main source for carbohydrates of the Gajiganna people.

Impressions of domesticated *Pennisetum* first appear in the potsherds of phase IIb, around 1200 cal BC, but they make up only 10% of the identifiable

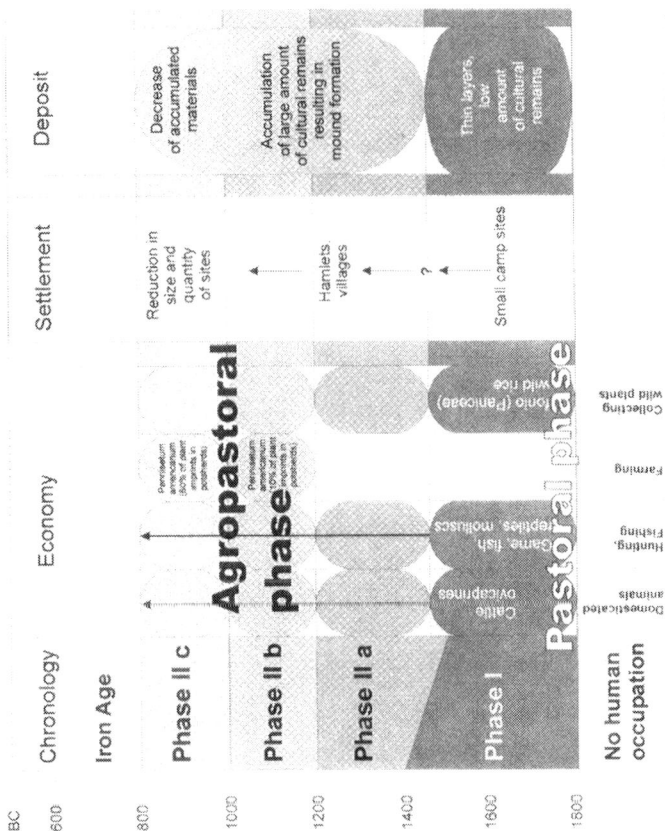

Figure 9.9. Economic and cultural appearance of the Gajiganna Culture.

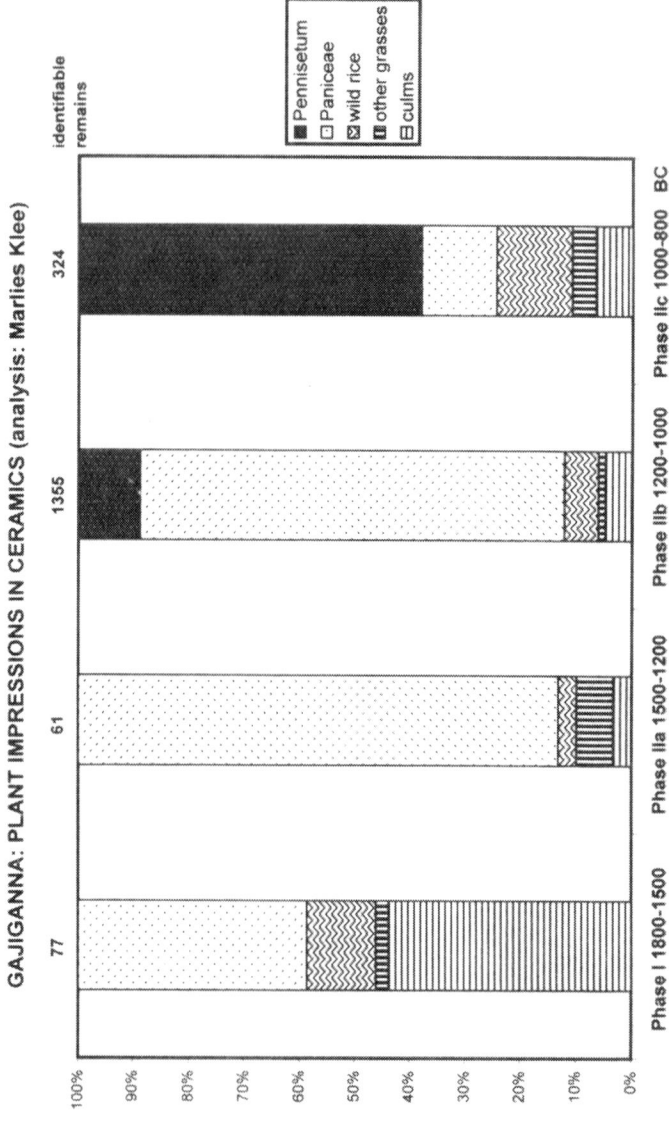

Figure 9.10. Gajiganna: Percentage values of plant impressions in potsherds from phase I to phase IIc.

remains. It is difficult to estimate the economic significance of millet by the amount of imprints in pottery. But 200 years later, in phase IIc, agriculture must have been well established, as 60% of all impressions were from *Pennisetum*. The basic question is: where were the domesticated plants from? Did they derive from a reactivation of an old tradition which had been brought to the Chad Basin but not used for centuries, because of abundant natural resources? Were they an introduction from outside by further immigrations which left no visible traces among the archaeological material? Or do we have to consider an independent local domestication of *Pennisetum* by the people of the Gajiganna complex? The botanical evidence suggests that local domestication is improbable. In the impressions, the morphological features of domestication are clearly visible and wild forms are absent; these two factors would point to an introduction from outside.

Two different settlement types coincide with the two economic stages. The pastoral phase produced only very flat mounds with a low accumulation of cultural material, demonstrating high mobility. Some sites of phase I, consisting only of a few potsherds, are interpreted as short-term camps and underline this mobility. Although the well-watered clay plains provided plenty of fodder, inundation might have been a serious problem during the rainy season as the area for pasture then was limited to the elevated sand islands in the surroundings of the camp. Mobility is also inevitable if water becomes short during the dry season. The present data do not allow for a decision whether the groups moved as a whole or transhumance was practiced.

Temporary campsites like the one mentioned above have not been found in a phase II context. By contrast, the large mounds of the subsequent agropastoral phase point to sedentary and village-organized communities. It is unknown how the phase II people managed their domesticated animals. Either they practiced transhumance or a larger area suitable for pasture was available, due to decreasing surface water, which would have allowed all-year-round grazing in the vicinity of the sites.

The transition from a pastoral to an agropastoral way of life might be seen as a fundamental change to such an extent that the material culture should reflect it. Indeed, there is a clear change in the pottery with regard to quantity as well as typology and technology. The sedentary population of the agropastoral stage produced pottery in enormous quantities, which would have hampered the mobility of the pastoralists. But other find categories have not shown significant differences so far. Grinding equipment is common in both stages, although less frequent among the pastoralists for the same reason as mentioned in the case of pottery. Furthermore, stone artifacts are polished axes, flakes struck from broken axes, and bifacially retouched arrow-points of Saharan type (Figure 9.11).

Based on the lack of any knapping debris, it can be concluded that all stone tools must have been manufactured elsewhere. This looks reasonable as there are no sources of stone raw materials in the Chad Basin. All stone artifacts had to be imported into the stoneless basin from outside, providing us with models of how a unique shortage of specific resources can be managed, as discussed by Connah and Freeth (1989) in the case of Daima. Bone tools might have compensated for this lithic shortage. These tools comprise harpoons, chisels,

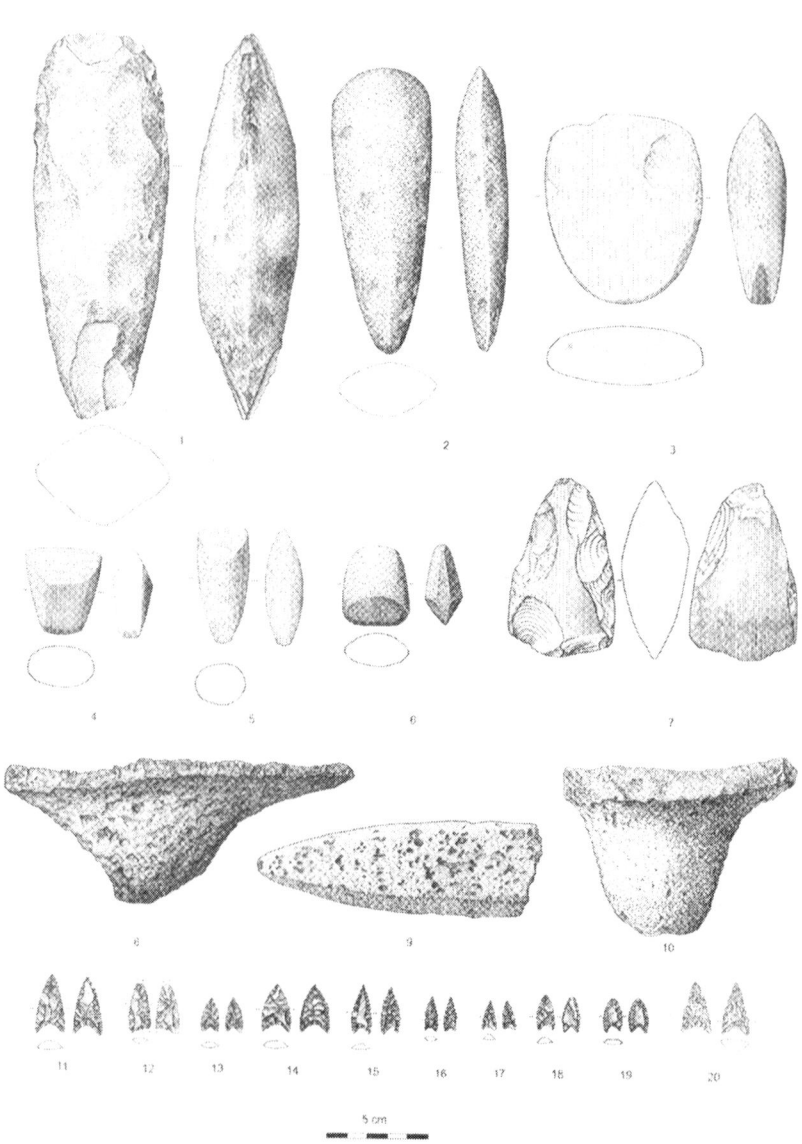

Figure 9.11. Stone artifacts of the Gajiganna Culture.

bone points or gorge-hooks for fishing and bone beads as ornament (Figure 9.12). Finally, there are art objects such as baked clay figurines of anthropomorphic and zoomorphic shapes (Breunig, 1994).

In the same way as the transition from pastoral to agropastoral economy becomes visible in the material culture, one might also expect a difference in the environmental setting and location of the sites. To verify this expectation, a site catchment analysis has been carried out with sites of the northwest group between Gajiganna and Dikwa (Figure 9.7). It can be shown that more than 80% of all pastoral sites had more water than land within a radius of 5 km—a distance of an hour's walk. This is not surprizing as watering their livestock is a major part of the labor required by pastoralists. The environment of the region under consideration is rather similar to the one inhabited by the Nuer in Southern Sudan. The Nuer are pastoralists and preferably stay with their livestock as close as possible to open waters in the flooded clay plains (Evans-Pritchard, 1940). The relation between water and land changes in the agropastoral phase, when only half of the sites have more water than land in their surrounding, which most probably reflects the increased need for sandy areas to grow millet.

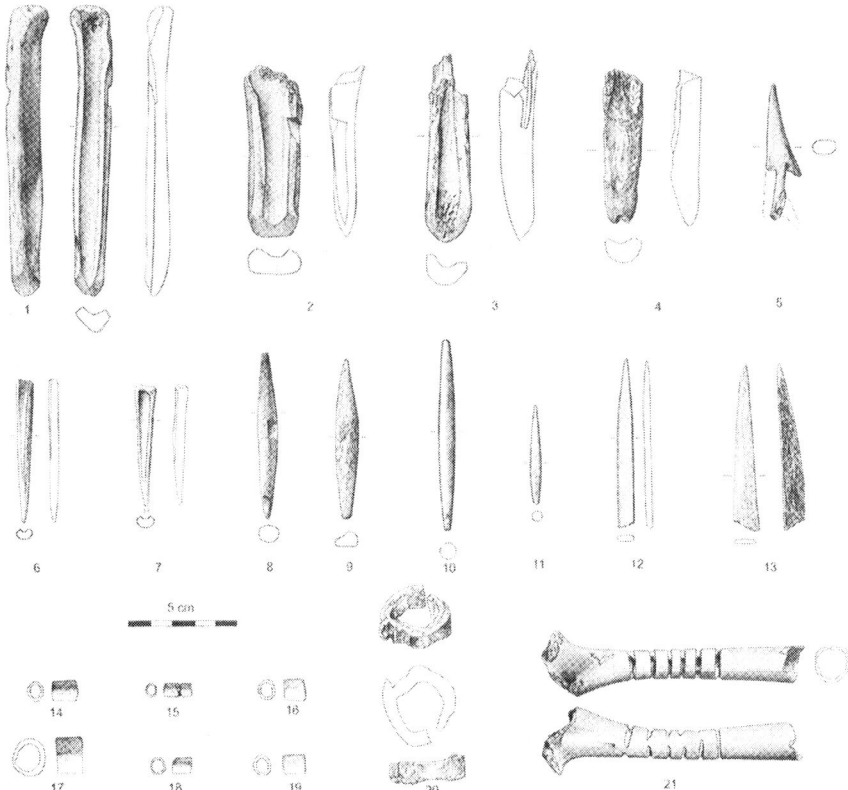

Figure 9.12. Bone tools of the Gajiganna Culture.

The Origin of the Gajiganna Culture

One of the principal questions is the origin of the Gajiganna complex and its migration into the Chad Basin. The distribution of sites already indicates the direction to which one should look. As Figure 9.7 shows, the pastoral sites have a smaller spatial distribution than the agropastoral ones. That means that the farmers of phase II penetrated into areas which had not been occupied by the pastoralists before. As the earlier pastoral sites are concentrated in the northwest, the whole complex might have originated from an area west or northwest of the Chad Basin. The east was completely uninhabited during the time of the Gajiganna Culture. During the whole period, apparently no Gajiganna people lived beyond the solid line indicated in Figure 9.7. The land east of this line—the firki—was not accessible until about 1000 cal BC, when human occupation commenced here as a result of increasing drought, the same drought which probably terminated the Gajiganna Complex. We suppose that the firki area was colonized from the east. The eastern migration on the one hand and the Gajiganna migration on the other is reflected in a distinct frontier or contact zone between Chadic speaking and Niger-Congo speaking people which existed around 1000 cal BC—the archaeological support for a linguistic model (Jungraithmayr and Leger, 1993). As shown by the excavations of Connah (1981) and Gronenborn (1996a) the eastern people were living here on small sandy spots from about 1000 cal BC onwards throughout the Iron Age until today.

       To locate the origin of the Gajiganna Complex it is necessary to consider the evidence beyond the area discussed here. The occupation of the Chad Basin commenced in times of increasing drought in the Sahara. Pastoral communities who had lived there for a long time must have been affected by the drought. The first colonists in the Chad Basin were also pastoralists. Their appearance fits well with an isochronic model depicting the emergence of cattle in the Eastern Sahara and tracing their spread over the Central Sahara down into the Sahel Zone of West Africa (Shaw, 1981, p. 224).

       The migration route of the Gajiganna pastoralists from the Sahara to the Chad Basin might be traced by sites located in the Manga Grasslands, close to the border between Niger and Nigeria (Figure 9.13). These sites did not provide economic data, but the pottery clearly belongs to the Gajiganna type.

       Pottery with decoration similar to the Gajiganna type can be traced even further north into the Central Sahara (Wendt, 1997). Final Neolithic and early Iron Age pottery has been recorded by Treinen-Claustre (1982, p. 62) from Borkou, Chad. Triangular stamping forming a zigzag pattern in combination with horizontal bands of comb impressions as well as grooved lines have parallels among the Gajiganna materials. Courtin (1966, p. 280) published final Neolithic potsherds with decoration of triangles filled with comb impression. They have almost completely identical counterparts among the Gajiganna finds, as already noticed by Connah (1981, p. 91) in connection with Borno 38, a site belonging to the Gajiganna Complex, re-visited by us in 1996. Further east, in the Ennedi, there might also be parallels (Camps-Fabrer 1966, pl. 56, fig. 4). Some potsherds decorated by comb impression and triangular stamping from northeast of the Hoggar (Camps-Fabrer, 1966, pl. 51, fig. 2) also would not be separated as an

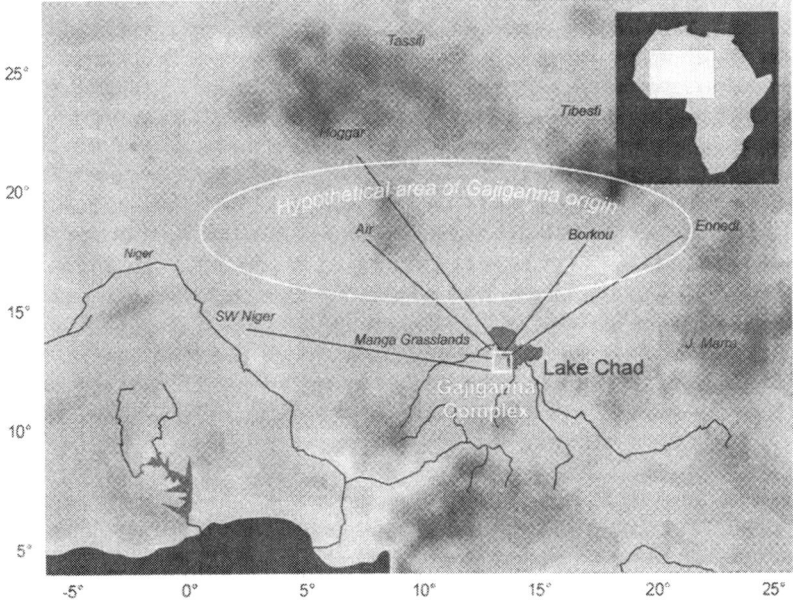

Figure 9.13. Map of sites with Gajiganna related pottery.

alien element in Gajiganna context. The most striking parallels, as much as one can recognize from drawings, are from the final Neolithic site of In Tuduf (Vallée de l'Azawagh), in the west of the Aïr (Paris, 1995, fig. 5; 1996, p. 80, fig. 22a). The indicated age of 4000 bp fits well with the supposed relation between Gajiganna people and their Saharan ancestors. The westernmost comparison of horizontal comb decoration is from Southwest Niger, dated to 3600-3500 bp (Néolithique phase B; Vernet, 1996, fig. 100). As the material from the Sahel Zone of Burkina Faso, very close to Southwest Niger, is quite different, one might assume that the river Niger was a frontier in those times.

Comparison between Saharan material and Gajiganna clearly points to a cultural tradition or connection, which allows one to place the origin of the Gajiganna Complex somewhere between Ennedi in the east, the Hoggar in the north, and the river Niger in the west. The available radiocarbon dates for the Saharan predecessor are slightly older than the Gajiganna dates and support the migration model. However, the archaeological evidence from those areas shows that they were not completely depopulated by the emigration of the Gajiganna predecessors. It is conceivable that the Chad Basin was settled by the southern representatives of a large Saharan complex without affecting the whole. They just filled up a vacuum, which finally underwent an independent development.

THE ORIGINS OF AGRICULTURE IN THE SAHEL

There is no doubt that the origins of pastoralism in the Sahel have to be sought in the Sahara, as wild cattle, sheep and goat do not belong to the natural Sahelian fauna. For the beginning of agriculture things are not so easy. In his often cited paper on centers and non-centers of agriculture, Harlan (1971) indicated the hypothetical domestication areas of the two most important African grain crops, *Pennisetum americanum* and *Sorghum bicolor,* stretching as a broad band in a west-eastern direction, more or less coinciding with the modern Sahel. When Harlan's map was published, direct archaeobotanical data were not available, and he had to base his domestication areas entirely on the modern distribution of the supposed wild ancestors of the crops. Wild races of *Sorghum* and *Pennisetum* are still found in the Sahel, hence it is not necessary to claim an introduction from the Sahara, and a local development cannot be excluded *a priori.*

On Harlan's map the domestication area of the two crops overlap in the Chad Basin, *Pennisetum* extending from there to the west and *Sorghum* to the east. Archaeobotanical evidence roughly confirms this general separation, but the origin of *Sorghum* has to be sought further to the east. In most West African archaeological sites, *Sorghum* is either missing completely or only appears quite late. Even in the firki plains around Lake Chad, where *Sorghum* is grown today on a large scale, this plant never played an important role in prehistoric times and has only been found in historical times after cal 1000 AD (Connah, 1981; Klee and Zach, 1999; Klee *et al.,* 2000; Zach *et al.,* 1996). On the other hand, wild *Sorghum* is present in several sites of the Eastern Sahara and the Central Sudan, dated to 8000-4000 bp (Barakat and Fahmy, 1999; Magid, 1989; Magid and Caneva, 1998; Wasylikowa, 1997; Wasylikowa and Dahlberg, 1999). The earliest finds showing distinct morphological features of domestication have been dated around the beginning of the Christian era: Jebel Tomat in Sudan (Clark and Stemler, 1975) and Qasr Ibrim in southern Egypt (Rowley-Conwy, 1991). At the same time domesticated Sorghum also appears in West Africa, at the site of Jenné-Jeno (McIntosh, 1995). The quite late occurrence of the distinctly domesticated Sorghum races in Northeast Africa is striking. Further confusion comes from remains of domesticated Sorghum from Saudi-Arabia and India— where wild Sorghum is absent in the natural flora—which are up to 2500 calendar years older than the earliest African domesticated forms (Cleuziou and Constantini, 1982; Fuller, 2000; Kajale, 1991). If identification and dating of the Indian and Saudi-Arabian *Sorghum* is correct, then cultivation and domestication should have started somewhere in the Southeast Sahara prior to 4500 bp. Hence the hypothetical centre might have extended somewhat further to the northeast than supposed by Harlan if climatic conditions before 4500 bp allowed for an extended distribution range of the wild ancestors. Recently, Harlan's theory has been substantially questioned and two other hypotheses have been brought into discussion: (1) *Sorghum* was transferred from Africa to the Arabian Peninsula at an early date, domesticated there and brought back to Africa around 2000 bp (Haaland, 1999); and (2) *Sorghum* was domesticated in Africa quite late, around 2000 bp, implying that either identification or dating of the Indian and Saudi-

Arabian finds must be wrong (Rowley-Conwy *et al.*, 1997). So far, all three hypotheses are waiting for their archaeobotanical confirmation or refutation.

As the *Pennisetum* remains found in the Sahelian sites of Burkina Faso and Nigeria are all fully domesticated (Klee and Zach, 1999; Neumann, 1999a), the initial steps of domestication probably did not take place in this zone. So far, the oldest domesticated *Pennisetum* finds come from Dhar Tichitt, Mauritania, where potsherds with impressions have been dated to 3500 bp (ca. 1900-1500 cal BC; Amblard, 1996). Interpretations concerning the development at Dhar Tichitt are contradictory. Munson (1976) proposed a shift from plant collecting to agriculture, which he supposed to have taken place around 1000 BC as a response to climatic deterioration. By contrast, Amblard (1996) claims that agriculture had been already established from the beginning of the occupation. However, the presence of domesticated as well as wild forms in the impressions of the Tichitt ceramics (Amblard and Pernès, 1989) indicates that in the second millenium BC the distribution area of wild *Pennisetum* extended farther to the north than today, and therefore a local domestication cannot be ruled out. The available data lead us to assume that *Pennisetum* domestication started somewhere in the Central or Westcentral Sahara, i.e. further to the north than supposed by Harlan, at some time before 3500 bp. The connections of the Gajiganna Culture with the Central Sahara, as discussed above, make it probable that the first domestication steps of *Pennisetum* took place in the Central Saharan mountains or their forelands. Remains of wild *Pennisetum* have been found in the early to middle Holocene sites of Ti-n-Torha and Uan Muhuggiag, Libya (Wasylikowa, 1992). Although they do not belong to *P. americanum* subsp. *monodii*, the wild ancestor of domesticated millet, they indicate that *Pennisetum* species were among the wild grasses collected for food in the Central Sahara during the moister phases of the Holocene.

The crucial point both for *Pennisetum* and *Sorghum* is that there are no archaeobotanical data for the areas and periods under discussion. Either there have been no excavations at all, or no archaeobotanical samples have been taken, or the sites were so badly preserved that sampling failed. We strongly advocate the need for more detailed archaeobotanical data and a closer cooperation between archaeologists and archaeobotanists to solve these problems.

Apart from information about the domestication history of *Pennisetum* and *Sorghum*, recent archaeobotanical data from the Sahel and other parts of northern Africa have revealed the important role of wild plants in the economies up to modern times. In this regard, Paniceae play a central role. Paniceae have been recovered from many archaeological sites in the Northern and Eastern Sahara as well from Lake Chad area. The Sahelian savannas, which extended far into the Sahara during the early and middle Holocene, were so rich in wild Paniceae, that for thousands of years there was no need to cultivate.

CONCLUSION

In the fourth millenium bp, increasing aridity in the Sahara resulted in large-scale population movements to the south, archaeologically traceable from Mauritania to

Niger (McIntosh, 1994). Migration seems to have been a successful response to the challenges of environmental deterioration. Under the assumption that there was a gradual southward retreat of those vegetation types which offered good pastures and enough open water, the southern Saharan populations simply followed their resources. However, apart from the general statement that there was a southward movement, a strong local and regional variability of the economic strategies can be observed.

In the second millenium BC, the first settlers of the Chad Basin in Nigeria entered a 'virgin' area which had been previously flooded. Their economy was strongly based on cattle keeping and to a lesser extent on hunting and the exploitation of aquatic resources; in addition, they collected wild grasses. In the dune areas of Burkina Faso, by contrast, there is no evidence for livestock keeping during the second millenium, and hunting, gathering and fishing seem to have been the main activities of a highly mobile population. Both in Nigeria and Burkina Faso, there are no indicators for agriculture before 1200 cal BC. Contemporary sites of the Windé Koroji complex in Mali are interpreted as "permanent agricultural or fisher-gatherer settlements possessing a migratory pastoral component" (MacDonald, 1996).

Indirect evidence from the Oursi pollen diagram, domesticated *Pennisetum* from Ti-n-Akof and plant impressions from Gajiganna indicate that agriculture started in the Sahel of Nigeria and Burkina Faso around 1200 cal BC (3000 bp) and was most probably introduced from the Sahara. So far, we cannot decide if this contemporaneity is mere accident or if 1200 BC marks a general boundary for the introduction of agriculture into the Sahel, but a correlation with climatic events seems reasonable. Severe aridity after 3300 bp, as visible in the pollen diagrams from the Manga Grasslands, must have seriously affected the environmental resources of the Sahel, even to the extent that the yields of wild grasses might have decreased and cultivation became an alternative. However, plant food production was not necessarily the preferred strategy all over the Sahel from its first appearance onwards. In places where wild grasses were abundant, their harvesting should retain its important role up to modern times, as has been shown for the sites Mege and Kursakata in the firki area south of Lake Chad (Klee and Zach, 1999; Klee *et al.*, 2000).

For the Sahara, the statement that ceramics, livestock keeping, sedentism and agriculture do not appear simultaneously is widely accepted. Our data show that even the comparatively late introduction of these cultural elements into the Sahel after 2000 cal BC did not occur in a 'Neolithic package'.

ACKNOWLEDGMENTS

We are indebted to the Deutsche Forschungsgemeinschaft (German Research Foundation) for funding all studies reported here. We also feel obliged to the National Commission for Museums and Monuments of Nigeria and the C.N.R.S.T. in Burkina Faso for excellent co-operation. Thanks are due to Barbara Voss and Monika Heckner for the drawings of archaeological materials. We are grateful to all scientific members of the Frankfurt project and our counterparts from the Universities of Maiduguri and Ouagadougou, who contributed to the results presented here. Special thanks are adressed to M. Adebisi Sowunmi for corrections and critical remarks. However, the authors are fully responsible for any shortcomings in the text.

REFERENCES

Amblard, S. (1996). Agricultural evidence and its interpretation on the Dhar Tichitt and Oualata, south-eastern Mauritania. In Pwiti, G., and Soper, R. (eds.), *Aspects of African Archaeology. Papers from the 10th Congress of the PanAfrican Association for Prehistory and Related Studies*, University of Zimbabwe Publications, Harare, pp. 421-427.

Amblard, S., and Pernès, J. (1989). The identification of cultivated pearl millet (*Pennisetum*) amongst plant impressions on pottery from Oued Chebbi (Dhar Oualata, Mauritania). *African Archaeological Review* 7: 117-126.

Andah, B. W. (1987). Early food producing societies and antecedents in Middle Africa. In Andah, B. W., and Okpoko, A. I. (eds.), Foundations of Civilization in Tropical Africa. *West African Journal of Archaeology* 17: 129-170.

Andres, W., Ballouche, A., and Müller-Haude, P. (1996). Contribution des sédiments de la Mare d'Oursi à la connaissance de l'évolution paléo-écologique du Sahel du Burkina Faso. *Berichte des Sonderforschungsbereichs 268* 7: 5-15.

Ballouche, A., and Neumann, K. (1995). Pollen from Oursi/Burkina Faso and charcoal from NE Nigeria: a contribution to the Holocene vegetation history of the West African Sahel. *Vegetation History and Archaeobotany* 4(1): 31-39.

Barakat, H. N., and Fahmy, A. G. (1999). Wild grasses as "neolithic" food resources in the Eastern Sahara: a review of the evidence from Egypt. In van der Veen, M. (ed.), *The Exploitation of Plant Resources in Ancient Africa*, Kluwer Academic/Plenum Publishers, New York, pp. 33-46.

Breunig, P. (1994). Early prehistoric art in Borno (NE Nigeria). *Sahara* 6: 98-102.

Breunig, P. (1995). Gajiganna und Konduga - zur frühen Besiedlung des Tschadbeckens in Nigeria. Bericht über die Ausgrabungen des Frankfurter Sonderforschungsbereichs 268 ("Westafrikanische Savanne") in Borno, Nordost-Nigeria. *Beiträge zur Allgemeinen und Vergleichenden Archäologie* 15: 3-48.

Breunig, P. (1996). The 8000-year-old dugout canoe from Dufuna (NE Nigeria). In Pwiti, G., and Soper, R. (eds.), *Aspects of African Archaeology. Papers from the 10th Congress of the PanAfrican Association for Prehistory and Related Studies*, University of Zimbabwe Publications, Harare, pp. 461-468.

Breunig, P. (ed.). *Archaeological Map of NE Nigeria*. (in preparation).

Breunig, P., and Neumann, K. (1996). Archaeological and archaeobotanical research of the Frankfurt University in a West African context. *Berichte des Sonderforschungsbereichs 268* 8: 181-191.

Breunig, P., and Neumann, K. (1999). Archäologische und archäobotanische Forschung in Westafrika. *Archäologisches Nachrichtenblatt* 4: 336-357.

Breunig, P., Garba, A., and Waziri, I. (1992). Recent archaeological surveys in Borno, Northeast Nigeria. *Nyame Akuma* 37: 10-17.

Breunig, P., Ballouche, A., Neumann, K., Rösing, F. W., Thiemeyer, H., Wendt, K. P., and Van Neer, W. (1993a). Gajiganna - new data on early settlement and environment in the Chad Basin. *Berichte des Sonderforschungsbereichs 268* 2: 51-74.

Breunig, P., Garba, A., Gronenborn, D., Van Neer, W., and Wendt, K. P. (1993b). Report on excavations at Gajiganna, Borno State, Northeast Nigeria. *Nyame Akuma* 40: 30-41.

Breunig, P., Neumann, K., and Van Neer, W. (1996). New Research on the Holocene Settlement and Environment of the Chad Basin in Nigeria. *African Archaeological Review* 13: 111-145.

Camps-Fabrer, H. (1966). *Matière et Art Mobilier dans la Préhistoire Nord-africaine et Saharienne*, CRAPE Mémoires 5, Art et Métiers Graphiques, Paris.

Casey, J. (1998). The ecology of food production in West Africa. In Connah, G. (ed.), *Transformations in Africa. Essays on Africa's Later Past*, Leicester University Press, London and Washington, pp. 46-70.

Clark, J. D., and Stemler, A. (1975). Early domesticated Sorghum from Central Sudan. *Nature* 254: 588-591.

Cleuziou, S., and Constantini, L. (1982). A l'origine des oasis. *La Recherche* 137(13): 1180-1182.

Connah, G. (1976). The Daima sequence and the prehistoric chronology of the Lake Chad region of Nigeria. *Journal of African History* 17: 321-352.

Connah, G. (1981). *Three Thousand Years in Africa: Man and his Environment in the Lake Chad Region of Nigeria*, Cambridge University Press, Cambridge.

Connah, G. (1984). An archaeological exploration in southern Borno. *African Archaeological Review* 2: 153-171.

Connah, G., and Freeth, S. J. (1989). A commodity problem in prehistoric Borno. *Sahara* 2: 7-20.

Courtin, J. (1966). Le Néolithique du Borkou, Nord-Tchad. *L'Anthropologie* 70: 269-281.

Davies, O. (1968). The origins of agriculture in West Africa. *Current Anthropology* 9(9): 1-5.

Durand, A. (1982). Oscillations of Lake Chad over the past 50,000 years: new data and new hypothesis. *Palaeogeography, Palaeoclimatology, Palaeoecology* 39: 37-53.

Evans-Pritchard, E. E. (1940). *The Nuer*, Clarendon Press, Oxford.

Fuller, D.Q. (2000). Fifty years of archaeobotanical studies in India: laying a solid foundation. In Settar, S., and Korisettar, R. (eds.), *Indian Archaeology in Retrospect*, Vol. III, Archaeology and Interactive Disciplines, Publications of the Indian Coucil for Historical Research, Manohar, New Delhi, pp. 247-363.

Garba, A. (1997). History of archaeology in the Lake Chad region of North-east Nigeria. *Nyame Akuma* 47: 38-41.

Gronenborn, D. (1996a). Kundiye: archaeology and ethnoarchaeology in the Kala-Balge area of Borno State, Nigeria. In Pwiti, G., and Soper, R. (eds.), *Aspects of African Archaeology. Papers from the 10th Congress of the PanAfrican Association for Prehistory and Related Studies*, University of Zimbabwe Publications, Harare, pp. 449-459.

Gronenborn, D. (1996b). Beyond Daima: Recent excavations in the Kala-Balge region of Borno State. *Nigerian Heritage* 5: 34-46.

Gronenborn, D. (1997). Bauern-Fischer-Fürsten. Ethnohistorische, archäologische und archäobotanische Arbeiten im Sonderforschungsbereich 268 zur Besiedlungs- und Kulturgeschichte des südlichen Tschad-Beckens (Borno State, Nigeria). *Archäologisches Nachrichtenblatt* 2(4): 376-390.

Gronenborn, D., Van Neer, W., and Skorupinski, T. (1995). Kleiner Vorbericht zur archäologischen Feldarbeit südlich des Tschad-Sees. *Berichte des Sonderforschungsbereich 268* 5: 27-39.

Gronenborn, D., Wiesmüller, B., Skorupinski, T., and Zach, B. (1996). Settlement history of the Kala-Balge region of Borno State, Nigeria. *Berichte des Sonderforschungsbereichs 268* 8: 201-213.

Haaland, R. (1999). The puzzle of the late emergence of domesticated sorghum in the Nile Valley. In Hather, J., and Gosden, C. (eds.), *The Prehistory of Food*, Routledge, London, pp. 397-418.

Hallier, M. (1999). Recherches archéologiques en hiver 1997/1998 au nord du Burkina Faso: les collines d'occupation de l'âge du fer. *Nyame Akuma* 51: 2-5.

Hambolu, M. O. (1996). Recent excavations along the Yobe Valley. *Berichte des Sonderforschungsbereichs 268* 8: 215-229.

Harlan, J. R. (1971). Agricultural origins: centers and non-centers. *Science* 174: 468-474.

Harlan, J. R. (1989). Wild-grass seed harvesting in the Sahara and Sub-Sahara of Africa. In Harris, D. R., and Hillman, G. C. (eds.), *Foraging and Farming: the Evolution of Plant Exploitation*, Unwin Hyman, London, pp. 79-98.

Harris, D. R. (1976). Traditional systems of plant food production and the origins of agriculture in West Africa. In Harlan, J. R., de Wet, J. M. J., and Stemler, A. B. (eds.), *Origins of African Plant Domestication*, Mouton, The Hague, pp. 311-356.

Harris, D. R. (ed.) (1996). *The Origins and Spread of Agriculture and Pastoralism in Eurasia*, University College London Press, London.

Hassan, F. A. (1996). Abrupt Holocene climatic events in Africa. In Pwiti, G., and Soper, R. (eds.), *Aspects of African Archaeology. Papers from the 10th Congress of the PanAfrican Association for Prehistory and Related Studies*, University of Zimbabwe Publications, Harare, pp. 83-89.

Hassan, F. A. (1997). Holocene palaeoclimates of Africa. *African Archaeological Review* 14(4): 214-230.

Holl, A. (1998). Livestock husbandry, pastoralism, and territoriality: the West African record. *Journal of Anthropological Archaeology* **17**: 143-165.

Jungraithmayr, H., and Leger, R. (1993). The Benue-Gongola-Chad Basin – zone of ethnic and linguistic compression. *Berichte des Sonderforschungsbereichs 268* **2**: 161-172.

Kajale, M. (1991). Current status of Indian palaeoethnobotany: introduced and indigenous food plants with a discussion of the historical and evolutionary development of Indian agriculture and agricultural systems in general. In Renfrew, J. M. (ed.), *New Light on Early Farming: Recent Developments in Palaeoethnobotany*, Edinburgh University Press, Edinburgh, pp. 155-189.

Klee, M., and Zach, B. (1999). Crops and wild cereals of three settlement mounds in NE-Nigeria - charred plant remains and impressions in ceramics from the last 4000 years. In van der Veen, M. (ed.), *The Exploitation of Plant Resources in Ancient Africa*, Kluwer Academic/Plenum Publishers, New York, pp. 81-88.

Klee, M., Zach, B., and Neumann, K. (2000). Four thousand years of plant exploitation in the Chad Basin of NE Nigeria. I: the archaeobotany of Kursakata. *Vegetation History and Archaeobotany* **9**: 223-237.

Lanting, J. N. (1997/98). Dates for origin and diffusion of the European logboat. *Palaeohistoria* **39/40**: 627-650.

Lézine, A. M. (1989). Late Quaternary vegetation and climate of the Sahel. *Quaternary Research* **32**: 317-334.

MacDonald, K. C. (1996). The Windé Koroji Complex: Evidence for the peopling of the eastern Inland Niger Delta (2100-500 BC). *Préhistoire Anthropologie Méditerranéennes* **5**: 147-165.

Magid, A. A. (1989). *Plant Domestication in the Middle Nile Basin*. Cambridge Monographs in African Archaeology 35, British Archaeological Reports International Series 523, Cambridge.

Magid, A. A., and Caneva, I. (1998). Economic strategy based on food-plants in the early Holocene Central Sudan: a reconsideration. In di Lernia, S., and Manzi, G. (eds.), *Before Food Production in North Africa. Proceedings of the Homonimous Workshop held in Forlì, September 1996, within the XIII World Congress of the International Union of the Prehistoric and Protohistoric Sciences*, ABACO Edizioni, Rome, pp. 79-90.

McIntosh, S. K. (1994). Changing perspectives of West Africa's past: archaeological research since 1988. *Journal of Archaeological Research* **2**(2): 165-198.

McIntosh, S. K. (ed.) (1995). *Excavations at Jenné-Jeno, Hambarketolo, and Kaniana (Inland Niger Delta, Mali), the 1981 Season*, University of California Press, Berkeley, Los Angeles, London.

McIntosh, S. K., and McIntosh, R. J. (1988). From stone to metal: new perspectives on the later prehistory of West Africa. *Journal of World Prehistory* **2**(1): 89-133.

Munson, P. J. (1976). Archaeological data on the origins of cultivation in the Southwestern Sahara and their implications for West Africa. In Harlan, J. R., de Wet, J. M., and Stemler, A. B. L. (eds.), *Origins of African Plant Domestication*, Mouton, The Hague, pp. 187-210.

Neumann, K. (1999a). Early plant food production in the West African Sahel: new evidence from the Frankfurt project. In van der Veen, M. (ed.), *The Exploitation of Plant Resources in Ancient Africa*, Kluwer Academic/Plenum Publishers, New York, pp. 73-80.

Neumann, K. (1999b). Charcoal from West African savanna sites - questions of identification and interpretation. In van der Veen, M. (ed.), *The Exploitation of Plant Resources in Ancient Africa*, Kluwer Academic/Plenum Publishers, New York, pp. 205-220.

Neumann, K., and Vogelsang, R. (1996). Paléoenvironnement et préhistoire au Sahel du Burkina Faso. *Berichte des Sonderforschungsbereichs 268* **7**: 177-186.

Neumann, K., Breunig, P., and Kahlheber, S. (2000): Early food production in the Sahel of Burkina Faso. *Berichte des Sonderforschungsbereichs 268* **14**: 327-334.

Neumann, K., Kahlheber, S., and Uebel, D. (1998). Remains of woody plants from Saouga, a medieval west African village. *Vegetation History and Archaeobotany* **7**: 57-77.

Paris, F. (1995). Le bassin de l'Azawagh: peuplement et civilisations, du néolithique à l'arrivée de l'islam. In Marliac, A. (ed.), *Milieux, Sociétés et Archéologues*, Éditions Karthala et Éditions de l'Orstom, Paris, pp. 227-257.

Paris, F. (1996). *Les Sépultures du Sahara Nigérien du Néolithique à l'Islamisation*, Orstom Éditions, Collection Études et Thèses, Paris.

Rowley-Conwy, P. (1991). Sorghum from Qasr Ibrim, Egyptian Nubia, c. 800 BC - AD 1811: a preliminary study. In Renfrew, J. (ed.), *New Light on Early Farming: Recent Developments in Palaeoethnobotany*, Edinburgh University Press, Edinburgh, pp. 191-212.

Rowley-Conwy, P., Deakin, W. J., and Shaw, C. H. (1997). Ancient DNA from archaeological Sorghum (*Sorghum bicolor*) from Qasr Ibrim, Nubia. Implications for domestication and evolution and a review of the archaeological evidence. *Sahara* 9: 23-34.

Salzmann, U. (1996). Holocene vegetation history of the Sahelian zone of NE-Nigeria: preliminary results. *Palaeoecology of Africa* 24: 103-114.

Salzmann, U., and Waller, M. (1998). The Holocene vegetational history of the Nigerian Sahel based on multiple pollen profiles. *Review of Palaeobotany and Palynology* 100(1-2): 39-72.

Shaw, T. (1981). The Late Stone Age in West Africa and the beginnings of African food production. In Roubet, C., Hugot, H. J., and Souville, G. (eds.), *Préhistoire Africaine. Mélanges Offerts Au Doyen Lionel Balout*, Editions A.D.P.F., Recherche sur les Grandes Civilisations 6, Paris, pp. 213-235.

Smith, A. B. (1980). Domesticated cattle in the Sahara and their introduction into West Africa. In Williams, M. A. J., and Faure, H. (eds.), *The Sahara and the Nile*, Balkema, Rotterdam, pp. 489-501.

Sowunmi, M. A. (1985). The beginnings of agriculture in West Africa: Botanical evidence. *Current Anthropology* 26(1): 127-129.

Sowunmi, M. A. (1999). The significance of the oil palm (*Elaeis guineensis*) in the late Holocene environments of west and west central Africa: a further consideration. *Vegetation History and Archaeobotany* 8:199-210.

Stahl, A. B. (1986). Early food production in West Africa: rethinking the role of the Kintampo culture. *Current Anthropology* 27(5): 532-536.

Stahl, A. B. (1993). Intensification in the West African Late Stone Age: a view from central Ghana. In Shaw, T., Sinclair, P., Andah, B., and Okpoko, A. (eds.), *The Archaeology of Africa: Food, Metals and Towns*, Routledge, London and New York, pp. 261-273.

Thiemeyer, H. (1992). On the age of the Bama Ridge - A new 14C-record from Konduga area, Borno State, NE-Nigeria. *Zeitschrift für Geomorphologie* N.F. 36(1):113-118.

Thiemeyer, H. (1997). *Untersuchungen zur Spätpleistozänen und Holozänen Landschaftsentwicklung im Südwestlichen Tschadbecken (NE-Nigeria)*, Jenaer Geographische Schriften 5, Friedrich-Schiller-Universität, Jena.

Thiemeyer, H., and Buschbeck, H.-M. (1993). Thermoluminescence dating of paleodunes in NE Nigeria. *Berichte des Sonderforschungsbereichs* 268 2: 221-226.

Thom, S. (1998). *Die Eisenzeitliche Nekropole von Kissi, Prov. Oudalan, Burkina Faso*, M.A. Thesis, University of Frankfurt, Germany.

Treinen-Claustre, F. (1982). *Sahara et Sahel à l'Age du Fer. Borkou, Tchad*, Mémoires de la Société des Africanistes, Paris.

Vernet, R. (1996). *Le Sud-ouest du Niger, de la Préhistoire au Début de l'Histoire*, Études Nigériennes 56, IRSH (Niamey), Sépia (Paris).

Vogelsang, R. (1995). Recherches archéologiques concernant l'histoire de l'occupation de la région sahélienne au nord du Burkina Faso: Campagne de fouille de 1994. *Nyame Akuma* 44: 16-20.

Vogelsang, R. (1996). Continuation des recherches archéologiques au nord du Burkina Faso: campagne de 1995. *Nyame Akuma* 46: 6-10.

Vogelsang, R. (1997). Etudes sur l'histoire de l'occupation de la région sahélienne du Burkina Faso: rapport des recherches sur le terrain (année 1996). *Nyame Akuma* 47: 2-6.

Vogelsang, R., Albert, K.-D., and Kahlheber, S. (1999). Le sable savant: les cordons dunaires sahéliens au Burkina Faso comme archive archéologique et paléoécologique du Holocène. *Sahara* 11: 51-68.

Wasylikowa, K. (1992). Holocene flora of the Tadrart Acacus area, SW Libya, based on plant macrofossils from Uan Muhuggiag and Ti-n-Torha / Two caves archaeological sites. *Origini* 16: 125-152.

Wasylikowa, K. (1997). Flora of the 8000 year old archaeological site E-75-6 at Nabta Playa, Western Desert, southern Egypt. *Acta Palaeobotanica* 37(2): 99-205

Wasylikowa, K., and Dahlberg, J. (1999). Sorghum in the economy of the early Neolithic nomadic tribes at site E-75-6, Nabta Playa, south Egypt. In van der Veen, M. (ed.), *The Exploitation of Plant Resources in Ancient Africa*, Kluwer Academic/Plenum Publishers, New York, pp. 11-32.

Wendt, K. P. (1995). Magerung und Oberflächenbehandlung zur chronologischen Interpretation technischer Merkmale in der Keramikentwicklung in Nordost-Nigeria. *Berichte des Sonderforschungsbereich 268* 5: 41-47.

Wendt, K. P. (1997). *Beiträge zur Entwicklung der Prähistorischen Keramik des Inneren Tschadbeckens in Nordost-Nigeria.* Ph. D. dissertation, University of Frankfurt, Germany.

Zach, B., Kirscht, H., Löhr, D., Neumann, K., and Platte, E. (1996). Masakwa dry season cropping in the Chad Basin. *Berichte des Sonderforschungsbereichs 268* 8: 349-356.

10

HOLOCENE CLIMATIC CHANGES IN THE EASTERN MEDITERRANEAN AND THE SPREAD OF FOOD PRODUCTION FROM SOUTHWEST ASIA TO EGYPT

M. Rossignol-Strick

INTRODUCTION

Whereas agriculture independently appeared in a few places in the world during the Holocene, its earliest beginning involving wheat, barley and pulses took place in Southwest Asia (Helbaek, 1959a,b; Zohary, 1969; Van Zeist, 1986; Zohary and Hopf, 1988; Moore, 1991; Moore and Hillman, 1992). Two aspects of this major advance in human history emphasize its exceptional character. Only in this area the wild progenitors of these primary domesticates are extant (Harlan and Zohary, 1966). These earliest domesticates still constitute a major part of the staple food of the world population. These founder crops entered Africa via the Nile Valley, the fertile oasis closest to Southwest Asia. The transfer occurred about two thousand and seven hundred years later than the appearance of domesticates in the Levant, where it defines the Neolithic. Here we investigate the time-scale of the regional climatic evolution in the Eastern Mediterranean basin in relationship to these protohistoric events.

THE CLIMATE HISTORY OF THE 11,000 TO 6000 YR BP INTERVAL IN THE EASTERN MEDITERRANEAN REGION

A period of drastic climate change accompanied the wasting of the continental ice sheets of the northern Hemisphere that had reached their maximum volume from 22,000 to 18,000 radiocarbon years ago. The final part of the last Ice Age imposed on the whole Earth extreme climate conditions that in the high latitudes and altitudes inhibited the potential arboreal vegetation by too low temperatures and too short summers, and by a deficit of moisture in the low latitudes and altitudes (CLIMAP 1976, COHMAP, 1988; Wright et al., 1993). In the Old World, an herbaceous ground cover of prairie and semi-desert was the dominant and very seasonal vegetation of non-glaciated Europe and Africa beyond the vastly expanded subtropical desert and the equatorial zone. In Europe, trees were restricted to refuges in discrete sites made favorable by edaphic moisture and protection from cold winds. It is under these not too hospitable conditions that the scarce human population carried out the long and fairly static Palaeolithic stage of its cultural course.

As in the previous glacial cycles, the wasting of the last ice sheets was, in the geological time-scale, a relatively rapid event. It took seven thousand years, from 15,000 bp to 8000 cal BP, to dissipate an ice volume that required about a hundred thousand years to accumulate. The climate first improved in the Bölling-Allerød interstadial from 14,450 to 12,700 cal BP (Johnsen *et al.*, 1992), which was interrupted by a cooling event observed in the GRIP ice core in Greenland, that corresponded to a first phase of rapid ice melt shortly before 13,800 cal BP (Bard *et al.*, 1996). A subsequent sharp reversal to glacial conditions, the Younger Dryas, took place from 12,700 to 11,550 cal BP (Johnsen *et al.*, 1992), or 11,000 to 10,000 radiocarbon bp. The Holocene begins with the transition to the Preboreal at 11,550 cal BP. A second phase of rapid ice melt took place early in the Preboreal, around 11,300 cal BP (9800 radiocarbon bp; Bard *et al.*, 1996).

In the Eastern Mediterranean region, these global events have been recorded in the isotopic and pollen records of marine sediments (Rossignol-Strick, 1993, 1995, 1997). Their time-scale rests on AMS radiocarbon dates, mainly in the Adriatic Sea. The time interval of the Younger Dryas displays a large expansion of the Chenopodiaceae pollen abundance. These herbs almost alone tolerate the saline soils developed in coastal marshes and also inland under very arid, desert conditions of less than 150 mm annual precipitation. The sage-brush, *Artemisia*, generally accompanies the Chenopodiaceae, but may also be displaced by them when aridity is too extreme, since its minimum requirement is 150-200 mm. Trees are reduced to minimal values. Rapidly thereafter, these drought-tolerant plants abruptly decrease, and the changing vegetation suggests improving climate. Increase in moisture, well above 650 mm annual precipitation, and possibly as high as 1200 mm, and also in temperature, is revealed by the onset and peak value of the deciduous oak, which signals the absence of summer drought in the relevant climate belt of mid-altitudes (ca. 300-1200 m asl). No-frost winters in low altitudes are revealed by *Pistacia* trees, which produce few pollen grains. Hence, even the low pollen abundance of *Pistacia* signals an important role of the trees in the savanna vegetation. The *Pistacia* phase is coeval with the deposition, in the Eastern Mediterranean sea of the most recent sapropel, a black organic-rich sediment layer, which took place from 9000 to 6000 AMS radiocarbon bp.

Terrestrial pollen records in Greece, Turkey, Iran and Syria display the same pollen succession that identifies the Younger Dryas and the early Holocene Climate Optimum. This successful comparison allows us to apply to them the time-scale of the marine record, which often results in some discrepancy with their few radiometric dates, mainly for the older part of the records (Rossignol-Strick, 1993, 1995, 1997). One site in northern Israel, Lake Hula (Figure 10.1), within the same Mediterranean zone as the other sites, has yielded a quite similar pollen record (Baruch and Bottema, 1991), but its own radiometric dates reveal a large discrepancy with the marine time-scale. It is consistently older by 3000 years than indicated by the comparative method.

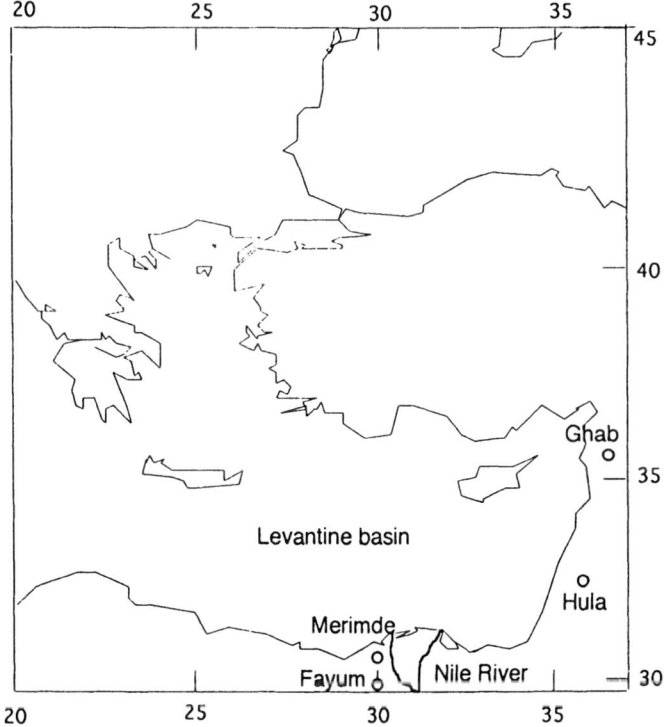

Figure 10.1. The Eastern Mediterranean (Levantine Basin), Ghab Valley and Lake Hula: pollen records, Merimde and the Fayum: first agricultural sites in Africa.

Consequently, the peak of deciduous oak pollen abundance, signaling the wettest time, would have occurred during the Bölling-Allerød, not in the early Holocene. Moreover, contrary to expectation, the time interval of the Younger Dryas, which is delimited by a 10,440 radiocarbon date (Bottema, 1995), but should reach 10,000 yr bp, displays not only the lowest pollen abundance for Chenopodiaceae and *Artemisia*, but also a peak for *Pistacia* and, after 10,440 yr bp, a minor peak for all Quercus, including its evergreen, thermophilous Mediterranean species. After 10,000 yr bp, not only is the expected improvement of the early Holocene, at least until 9270 bp, not seen (Bottema, 1995), but the abundance of all Quercus pollen decreases. These serious inconsistencies with the pollen record of the Ghab Valley, in the same vegetation and climatic zone and only 300 km to the north of Lake Hula, clearly suggest that the Hula chronology is incorrect.

This is supported by the isotopic study of speleothems in a cave near Jerusalem (Bar-Matthews and Ayalon, 1997; Kaufman *et al.*, 1998). The 95,000 yr continuous record, dated by the U-series method, shows the lowest $\delta^{18}O$ values, interpreted as corresponding to the wettest period (~1000 mm), from 10,000 to 7000 U/Th bp (more narrowly 8.5 to 7.0 kyr bp). They are preceeded by the drier Younger Dryas with ages 13,200 to 11,400 bp, and the Bölling-Allerød, which has heavier $\delta^{18}O$ values and is seen as less rainy than the early Holocene. This record is fully consistent with the timescale of the Hula pollen record obtained by the comparative method, in contrast with its radiocarbon-derived timescale. We suggest that the Rift of the upper Jordan Valley, where Lake Hula is located, may have been, and still be, the site of degassing of volcanic carbon dioxide that naturally introduces old, 'dead' carbon into the aquatic organic matter subjected to radiocarbon measurements. Indeed, volcanism was active as recently as the Bronze Age on the Damascus Plateau, a short distance from the Rift (Dubertret and Dunand, 1954-1955), and the hydrothermal sources of the Roman city of Tiberias on the Lake of the same name are still active, as well as sulfurous sources near Capharnaum on its northwestern shore. This circumstantial evidence suggests that degassing of dead radiocarbon is the probable cause of the radiometric discrepancy of the Lake Hula record. Solving this problem is important for reconstructing how agriculture was introduced in Africa (note: as of June 7, 2000, I have been told that the authors Baruch and Bottema have indeed accepted my reading of the time-scale of the Lake Hula pollen diagram).

That the early Holocene, not the Bölling-Allerød, was the wettest interval is also confirmed in the most desert part of Arabia by the 9000 to 6000 AMS radiocarbon bp dating of lacustrine sediments indicating high water level in fossil Lake Mundafan and other sites in the Rub Al Khali between 18° and 23°N (McClure, 1976). Also, in the northwestern Sudan between 19° and 22°N, the pollen records of fossil lake muds reveals a major pluvial episode between 9500 and 4500 bp (Ritchie *et al.*, 1985; Ritchie and Haynes, 1987). In other sites in the Eastern Sahara there were permanent fresh-water lakes between 8.5 and 7 kyr bp (Yan and Petit-Maire, 1994). All these data indicate that the Sahelo-Saharan boundary for the African monsoon rains shifted at least 500 km further north than presently, at 22°-23°N.

Thus, the early Holocene was unquestionably the period of the 'Climate Optimum', expressed in the Eastern Mediterranean domain by abundant annual moisture, in the 800-1200 mm range, without summer drought, in the widened belt of the oak forest at mid-elevation, and warm winters in the lower elevation of the *Pistacia* savanna. This summer moisture in the Eastern Mediterranean and the Near East was conveyed by the extra-tropical westerlies, not the southwesterlies of the Indian Ocean monsoon, as proposed by Bar-Yosef and Meadow (1995) according to an extended reading of the COHMAP (1988) model at 9 kyr bp. Over the low-lying African and Arabian land-mass, the southwesterly monsoonal flow from the south Atlantic is steered by the trough of low pressure on the ground generated by the course of the sun, and its penetration is limited by the Tropic of Cancer. In the early Holocene, two aspects of the insolation variation influenced the changes in the intensity and extension of the monsoon. The caloric summer radiation in the northern hemisphere was maximal at 10-11 kyr bp (Berger, 1978), and at 9 kyr bp, the Tropic of Cancer was at 24°14' (Berger, 1978), which is 47' north of its present latitude. This increased the latitude of the pressure trough and the northward penetration of the monsoonal westerlies by about 87 km. This latitude is that of the northernmost fossil lakes in Africa and Arabia. The simulation of the intensification and extension of the African monsoon under the increased radiation of the early Holocene has been improved by the introduction of the vegetation and soil feedbacks (Kutzbach et al, 1996).

THE TRANSITION FROM FOOD GATHERING TO AGRICULTURE IN SOUTHWEST ASIA

It is impossible not to infer that the precipitous climate changes from 11,000 to 6000 bp in the Eastern Mediterranean had an overwhelming influence on culture change. In Southwest Asia, the critical shift from wild food plant gathering to progressively intentional production of the domesticates occurred at the beginning of this climatic shift, along with the ongoing gathering and perhaps even cultivation of the wild plants. It is only in that region that the wild progenitors of wheat and barley, the founder crops, are presently living. In the Levant and Southeast Turkey, Iraq and Iran, they are found, in primary habitats, in the open oak-woodlands of the foothills of the Taurus-Zagros arc (Helbaek, 1959a,b; Harlan and Zohary, 1966).

The earliest evidence of these cereals displaying the tough rachis phenotype deemed domesticate has been discovered in archaeological sites situated in the northern part of the Fertile Crescent, mainly in Northeast and Central Syria. Among seeds of cereals and pulses found in archaeological sites, the wild and domesticate variants mainly differ by morphological and anatomical characters that are linked to the mode of seed dispersal (Zohary, 1969; Van Zeist, 1986; Zohary and Hopf, 1988). The wild barley, *Hordeum spontaneum*, has a brittle ear rachis that at maturity disarticulates and leaves a scar at a pre-determined site below the insertion of each spikelet. The dispersal unit is the arrow-shaped spikelet attached to its rachis section that anchors in the ground. In the domesticated barley, *H. vulgare*, at maturity the ear rachis does not dis-articulate, so that this tough rachis keeps the ear whole on top of the stem until the next autumn rains or until laying of the stems by overgrowth. In archaeological material of the early Neolithic sites, identification of wild versus domesticated barley rests mainly, but not even always unequivocally (Kislev, 1989), on the study of the fragments of the rachis. The fragments of the brittle rachis of the wild barley are limited by a clean scar whereas those of the tough rachis of the domesticated form are irregularly broken by the threshing performed by the farmers (Kislev, 1989, 1992; Hillman and Davies, 1990; Zohary and Hopf, 1988). The wheat variety most frequently found is the tetraploid emmer. The distinction between wild emmer (*Triticum dicoccoides* Körn.) and domesticated emmer (*T. dicoccum* Schübl; Zohary and Hopf, 1988) in early archaeological sites is more uncertain than for barley because the rachis breaks similarly at maturity in natural conditions and during threshing (Kislev, 1989). The two wheat forms are fully interfertile and appear to differ only by one major gene (Zohary and Hopf, 1988). However, the genes that control the rachis brittleness and seed dormancy are still unidentified (Blumler and Byrne, 1991). For the diploid einkorn wheat (wild: *T. boeticum*; domesticate: *T. monococcum*), it seems that the seeds identified as *T. monococcum* that were found in some early Neolithic sites such as Jericho, well outside their present distribution area, may be interpreted as the one-seeded basal or terminal spikelets of emmer (Kislev, 1992).

The emergence of agriculture is identified according to the opinion that the tough-rachised phenotype of cereals cannot presently survive without the human intervention of reaping, threshing and sowing (Zohary, 1969, 1992; Harlan *et al.*, 1973; Kislev, 1984; Zohary and Hopf, 1988; Hillman and Davies, 1990). Thus, the presence of this phenotype in archaeological layers would reveal agricultural practice, although it does not imply the conscious selection of this variety, with the corollary that cultivation would have been initiated with wild cereals, possibly during the Epipalaeolithic (the Natufian in the Levant). Recognition of that cultivation stage could be based on the observation of the wear on flint sickle-blades (Unger-Hamilton, 1992). The emergence of agriculture is best documented at Tell Abu Hureyra in Northeast Syria on the Euphrates (Lewin, 1988; Moore, 1991, 1992). There, the late Epipalaeolithic stage (Hureyra 1, coeval with the Natufian of the Levant) from 11,550 to 10,000 AMS radiocarbon bp has yielded only wild brittle rachis cereals, mainly einkorn and also barley and rye, which have been gathered by foraging. It is succeeded by a hiatus of occupation of three hundred years, until 9700 AMS radiocarbon bp, which corresponds to the end of the Younger Dryas, and by a second phase of occupation (Hureyra 2) from 9700 to 7000 AMS radiocarbon bp. In this early Neolithic phase, the earliest tough-rachised phenotypes of crops are identified, thus revealing that, since domestication had occurred, intentional cultivation was practiced (Moore and Hillman, 1992). In different abundances, they are cereals: emmer, barley and einkorn, and pulses: peas, chickpeas and lentils (Hillman, 1975). Other early Neolithic sites in the Levant have also revealed these founder crops, but they do not encompass the continuity observed in Tell Abu Hureyra from Epipalaeolithic to early Neolithic. In Mureybit, 20 km from Hureyra on the Euphrates, the Epipalaeolithic establishment, dated 10,450-8850 radiocarbon bp, yielded only wild cereals and pulses: einkorn, barley and lentils (Van Zeist and Bakker Heeres, 1984). On the Middle Euphrates, two Neolithic preceramic sites, Dja'de and Jerf al Ahmar, have revealed, between 9800 and 9200 bp, only wild cereals (einkorn, rye and barley) and pulses, while at a third site, Halula, domesticated emmer and barley were observed in ninth millenium bp layers (Willcox, 1996). In three early Neolithic sites in the Damascus basin, Aswad, Ghoraïte and Ramad, no wild cereals are found, and in Aswad I at 9800 radiocarbon bp, domesticated emmer is abundantly found, as well as einkorn, but the status of the two-rowed hulled barley is not certain (Van Zeist and Bakker Heeres, 1982). In the Jordan Valley north of Jericho, the site of Netiv Hagdud has yielded barley seeds and enough non-brittle rachis fragments to see them as domesticated in Pre-Pottery Neolithic layers dated 7775 BC (Kislev *et al.*, 1986) and in Gilgal at 8000-7800 BC (Noy in Zohary, 1992). In the oldest layers at Jericho, 10,000-9300 radiocarbon bp, emmer and barley are found in small amounts (Hopf, 1983 in Zohary and Hopf, 1988). In the southern Levant, the earliest domesticates appear later (Zohary and Hopf, 1988). At Beidha, in Jordan near Petra, emmer and barley are dated 6780-6600 BC (8730-8550 radiocarbon bp, ca. 9600 cal BP). In Arad, Northern Neguev, emmer, barley and some einkorn form rich remains in the early Bronze Age at 2770 BC (4720 radiocarbon bp, ca. 5650 cal BP), and also at the same time in Bab edh-Dhra, Southeast of the Dead Sea.

It has been suggested that the stress exerted on human behavior by the arid and cold conditions of the Younger Dryas significantly contributed to modifications in the plant gathering pattern of the Epipalaeolithic population, and led to the emergence of agriculture (Moore and Hillman, 1992; Bar-Yosef and Meadow, 1995). However, a sound basis for this suggestion was lacking at the time, because the radiometric timescales of the pollen records of Lake Hula and even of the Ghab Valley, which were assumed valid, did not support the notion that the Younger Dryas was regionally arid and cold.

In contrast, having established, by comparison with the marine record, that the Younger Dryas was actually the most arid time of the late Quaternary in the Eastern Mediterranean domain (Rossignol-Strick, 1993, 1995, 1997), and noting that the earliest domesticates in Syria appear immediately after the Younger Dryas, we now propose that the aridity and cold themselves had a direct role in the selection of tough rachised cereals and of pulses with non-dehiscent pods. The mutation for this new mode of seed dispersal would have randomly appeared in one individual shortly before or during the Younger Dryas, and the early generations of mutants would have been naturally selected by the extreme environmental conditions of that period. Genetic evidence supports the view that for emmer, pea and lentil, and perhaps also for einkorn, a single domestication event, or at most very few, for barley, has been at the origin of agriculture in the Neolithic of Southwest Asia (Zohary, 1996). Although it is widely believed that these recessive mutants cannot survive except under cultivation (Zohary, 1969; Zohary and Hopf, 1988; Hillman, 1996; Hillman and Davies, 1990, 1992), the possibility that the tough (or semi-tough) rachis plants appeared before cultivation is also raised: "The domesticated phenotypes may have become locally abundant for a very short time under special conditions" (Blumler and Byrne, 1991). These special conditions may precisely have been those of the Younger Dryas. Indeed, it is hard to believe that this phenotype, as it first appeared in one individual plant, according to the monophyletic view of Zohary implied by the one domestication event, was most opportunely protected from its own vulnerability in the wild by human harvesting. Under the more favorable conditions of the early Holocene, which included warm winters and summer precipitation (Rossignol-Strick, 1993, 1995, 1999), harvesting did ensure the diffusion of the mutant. Nevertheless, the Younger Dryas was the initial period during which the mutant, fending for itself in a favorable niche amid the arid environment, had established some sizeable population. Eventually, in the early Holocene, this population fell under the sickle or the beating stick of humans, who may or may not have noticed it among the wild type plants. Since the present distribution of wild cereals was initiated in the early Holocene Climate Optimum, we do not know how restricted their distribution was under the aridity of the Younger Dryas. The necessary moisture may well have been edaphic, since rains were so scarce and irregular that no location could consistently benefit from them on an interannual pattern. Thus, it could have become a survival asset for the mutant to keep together the seeds of a tough rachis ear, because at least some of them would have a chance to find the ground moisture necessary for their germination. Moreover, the tough rachis trait may have protected the seeds from predation by ants and rodents by keeping them above ground until their fall and germination. Also, the harshness of the Younger Dryas must have much reduced the population of rodents and other predators. In contrast, it has been noticed in cultivation experiments of wild cereals that mild winters considerably increased the number of rodents, resulting in the loss of much of the crop (Willcox, 1992). We suggest that only the character of aridity and cold of the Younger Dryas can have selected the early generations of recent mutants and thus account for the 'astonishing synchroneity' (Blumler and Byrne, 1991) of their archaeological appearance in the first third of the tenth millenium bp.

THE EARLIEST AGRICULTURE IN AFRICA

From Southwest Asia, agriculture based on those crops radiated in three major directions: toward the west, the east, and Africa.

Significantly, it is in the corner of Africa closest to Southwest Asia that the oldest traces of wheat and barley agriculture have been discovered. It is not happenstance that the Nile Valley, the thoroughfare that links the wet tropics across the Saharan desert to the Mediterranean, opens to the sea near to the land-bridge to Asia. The main Nile north of Khartoum flows in the tectonic low behind the uplifted eastern rim of the continent, the Red Sea hills that rose in the Upper Eocene as the Red Sea was formed (Shahin, 1985). The main Nile is therefore a by-product of the Red Sea Rift that extends northward into Southwest Asia as the Arava- Dead Sea- Jordan Valley. The spread of the founder crops into Africa was steered by the geomorphology that evolved in the Eocene.

In the Fayum, the early Neolithic, ca. 5200-4500 cal BC (7150-6500 cal BP, 6300-5700 radiocarbon bp; Hassan, 1985, this volume) has produced rich remains of mainly emmer wheat and two- and six-row barley (Zohary and Hopf, 1988). In the western Nile Delta, about 600 km from Jericho, the early Neolithic site of Merimde has revealed, ca. 4800 cal BC (6750 cal BP, 5900 radiocarbon bp), rich remains of emmer wheat, six-row barley, lentils and peas (Wetterstrom, 1993, 1996; Zohary and Hopf, 1988). Wheat and barley would have been introduced from Southwest Asia perhaps a millennium before the early farming communities in Merimde. Later, in the Nagada region of Upper Egypt, in site KH3 at Khattara ca. 3800 radiocarbon BC, and in South Town, at 3400 radiocarbon BC, emmer wheat dominates, with six-row barley, and pea (Zohary and Hopf, 1988). In Lower Egypt, in Saqqara, the Zoser pyramid of the third

Dynasty, 2900 BC, has yielded rich remains of emmer and barley, and rare lentil, while emmer wheat also prevails in Queen Icheti's tomb of the sixth Dynasty, 2550 BC (Zohary and Hopf, 1988).

Thus, in Lower Egypt, domesticated emmer, barley and pulses (but not einkorn) were cultivated in the late eighth millenium BP, and from there they spread into the rest of Africa. The entry into Africa occurred about three thousand and five hundred years later than its onset in Northeast Syria, and two thousand and one hundred years later than in Beidha in southern Jordan, but earlier than in Arad in the northern Neguev. Whereas in Southwest Asia the earliest tough-rachis cereals are observed—and perhaps cultivated—as soon as the Younger Dryas ended, in Egypt this stage was reached at the end of the Climate Optimum of the early Holocene. We suggest that the cause of that significant delay was that the transfer of crops has been dependent upon the domestication of sheep.

The Holocene Climate Optimum may have steered the spread of plant cultivation to Egypt because at this time the sheep has been domesticated in Southwest Asia. The change in animal use by farmers, from gazelle hunting to sheep herding, occurred in Tell Abu Hureyra at 8300 AMS radiocarbon yr bp (Moore, 1992). At that time, the Syrian desert between the Euphrates and the Mediterranean zone of the Levant would have been restricted by the concomitant spread of a grass steppe favored by the relatively wet climate, particularly in summer. Such a vegetation is indicated in the pollen record of the Ghab Valley in Northwest Syria, where the abundance of Gramineae steadily increases from the end of the Younger Dryas to maximum value around 7000 bp, when the oak, past its maximum at 135 cm core depth around 8000 bp, has decreased to a minimum (Niklewski and Van Zeist, 1970, zone Z3, 73 cm core depth). This signals that moisture has peaked and then decreased, although to a level significantly higher than during the Younger Dryas, around 350-450 mm in the present margin of the Syrian desert. Whereas the core of the Sinai desert probably persisted during the Holocene Climate Optimum, this wet period caused the desert to shrink at its periphery. This provided pasture and made possible the southwestward crossing of the desert by the herders of sheep, who were also farmers, eventually settling in Merimde ca. 7000 bp. Thus, the onset of agriculture in Africa would have been consequent to the domestication of sheep in Southwest Asia at the end of the wettest period.

Could a brief arid and cold event at that time also have influenced the transfer of agriculture to Egypt (see Hassan, 2000a,b)? Such a spell has been documented ca. 8.2 kyr bp in the Greenland ice-cores (Alley et al., 1997) and identified in many other sites worldwide. In the Eastern Mediterranean domain, it is most clearly documented by a pulse of Artemisia in the pollen record of Tenaghi Philippon in Macedonia, by an interruption of the Holocene sapropel in the Adriatic Sea (Fontugne et al., 1989; Rossignol-Strick et al., 1992; Rohling et al., 1997, this volume) and by a brief decrease of the Nile flood plume off the coast of Israel (Luz, 1979). Although this arid event may have played some role, our opinion is that the overall favorable climate of the early Holocene during three thousand years exerted the main influence in the transfer of domesticated crops to Africa and in the demic expansion which may have achieved it, as occurred in Europe (Sokal et al., 1991).

CONCLUSION

The importance of the cultural stage of the primordial agriculture emergence cannot be overestimated. It occurred in Southwest Asia in the early Holocene which, in the Eastern Mediterranean basin, enjoyed the most favorable, warm and wet climate, and immediately followed the most arid and cold period of the Younger Dryas. In the early tenth millenium before present in the radiometric time-scale, or ca. 11,500 calendar years before present, wheat, barley and pulses for the first time display the anatomical variant deemed domesticate. There are many prerequisites for this emergence which are now becoming better understood. They include the natural environment, geology, climate and vegetation, and the human population present in that area at that time.

In contrast with Southwest Asia, Africa does not possess the first of the necessary natural conditions. It does not harbor the wild progenitors of wheat, barley and pulses. Presently they are extant only in Southwest Asia. Insofar as the widely held view is correct that for each cereal the variant called domesticate cannot subsist without human intervention, its first appearance in archaeological layers would mark the unmistakable starting point for agriculture, defined as reaping, threshing and sowing of these domesticates (Harris and Hillman, 1989; Harris, 1996). The same activities for the wild progenitors could be called cultivation (Harris, 1996), and are less accessible to identification, although it is suspected that they might have been initiated during the Bölling-Allerød interstadial prior to the Younger Dryas, with the Natufian or Epipalaeolithic cultures.

We are impressed by the simultaneity of the first appearance of domesticates in several archaeological sites of Southwest Asia immediately after the end of the Younger Dryas period at 10,000/9700 bp. We here contend that the timing and the simultaneity strongly suggest that the natural conditions of cold and aridity selected in the wild the recent and monophyletic mutant forms of cereals and enabled them to achieve a population size large enough not to be missed in the gathering yield of the human foragers.

In Africa, the earliest evidence of agriculture based on these crops is found in the Nile Valley, whose origin was a consequence of the formation of the Red Sea and Near Eastern Rift, and therefore was intrinsically linked to Southwest Asia. Thus, it is no surprise that the earliest agriculture in Africa had for its cradle the fertile alluvium of the Nile River that soon thereafter became the land of Egypt. There was a 3500 year delay between the appearance of domesticates in Southwest Asia and in the Nile Valley. This can be accounted for by the fact that, in Southwest Asia, the Pre-Pottery Neolithic farmers, after about 1000 years of agriculture, became sheep herders. As their population density increased with their economic success, their spread toward the southwest in the direction of Africa became possible when pasture for their flocks became available along the way. They crossed the Sinai desert, most probably along the coast, during the wet early Holocene as the desert margins receded. This view is supported by the earliest and simultaneous presence of sheep, wheat and barley around seven thousand years ago in the Fayum and the Delta in Merimde. This early initial stage was soon followed by the rise of the Predynastic period of the Egyptian civilization.

REFERENCES

Alley, R.B, Mayewski, P.A., Sowers, T., Stuiver, M., Taylor, K.C., and Clark, P.U., (1997). Holocene climatic instability: a prominent, widespread event 8200 yr ago. *Geology* 25: 483-486.

Bard, E., Hamelin, B., Arnold, M., Montaggioni, L., Cabioch, G., Faure, G., and Rougerie, F. (1996). Deglacial sea-level record from Tahiti corals and the timing of global meltwater discharge. *Nature* 382: 241-244.

Bar-Matthews, M., and Ayalon, A. (1997). Late Quaternary paleoclimate in the Eastern Mediterranean region from stable isotope analysis of speleothems at Soreq Cave, Israel. *Quaternary Research* 47: 155-168.

Baruch, U., and Bottema, S. (1991). Palynological evidence for climatic changes in the Levant ca. 17,000-9,000 B.P. In Bar-Yosef, O., and Valla, F.R. (eds.), *The Natufian Culture in the Levant*, International Monographs in Prehistory, Archaeological Series 1, Ann Arbor, Michigan, pp. 11-20.

Bar-Yosef, O and Meadow, R.H. (1995). The origins of agriculture in the Near East. In Price, T.D., and Gebauer, A.B. (eds.), *Last Hunters, First Farmers: New Perspectives on the Prehistoric Transition to Agriculture*, School of American Research Press, Santa Fe, pp. 39-94.

Berger, A. (1978). Long-term variations of caloric insolation resulting from the Earth's orbital elements. *Quaternary Research* 9: 139-167.

Blumler, M.F., and Byrne, R. (1991). The ecological genetics of domestication and the origin of agriculture. Current Anthropologist 32: 23-54.

Bottema, S. (1995). A short review of the Younger Dryas in the Eastern Mediterranean area (extended abstract). *Geologie en Mijnbouw* 74: 271-273.

CLIMAP project members. (1976). The surface of the ice-age Earth. *Science* 191: 1131-1137.

COHMAP members. (1988). Climatic changes of the last 18,000 years: observations and model simulations. *Science* 241: 1043-1052.

Dubertret, L., and Dunand, M. (1954-1955). *Les Gisements Ossifères de Khirbet el-Umbachi et de Hebariye (Safa)*, Annales Archéologiques de Syrie, tome IV et V, pp. 59-76.

Fontugne, M., Paterne, M., Calvert, S., Murat, A., Guichard, F., and Arnold, M. (1989). Adriatic deep water formation during the Holocene: implication for the reoxygenation of the deep Eastern Mediterranean sea. *Paleoceanography* 4: 199-206.

Harlan, J., and Zohary, D. (1966). Distribution of wild wheats and barley. *Science* 153: 1074-1080.

Harlan, J. R., de Wet, J.M.M., and Price, E.G. (1973). Comparative evolution of cereals. *Evolution* 27: 311-325.

Harris, D.R. (1996). Introduction: themes and concepts. In Harris, D.R. (ed.), *The Origin and Spread of Agriculture and Pastoralism in Eurasia*, University College London Press, London, pp. 1-9.

Harris, D.R., and Hillman, G.C. (1989). *Foraging and Farming: the Evolution of Plant Exploitation*, Unwin Hyman, London.

Hassan, F. A. (1985). Radiocarbon chronology of Neolithic and Predynastic sites in Upper Egypt and the Delta. *African Archaeological Review* 3: 95-116.

Hassan, F. A. (2000a). Climate and cattle in north Africa. In Blench, R., and MacDonald, K. C. (eds.), *The Origins and Development of African Livestock: Archaeology, Genetics, Linguistics, and Ethnography*, University College London Press, London, pp. 61-86.

Hassan, F. A. (2000b). Holocene environmental change and the origins and spread of food production in the Middle East. *Adumatu* 1: 7-28.

Helbaek, H. (1959a). Domestication of food plants in the Old World. *Science* 130: 365-372.

Helbaek, H. (1959b). How farming began in the Old World. *Archaeology* 12: 183-198.

Hillman, G.C. (1975). The plant remains from Abu Hureyra: a preliminary record. In Moore, A.T.M. (ed.), The excavations of Tell Abu Hureyra in Syria: a preliminary record. *Proceedings of the Prehistoric Society* 41: 70-73.

Hillman, G C (1996). Late Pleistocene changes in wild plant-foods available to hunter gatherers of the northern Fertile Crescent: possible preludes to cereal cultivation. In Harris, D.R. (ed.),

The origin and spread of Agriculture and Pastoralism in Eurasia, University College London Press, London, pp. 159-203.

Hillman, G.C., and Davies, M.S. (1990). Domestication rates in wild-type wheats and barley under primitive cultivation. *Biological Journal of the Linnean Society* 39: 39-78.

Hillman, G.C., and Davies, M.S. (1992). Domestication rates in wild wheats and barley under primitive cultivation: preliminary results and archaeological implications of field measurements of selection coefficient. In Anderson, P.C. (ed.), *Préhistoire de l'Agriculture: Nouvelles Approches Expérimentales et Ethnographiques*, Monograph CRA 6, CNRS, Paris, pp. 154-158.

Hopf, M. (1983). Jericho plant remains. In Kenyon, K.M., and Holland, T.A. (eds.), *Excavations at Jericho*, vol 5, British School of Archaeology in Jerusalem, London, pp. 576-621.

Johnsen, S.J., Clausen, H.B., Dansgaard, W., Fuhrer,K., Gundestrup, N., Hammer, C.U., Iversen, P., Jouzel, J., Stauffer, B., and Steffensen, J.P. (1992). Irregular glacial interstadials recorded in a new Greenland ice core. *Nature* 359: 311-313.

Kaufman, A., Wasserburg, G.J., Porcelli, D., Bar-Matthews, M., Ayalon, A., and Halicz, L. (1998). U-Th isotope systematics from the Soreq cave, Israel and climatic correlations. *Earth and Planetary Science Letters* 156: 141-155.

Kislev, M.E. (1984). Emergence of wheat agriculture. *Paléorient* 10: 61-70.

Kislev, M.E. (1989). Pre-domesticated cereals in the Pre-pottery Neolithic A period. In Hershkovitz, I. (ed.), *Man and Culture in Change*, British Archaeological Reports International Series 508 (1), Oxford, pp. 147-151.

Kislev, M.E. (1992). Agriculture in the Near East in the VIIth millenium BC. In Anderson, P. (ed.), *Préhistoire de l'Agriculture: Nouvelles Approches Expérimentales et Ethnographiques*, Monograph CRA 6, CNRS, Paris, pp. 87-94.

Kislev, M. E., Bar-Yosef, O., and Gopher, A. (1986). Early Neolithic domesticated and wild barley from the Netiv Hagdud region in the Jordan Valley. *Israel Journal of Botany* 35: 197-201.

Kutzbach, J., Bonan, G., Foley, J., and Harrison,S.P. (1996). Vegetation and soil feedbacks on the response of the African monsoon to orbital forcing in the early to middle Holocene. *Nature* 384: 623-626.

Lewin, R. (1988). A revolution of ideas in agricultural origins. *Science* 240: 984-986.

Luz, B. (1979). Palaeo-oceanography of the post-glacial Mediterranean. *Nature* 278: 847-48.

McClure, H.A. (1976). Radiocarbon chronology of late Quaternary lakes in the Arabian desert. *Nature* 263: 755-756.

Moore, A.T.M. (1991). Abu Hureyra 1 and the antecedents of agriculture on the middle Euphrates. In Bar-Yosef, O., and Valla, F. (eds.), *The Natufian Culture in the Levant*, International Monographs in Prehistory, Ann Arbor, Michigan, pp. 277-294.

Moore, A.T.M. (1992). The impact of accelerator dating at the early village of Abu Hureyra on the Euphrates. *Radiocarbon* 34: 850-858.

Moore, A.T.M., and Hillman, G.C. (1992). The Pleistocene to Holocene transition and human economy in southwest Asia: the impact of the Younger Dryas. *American Antiquity* 57: 482-494.

Niklewski, J., and Van Zeist, W. (1970). A late Quaternary pollen diagram from NW Syria, *Acta Botanica Neerlandica* 19: 737-754.

Ritchie, J.C., and Haynes, C.V. (1987). Holocene vegetation zonation in the eastern Sahara. *Nature* 330: 645-647.

Ritchie, J.C., Eyles, C.H., and Haynes, C.V. (1985). Sediment and pollen evidence for an early to mid-Holocene humid period in the eastern Sahara. *Nature* 314: 352-355.

Rohling, E.J., Jorissen, F.J., and de Stigter, H.C. (1997). 200 year interruption of Holocene sapropel formation in the Adriatic sea. *Journal of Micropalaeontology* 16: 97-108.

Rossignol-Strick, M. (1993). Late Quaternary climate in the Eastern Mediterranean region. *Paléorient* 19(1): 135-152.

Rossignol-Strick, M. (1995). Sea-land correlation of pollen records in the Eastern Mediterranean for the Glacial-Interglacial transition: biostratigraphy versus radiometric time-scale. *Quaternary Science Review* 14: 893-915.

Rossignol-Strick, M. (1997). Paléoclimat de la Méditerranée Orientale et de l'Asie du Sud-Ouest de 15000 à 6000 BP. *Paléorient* 23(2): 175-186.

Rossignol-Strick, M. (1999). The Holocene Climatic Optimum in pollen records of sapropel 1 in the Eastern Mediterranean, 9000-6000 BP, and the start of the Neolithic. *Quaternary Science Review* 18: 515-530.

Rossignol-Strick, M., Planchais, N., and Paterne, M. (1992). Vegetation dynamics and climate during the deglaciation in the south Adriatic basin from a marine record. *Quaternary Science Review* 11: 415-423.

Shahin, M. (1985). *Hydrology of the Nile Basin*, Elsevier, Amsterdam.

Sokal, R.R., Oden, N.L., and Wilson, C. (1991). Genetic evidence for the spread of agriculture in Europe by demic diffusion. *Nature* 351: 143-145.

Unger-Hamilton, R. (1992). Experiments in harvesting wild cereals and other plants. In Anderson, P. (ed.), *Préhistoire de l'Agriculture: Nouvelles Approches Expérimentales et Ethnographiques*, Monograph CRA 6, CNRS, Paris, pp. 211-244.

Van Zeist, W. (1986). Some aspects of early Neolithic plant husbandry in the Near East. *Anatolica* XV: 49-68.

Van Zeist, W., and Bakker-Heeres, J.A.H. (1979). Some economic and ecological aspects of the plant husbandry of Tell Aswad. *Paléorient* 5: 161-169.

Van Zeist, W., and Bakker-Heeres, J.A.H. (1982). Archaeobotanical studies in the Levant. 1. Neolithic sites in the Damascus basin: Aswad, Ghoraifé, Ramad. *Palaeohistoria* 24: 165-256.

Van Zeist, W., and Bakker-Heeres, J.A.H. (1984). Archaeobotanical studies in the Levant. 3. Late-Paleolithic Mureybit. *Palaeohistoria* 26: 171-199.

Wetterstrom, W. (1993). Foraging and farming in Egypt: the transition from hunting and gathering to horticulture in the Nile Valley. In Shaw, T., Sinclair, P., Andah, B., and Okpoko, A. (eds.), *The Archaeology of Africa: Food, Metals and Towns*, Routledge, London, pp. 165-226.

Wetterstrom, W. (1996). L'apparition de l'agriculture en Égypte. *Archéo-Nil* 6: 51-75.

Willcox, G. (1992). Archaeobotanical significance of growing Near Eastern progenitors of domestic plants at Jalès, France. In Anderson, P. (ed.), *Préhistoire de l'Agriculture: Nouvelles Approches Expérimentales et Ethnographiques*, Monograph CRA 6, CNRS, Paris, pp. 159-178.

Willcox, G. (1996). Evidence for plant exploitation and vegetation history from three Early Neolithic pre-pottery sites on the Euphrates (Syria). *Vegetation History and Archaeobotany* 5: 143-152.

Wright, H.E. Jr, Kutzbach, J.E., Webb III, T., Ruddiman, W.F, Street-Perrott, F.A.S., and Bartlein, P.J. (eds.) (1993). *Global Climates Since the Last Glacial Maximum*, University of Minnesota Press, Minneapolis.

Yan, Z., and Petit-Maire, N. (1994). The last 140 ka in the Afro-Asian arid/semi-arid transitional zone. *Palaeogeography, Palaeoclimatology, Palaeoecology* 110: 217-233.

Zohary, D. (1969). The progenitors of wheat and barley in relation to domestication and agricultural dispersal in the Old World. In Ucko, P.J., and Dimbleby, G.W. (eds.), *The Domestication and Exploitation of Plants and Animals*, Gerald Duckworth, London, pp. 47-66.

Zohary, D. (1992). Domestication of the Neolithic Near Eastern crop assemblage. In Anderson, P.C. (ed.), *Préhistoire de l'Agriculture, Nouvelles Approches Expérimentales et Ethnographiques*, Monograph CRA 6, CNRS, Paris, pp. 81-86.

Zohary, D. (1996). The mode of domestication of the founder crops of Southwest Asian agriculture. In Harris, D.R. (ed.), *The Origin and Spread of Agriculture and Pastoralism in Eurasia*, University College London Press, London, pp. 142-158.

Zohary, D., and Hopf, M. (1988). *Domestication of Plants in the Old World: The Origin and Spread of Cultivated Plants in West Asia, Europe and the Nile Valley*, Clarendon Press, Oxford.

11
SUSTAINABLE AGRICULTURE IN A HARSH ENVIRONMENT: AN ETHIOPIAN PERSPECTIVE

A. Butler

INTRODUCTION

Ethiopia is situated between 4° and 15°N, with altitudes ranging from below sea level in the Danakil depression and deserts in the east, through the tropical regions (*qolla*) below altitudes of 1600 m above sea level (asl), to the highlands reaching over 4000 m asl (*dega*). The altitude governs different subsistence strategies which range from pastoralism and hoe-cultivation of ensete (*Ensete ventricosum* (Welw.) E.E. Cheesm.), tropical roots and tubers in the lowlands to the ox-plow cultivation of predominantly grain crops which is practiced in the highlands.

   This article focuses upon crop production and utilization in the mid-altitude highland regions between 1600 and 2400 m asl (*woina dega*), and employs data from the preliminary phases of an ethnoarchaeological project in Tigray. The evidence for the beginnings of agriculture in the Ethiopian Highlands is first examined, following which current systems of traditional agronomy are described and strategies for sustainability are highlighted.

CURRENT VIEWS ON THE BEGINNINGS OF AGRICULTURE IN THE ETHIOPIAN HIGHLANDS

The beginnings of grain cultivation in highland Ethiopia are conjectural. The assemblage of Southwest Asian crops, together with the ox-plow technology, was introduced in antiquity, but the date and the route are uncertain. One view holds that the introduction was associated with the arrival of Semitic-speaking peoples from the Arabian Peninsula in the first millennium BC (Doggett, 1965; Westphal, 1975, pp. 74-75). However it appears more plausible that the agricultural package was brought in earlier from Egypt (Clark, 1962, 1970; Doggett, 1991; Simoons, 1965, 1970). This is supported by the linguistic evidence, which shows that much agricultural terminology is Cushitic rather than Semitic (Ehret, 1980). The earliest evidence for agriculture in Egypt comes from Merimde Beni Salama in the Delta and the Fayum A sites from the seventh millennium bp (Hassan, 1988, 1998) and from the site at Hemamieh in Upper Egypt of the later seventh millennium bp

(Wetterstrom, 1993). On this basis it has been postulated that the Southwest Asian crop assemblage could have arrived in Ethiopia via the Nile around the fifth millennium bp. The cultivation of the indigenous species such as teff (*Eragrostis tef* (Zucc.) Trotter), finger millet (*Eleusine coracana* (L.) Gaertn.), noog (*Guizotia abyssinica* Cass.) and chat (*Catha edulis* Forsk) may have already been established by that time, perhaps as much as a millennium earlier, a view propounded by both Clark (1976, p. 78) and Harlan (1992; Harlan *et al.*, 1976), based on the fact that these crops, particularly the very small-grained cereal, teff, are unlikely to have been taken into cultivation if the other grain crops were already available.

In 1986, Harlan (1993) noted the 'dearth of archaeobotanical evidence' in Africa. We still have no archaeological data on early cultivation in Ethiopia. Four occupation areas have been examined in Agordat in Eritrea, which have pottery and a figurine of an ox, typologically similar to artifacts of the C Group of Kerma in Nubia. This dates from the Second Intermediate Period of Egypt, 1567 BC. (Arkell, 1954; Phillipson, 1993, p. 150). From the excavations of a rock shelter at Gobedra, west of Aksum, dating from the mid seventh millennium to the fifth millennium bp (Phillipson, 1977, 1993a, p. 150, 1993b) and cave sites dating from around 2500 bp at Lalibela and Natchabiet (Dombrowski, 1970), plant remains have been recovered, but all are now known to be intrusive. Thus none of these sites has yielded evidence of early agriculture. The best evidence comes from the analyses of animal bones at Lake Besaka from the early fourth millennium bp and from the later levels at Gobedra dating from the second millennium bp which are said to include domestic cattle (Phillipson, 1993a, p. 150, 1993b).

During the 1990s two major archaeological projects have been undertaken at Aksum in northern Tigray. The archaeobotanical remains, which have been recovered from Aksumite domestic areas about a kilometer north of the town, date from between the fourth and the seventh centuries AD and include most of the crops that are grown today: barley, wheat, possibly teff, oat, sorghum, pea or faba bean, lentil, linseed, cotton, noog, grape and gourd (Boardman, 1996, 1999; Phillipson, 1998, p. 59; Phillipson and Reynolds, 1996). In pre-Aksumite levels of the first millennium BC, the plant remains are mainly teff, barley and linseed, with a remarkable absence of sorghum and finger millet (Boardman, unpublished). A detailed archaeobotanical report, including radiocarbon dates for the plant material from some earlier levels, is awaited. An interim report from the second project at Beta Giyorgis near Aksum lists emmer and free-threshing wheat, barley, teff, flax, lentil and grape from the Ona Nagast area, and includes emmer from a Proto-Aksumite level with pottery dating from ca. 90-70 BC (Bard and Fattovich, 1997). A recent summary of the linguistic, historical and archaeological evidence concludes that domesticated cattle were introduced on to the Tigrean Plateau from the western lowlands between about 3500-1500 cal BC, and that wheat and barley probably accompanied them (Fattovich, 1997).

Recently surveys have been made of two rock shelters near Aksum, which contain lithic material similar to that of the fifth millennium bp levels at Gobedra. Trial trenches have yielded plant remains which include the chaff of wheat and barley. The results of AMS dating are awaited (Finneran, personal communication; Phillipson and Reynolds, 1996). Further similar sites in north Tigray and Eritrea could be promising foci for future research.

TRADITIONAL FARMING IN HIGHLAND ETHIOPIA

Agriculture and food production in Ethiopia have been well-documented during the past seventy years under changing political circumstances and in different highland regions (for example Alemayehu Konde, 1993; Bahru Zewde 1991; Dessalegn Rahmato, 1991a; Holt and Lawrence, 1993; McCann, 1995; Simoons, 1960, 1970; Vavilov, 1992; Westphal, 1974, 1975). The descriptions of the crops and the techniques of cultivation and crop processing in the highlands are broadly similar to those witnessed in Tigray during this current project, and demonstrate the essential conservatism of traditional farming.

FARMING AT ADI AINAWALID, MEKELLE, TIGRAY: OBSERVATIONS BETWEEN 1996 AND 1997

This preliminary phase of an ethnoarchaeological study of traditional agriculture was carried out over two years mainly at Adi Ainawalid, a small village settlement (*kushet*) of 180 households, some 15 km northwest of Mekelle in Southcentral Tigray. Supplementary visits were made to four *kushets* in the same village group (*tabia*) and also to three further *kushets* near Adi Gudem about 30 km south of Mekelle (D'Andrea *et al.*, 1997, 1999; Butler *et al.*, 1999). Altogether interviews were recorded in 94 households and additional information was obtained in the fields. The study area is set on the edge of the Mekelle plain on the Giba Plateau at an altitude of 2000 m asl. The climate is temperate, with temperatures averaging 22°C.

The houses are largely circular, made of stone with timber supports and thatched or earthen roofs. Land-holdings vary between one eighth and two ha, divided into one or more plots per household, usually one adjacent to the settlement and others at a greater distance.

Soil and Water

The soils of the Giba Plateau are thin and stony, mainly of metamorphosed limestone, which weathers to lithosols, cambisols and vertisols (Hunting Technical Services, 1973/74). At Adi Ainawalid the soils are predominantly thin, sandy cambisols, with some patches of deeper vertisols (Mitiku Haile, personal communication). The surrounding hill slopes are steep and virtually treeless. Erosion due to the action of both water and wind is a major problem.

In the highlands, the rainfall is largely governed by the position of the Intertropical Convergence Zone (ITCZ; Griffiths, 1972); over 90% of the rain normally falls in the monsoon season (*meher* or *keremt*), between July and September, with a lighter *belg* period between March and April and showers at other seasons. The summer rains are commonly heavy and damaging, often with hailstorms. As is common in Tigray (Tsegay Fithanest, 1996, p. 34), slopes have been terraced with stone quarried from local outcrops to reduce run-off damage (Figure 11.1). While cross-contour ditches (Abate Tedla and Mohamed Saleem, 1992) and contour plowing (Dessalegn Rahmato, 1991b, p. 83) are often seen in the highlands, the field systems at Adi Ainawalid are small and set on relatively shallow slopes, and drainage is confined to shallow channels dug along some field boundaries. A clay-lined depression constructed downslope from the hills at the edge of the village serves as a reservoir for watering livestock. This was virtually empty during the periods of field study, which were in dry years. Until recently, water for human use has had to be fetched from the river twice a day, which was up to an hour's walk away; now a well has been constructed near the settlement area.

Figure 11.1. Treeless landscape with stone terracing near Adi Ainawalid, Tigray.

The farming seasons in the highlands are determined by the bimodal rainfall, thus the main season follows the *meher* and a secondary minor growing season is supported by the *belg*. However in much of Tigray for some years there have been no spring rains. Thus farm production is now dependent on a single harvest following the summer rains, with catch-cropping at other seasons when sufficient rain occurs. Tillage of the field plots is minimal. The plots are plowed by teams of two oxen between February and July (Figure 11.2). Stones and tree-stumps are retained in the soil to reduce erosion and help retain moisture. To

Figure 11.2. Minimum tillage: 2-oxen scratch plow, leaving stones on the fields, Adi Ainawalid, Tigray.

restore soil fertility, soil-burning (*guie*) to release minerals, kill weeds and disinfect, common in some highland areas of Ethiopia (Abate Tedla and Mohamed Saleem, 1992), is not practiced, but dung is applied to the cereal plots. The pulse crops are not fertilized, but when the concentration of soil nutrients appears to be low, pulses are themselves planted to 'refresh the soil' (Tadesse Desta and Tekle Haimanot Waldo, personal communication; Tsegay Fithanegest, 1996, pp. 23-28). The size of the holdings means that fallowing is seldom possible and fertility is generally maintained by crop rotation, with ratios of one year of pulses to three or four of cereals.

Crop Selection and Sowing

The main cereal crops at Adi Ainawalid are sorghum (*Sorghum bicolor* (L.) Moench, *mashella*), teff (*Eragrostis tef* (Zucc.) Trotter, *taff*), finger millet (*Eleusine coracana* (L.) Gaertn., *dagusha*), bread wheat (*Triticum aestivum* subsp. *vulgare* (Vill.) MacKey, *sindai*) and barley (*Hordeum vulgare* L., *segem*). The cool season grain legumes (pulses) cultivated are lentil, (*Lens culinaris* subspecies *culinaris* Medikus, *bersheem*), chickpea (*Cicer arietinum* L., *shimbra*) and grass pea (*Lathyrus sativus* L., *gwayya* or *sebero*). Some fenugreek (*Trigonella foenum-graecum* L., *abaka*) is grown and is regarded as a pulse. Small amounts of linseed (*Linum usitatissimum* L., *indarta*) arc also cultivated.

Commonly, the most-promising crop varieties are identified and are monitored throughout the growing season; then the seed for the following year's cultivation is selected on the threshing floor following the harvest. However, one farmer categorically stated that all selection is made after harvest, when seed is sometimes exchanged by neighbors for better quality (Hailu Asbala, personal communication). At the start of the following sowing period, the crop species and the variety of each crop are carefully chosen from amongst these special reserves for the optimal performance under the environmental conditions that are anticipated for that particular year. Most importantly, the traits mainly sought are those associated with the greatest security of production rather than the heaviest yield, a general practice throughout the highlands (Melaku Worede and Hailu Mekbib, 1993). Although no quantifications were made during the study, cereals predominate as the major type of cultigen, amongst which teff is the most highly valued crop.

To spread risk and intensify production on the small plots, intercropping is frequently practiced. At Adi Ainawalid mixtures of wheat and barley (*hanfetse*) are sown, pairing particular varieties of each crop with synchronous rates of development. Near Adi Gudem the crop mixtures also include pea with faba bean (*ater-abie*) and other combinations such as chickpea with sorghum. When a single species is monocropped it includes several varieties (Figure 11.3). The ratio of the different taxa usually differs at harvest from that sown. All grain crops are broadcast; the cereals are sown mainly between February and July, lentil is sown May to September, and teff and the other pulses are more usually sown later, between July and September. This staggering of cropping cycles exploits the shorter growing seasons of the pulses, which helps to avoid competition for labor.

Cultivation and Crop Management

A secondary plowing (*gamsa*) and weeding are carried out during or following the rainy season from July to September. Input is directed mainly towards the cereals: thus weeding is usual for cereals, but rare for pulses. Palatable weeds are fed to the livestock. Crop rotation is used to break infestations of parasitic weeds, particularly broomrape (*Orobanche minor* Sm., *m'andat tali*), which is found on legume crops, and striga (*Striga asiatica* (L.) Kunze., *selemi*), which infests sorghum. Insect pests, such as stalk borer, army worm and locusts, as well as rodents and birds can be a serious threat to the grain yield. Insecticides are rarely used, but children and scarecrows are employed as bird scarers.

All the traditional grain crops are totally rainfall-dependent. Irrigation is confined to a small area adjacent to the river in which small plots can be rented for the production of cash crops such as tomato, potato and maize.

Figure 11.3. Sorghum crop with mixed varieties. Chickpea in foreground. Adi Bakel, Tigray.

Harvesting and Crop-Processing

The ripe cereals are cut or uprooted by sickle around November and December; the wheat/barley mixture *hanfetse* is harvested and treated as a single crop throughout all the processing stages. The pulses are uprooted by hand in the semi-green state a month or so earlier, except for grasspea and chickpea which are harvested in November/December and January respectively. Great care is taken to minimize premature seed loss from pod shatter. Weeds are avoided and left standing in the fields. After the crops have been piled to dry briefly in the field they are then taken to the threshing floors, and the fields are gleaned. Threshing teams of between two and eight or more oxen trample the crop to release the grains.

Crop-processing is carefully carried out and various stages may be repeated to ensure maximum separation of all fractions (D'Andrea *et al.*, 1997, 1999; Butler *et al.*, 1999). The winnowing implements include 2-4-tined forks, paddles, basket trays, and brushes used in succession. They are largely homemade from local plant materials. The separated grain and crop residues are very carefully collected so that finally no remnants can be seen on the threshing floor. All the straw and chaff residues are highly valued as livestock feed.

Rare Crops

Some crops which were common in the past have recently become rare. The rare cereals include two varieties of barley, emmer (*Triticum turgidum* L. subsp. *dicoccum* (Schrank) Thell., *ares*) and durum wheat (*T.turgidum* conv.*durum* (Desf.) MacKey, *kinkinai*). The production of emmer has decreased, but it is still used as an invalid and infant food and is cultivated in the *kushets* near Adi Gudem. Durum wheat still accounts for over 60% of wheats grown in Ethiopia as a whole and a range of varieties are adapted to very specific localities (Tesfaye Tesemma, 1991). Today, there seems to be little experience of durum cultivation at Adi Ainawalid, but its value is acknowledged: it makes a heavy bread used as a sustaining harvest food. On two recent occasions one farmer at the adjacent *kushet* of Adi Bakel had attempted to reintroduce it, but the crop was sterile (Hailu Asbala, personal communication). Faba bean (*Vicia faba* L., *abie*), the common field pea (*Pisum sativum* subspecies *arvense* L., *ater*), and the Ethiopian variety of the field pea (*Pisum sativum* L. var. *abyssinicum* A.Br., *dekoko*), which were once common, are now very infrequently grown near Adi Ainawalid. They are important festival foods and, despite their high price, they are bought from Mekelle market in small quantities, four or five times a year. All these crops were said to have been common in 'the time of Haile Selassie' (prior to 1973). The main cause for their decline was given as the lack of rainfall, coupled with the small land holdings. Implicated also is the political disruption of traditional farming during the 1970s and 1980s, which caused much loss of germplasm. At a national level, attempts are now being made to conserve and reintroduce the landraces, which may include varieties better adapted to the current climatic conditions (Melaku Worede and Hailu Mekbib, 1993).

CROP PRODUCTION 1996 AND 1997

Production is taxable and a sensitive topic, thus during this study the yield was not recorded in detail, but it was noted that it was much higher in cereal grains than pulses.

The success of the different crops in the years 1996 and 1997 was compared. In 1996, the farmers of Adi Ainawalid described production as very low, but all species were yielding a harvest. In November lentil had already been processed, chickpea was still green in the fields, and the sorghum, teff, finger millet, wheat, barley, *hanfetse* and grasspea, fenugreek and flax crops were being processed. In the following year, the main rains had ceased two months early, in July. At Adi Ainawalid many crops had totally failed. Stands of sorghum and finger millet were rare and teff was reduced. During that November to December harvest period there were heavy showers ('asmara rains') which further destroyed many of the remaining standing crops, a widespread occurence that year (USAID, 1998). In order to make up some of the shortfall and exploit the late autumn rains, some of the land was plowed in November for catch crops of chickpea and barley ('asmara cropping').

Importantly, the distribution of the rainfall is uneven and produces a mosaic of local effects. This was illustrated in 1997 when some *kushets* immediately adjacent to Adi Ainawalid and near Adi Gudem were successfully producing a wider range of crops. However, food production appears generally to have been low in the area for some years. A recent Food Aid survey in Tigray has estimated that each household averaging 4.2 members would require a minimum of 10 quintals (1000 kg) of cereal grain and pulses a year, which includes one quintal for seed. While the annual production between 1988 and 1992 averaged 9.2 quintals per family, this was unequally spread and in most cases was less than the recommended amount: of the harvests between 1988 and 1992, 27% produced sufficient or surplus and 73% were deficient, averaging only 4 quintals annually (Holt and Lawrence, 1993, pp. 26-31).

Other Sources of Grain

Social mechanisms have been developed to balance the food supply. The local effects of crop failure and short-term food-shortage can be mitigated by community exchange networks both within and between villages (Dessalegn Rahmato, 1991a). Cooperation within kinship groups and between neighbors (*mahber*) generally helps provide food on the larger scale required for the numerous religious festivals (Alemayehu Konde, 1993, p. 27). However, these measures could only be effective in the short term and when the needs were not widespread. In this study they were mentioned but the extent of their use was not investigated.

The local and the regional markets draw in crops from a wider area, and can supply both food grains to counteract local shortages and seed corn when village grain is of poor quality; inter-regional exchange can also be a source of particular landraces (Melaku Worede and Hailu Mekbib, 1993). Mekelle market supplies the seeds of the crop species grown at Adi Ainawalid, and also a wider range of food grains. Prices fluctuate throughout the year; as might be expected, they tend to be highest at the end of spring and lowest at harvest time when the farmers are selling their surplus grains (Alemayehu Konde, 1993, pp. 41-42; Holt and Lawrence, 1993, p. 60).

GRAIN STORAGE

The harvested grain is stored in the houses in tall clay or bamboo vessels sealed with dung. The latter is believed to have some insect repellent action and is the only preventative against insect damage, although in other areas fumigants are used to deter insects. When grains are suspected of being infested, they may be heated lightly to kill larvae. Cats are kept to catch rodents. The yields from small harvests are stored in sacks. Some unthreshed crops are also stacked in yards adjacent to the houses.

Emergency stores of grain are said to be found in most villages. Grain, particularly sorghum, apparently can survive in underground pits (*goudguard*) for up to five years without spoilage when conditions are dry. These storage pits are said to be most prevalent in the dry areas of Southeastern Tigray (Abate Tedla, personal communication; Dessalegn Rahmato, 1991b, p. 31; McCann, 1995, p. 68) but, because of the secrecy attached to them, their presence was not investigated in this study, and it is not known to what extent they play a part in the subsistence of Adi Ainawalid.

Livestock and Feed

The livestock are mainly cattle, sheep, goats, donkeys and chickens. Of 77 of the households questioned in 1996 and 1997, 30% did not own any cattle or donkeys and 58% had fewer than four. Occasionally, camels and mules may be kept. It was difficult to ascertain numbers of small ruminants. A major limiting factor to the amount of farm work undertaken is the availability of oxen. This survey found that about half the farmers do not own an ox, about a third of them have one ox and the rest own two or more oxen, results similar to those published for other highland regions (FAO, 1986). Thus most farmers are dependent on borrowed animals for both traction and threshing.

Feed crops are not cultivated. Livestock graze the stubble, field edges and patches of rough pasture. Hay is rarely made, but palatable weeds are uprooted from the fields for feed following the harvest. Crop residues are the prime feed; leaf forage is gathered in times of shortage. Rainfall affects the animals as much as the human population: feed and forage tend to be short in the spring prior to the rains before the plowing period; thus the oxen are frequently undernourished when about to undertake their heaviest work.

Summary of Cultivation Observations

From the observations above, three major points are noted. Firstly, a spatial effect: crop production is governed by a fine mosaic of different environmental conditions; thus yields can differ in closely neighboring areas. Secondly, a temporal effect: the same area can produce very variable yields, both in quality and quantity in successive years. Both these types of variation in yield are said to be more marked in Tigray than in other highland Ethiopian provinces (Holt and Lawrence, 1993, pp. 34-35). Thirdly, social equilibrating systems: local exchange networks, communal village storage systems and regional markets can balance food resources. When grain from these extraneous sources is acquired in small quantities for immediate use, it may not reach the domestic storage systems, and thus may be virtually undetectable by the outsider. The observed production of a specific food grain does not necessarily represent the real availability of that resource or indicate the true extent of its utilization. Further, important plant food resources, even when imported from very near their point of production, may leave little or no trace of their use.

Food and Food-Processing

The community is virtually vegetarian and the diet is based on grains. The pulses in the unripe green stages and all grains when lightly toasted (*kollo*) are eaten as snacks. The cereals are mainly ground and made into a variety of breads to which may be added flour of ground pulses, fenugreek or linseed. The staple pancake bread of fermented batter (*injera*) is made two or three times a week from teff when available, but also from combinations of other cereals and pulse flour. This is eaten three times a day, when supplies allow. Beer is brewed every week, primarily from sprouted finger millet or barley. Pulses, today mainly grasspea, are soaked, roasted, hulled, spiced and boiled to a coarse purée (*totoco*); they are also finely ground, spiced and boiled to make a vegetarian sauce (*shiro wot*) to eat with *injera*, ideally at every meal. The spices include fenugreek, black cumin, basil and chili (*berbere*) with onion and garlic.

Grasspea contains a toxin which causes the irreversible paralytic condition lathyrism. However, it is a particularly favored resource because, as well as being highly palatable and protein-rich, it is the most drought-resistant and environmentally-tolerant of the crops. Nationally, as well as locally, its cultivation is currently being expanded and measures are being sought to minimize the resulting outbreaks of lathyrism (Adugna Negere and Shelemew Mariam, 1994). At Adi Ainawalid and all the other *kushets* studied, the preparation of pulses described above is thought to reduce the concentration of toxin and thus remove the risks to health; however, it was noted that as the water for soaking and the fuel for roasting are scarce resources, the minimum was used and the detoxification procedure did not appear very thorough.

Recent studies in a grasspea-growing region of the central highland province of Gonder record diets similar to those found at Adi Ainawalid. The combination of cereals, pulses and spices provides the carbohydrate, protein, vitamins and trace elements for a well-balanced diet; but generally the total intake is too low. With an average monthly intake per person of grasspea of 5 kg (Elizabeth Wuhib *et al.*, 1994), and despite efforts at detoxification, outbreaks of lathyrism are common (Redda Tekle-Haimanot *et al.*, 1994). In comparison, one family of two adults and five children in Adi Ainawalid estimated their consumption of grasspea at only 7 kg a month. Although today lathyrism is said to be unknown at Adi Ainawalid, the efficacy of the preparative techniques is uncertain; it is possible that the general paucity of food supply may limit the extent of toxin consumption to amounts below the dangerous level (Butler *et al.*, 1999). The concentration of toxin is currently being analyzed in samples of grasspea collected at each stage of preparation.

Trees and Fuel

There is little native woodland near Mekelle. Small trees of mainly leguminous species are often planted in the house compounds for shade and for emergency fodder. Large trees are mainly limited to single protected specimens used as meeting places, and conserved zones within church compounds. As fuel-wood is in very short supply, its collection is now licensed. *Eucalyptus* species and other fast-growing exotic tree species have been introduced to supply architectural timber, the raw material for farm implements, shade and fodder, erosion control and fuel. Planting schemes using native trees have begun to be implemented (Mitiku Haile, personal communication). Fuel is mainly dung with lesser amounts of crop residues (particularly sorghum stems), leaves, weeds and shrubs.

SUMMARY

A number of farming strategies are employed by the farmers in the area of Adi Ainawalid to overcome environmental difficulties largely incurred by the pattern of rainfall. These are summarized in Table 11.1. They appear to be similar to those found in other systems of subsistence farming (for example Riebsame, 1989, p. 14). Perhaps the survival strategy most specific to Ethiopia can be found in the crops themselves: the Ethiopian grain crops of Southwest Asian origin are particularly noted for the rich diversity of landraces (Vavilov 1992, pp. 154-155, 313, 319, 323-324), selected for an ability to tolerate a wide range of variable environmental conditions.

The Effects of Drought

It has been shown that, during the last millennium, the frequency of years of low rainfall in the northern Ethiopian Highlands has been increasing (Machado *et al.*, 1998; Mohammed and Bonnefille, this volume). Recently Tigray has been the scene of some devastating famines, and their effects have stimulated new research into the causes (Kiros, 1993, p. 380; Mesfin Wolde Mariam, 1986).

Lack of rainfall is identified as the primary problem facing farmers in the highlands (Simon Adebo, 1993, pp. 74-75). Crop failure and food shortage can result from what appears to be a minor and very localized decrease in annual precipitation. Prior to the regional famine in Tigray of 1966 the reduction in mean rainfall was only 6%, while the preceding year's rainfall had been as high as 129% (Mesfin Wolde Mariam, 1986, p. 127). It has been estimated in the highlands that an annual rainfall of 400-500 mm is sufficient to support agriculture (Alemayehu Konde, 1993, p. 21), thus the yearly average in Tigray which varies between 450 and 980 mm (Berhanu Gebremedin and Mitiku Haile, 1997, p. 6), with 630 mm in Aksum (Machado *et al.*, 1998), could be considered to be sufficient to sustain food production. At Adi Ainawalid in 1997 there had been reduced *meher* rains, but the unseasonably heavy autumn rainfall will have brought the total rainfall that year nearer to the annual average. The distribution of rainfall throughout the year is the more important determinant of food production.

It has been found that famine is not usually directly associated with drought. While it appears that two or more consecutive years of crop failure tend to result in famine (Holt and Lawrence, 1992, p. 34), in most such disasters

Table 11.1. Strategies for sustainable subsistence at Adi Ainawalid, Tigray

Aim	Strategy
Erosion control	stone-terracing of slopes
	tree planting
	drainage ditches around fields
	minimum tillage, retaining stones and tree stumps
Water conservation	reservoirs
Crop selection	choice of cultigen for performance not yield
	choice of landrace for performance not yield
	exchange optimal landraces with neighbors
	increased cultivation of grasspea
	rare cultivation of certain crops, e.g., durum wheat
Agronomic system	monocropping with mixed varieties
	intercropping to spread risk
Priority of input	priority of labor towards prime cereal staples
Optimal schedule	cease *belg* farming when spring rains fail
	plant catch crops with *asmara* rains
Best recovery of yield	harvest of pulses prior to ripeness and pod shatter
	complete collection of all crop fractions
	collection of palatable weeds for livestock feed
Longer-term reserves	? community grain storage for up to 5 years
Other food sources	? kinship group sharing
	? exchange networks between neighbors
	regional markets
Dietary adaptation	increasing consumption of grasspea
Income for purchases	sale of local resources, especially livestock
	sale of labor

recorded over the past two centuries, food shortage has been combined with a second factor; which can include diseases of either humans, such as the Spanish flu epidemic in 1918, or of livestock, such as rinderpest in cattle in 1888, and locust plagues in 1811 (McCann, 1995, pp. 89-91). The political upheavals between the 1970s and 1990s certainly exacerbated the effect of the changes in the pattern of rainfall to produce the famines of the 1980s.

Communities supported by subsistence farming are known have a high susceptibility to famine (Mesfin Wolde Mariam, 1986, p. 169). The criteria which categorize this vulnerability are said to include fragmented land holdings in small plots, production geared towards the household rather than to markets, the lack of seasonal opportunities for alternative employment and little or no reserves of cash or grain (Mesfin Wolde Mariam, 1986, p. 23). Farming at Adi Ainawalid fulfills most of these criteria. However income is generated by a number of means which mainly exploit local materials: these include the sale of livestock and their produce (such as eggs, dairy produce and skins), locally-quarried stone, fuel-wood collection, basketry, cotton spinning and weaving and making the week-long journey with donkeys and camels to the Danakil to dig out salt for sale. Also neighbors' livestock are herded, land rented out, and labor sold for construction work and domestic service in Mekelle.

Contemporary Farming Pattern

As a consequence of the political disturbances, it is difficult to judge to what extent agriculture in Tigray today represents the traditional practices of the past. As far as has been possible to ascertain, Adi Ainawalid demonstrates comparatively little long-term change in farming techniques; following the civil disruption of the 1970s and 1980s and the subsequent resettlement program, most people now occupy their original homes, which in a number of cases have been in the same family for at least three generations, and farming appears to have been restored to the traditional pattern. However, as we have seen, several traditional crops are disappearing and the conservation of the Ethiopian germplasm is now a major issue (Engels et al., 1991; Melaku Worede, 1996). Similarly in some areas there has been an erosion of rural people's knowledge. There is an equal concern for its retrieval and dissemination to speed and stabilize the agricultural recovery, and to which it is intended that this project will contribute.

CONCLUSION

Recent investigations indicate that periods of drought in Ethiopia may be associated with the El Niño Southerly Oscillation (ENSO), a view endorsed by local agronomists in Tigray (Amare Belay, personal communication). This is as yet unproven (Blench and Marriage, 1998), and it appears that additional climatic factors complicate the relationship (Fekadu Bekele, 1997). However, the rainfall data suggest that ENSO stimulates a drought frequency of every three to ten years, with regional variations, amongst which northern Ethiopia has the highest occurrence (Tsegay Wolde-Georgis, 1997). This contribution revealed that the farmers have devised means to withstand short-term cyclical droughts; but should climatic shifts endure for longer periods, further measures would be required. Accordingly, a greater understanding of longer-term climatic changes could facilitate the creation of an early warning system with the development and employment of further strategies for sustainable food production.

ACKNOWLEDGMENTS

The fieldwork was part of a project led by Dr A.C. D'Andrea and funded by a research grant from the Social Science and Humanities Research Council of Canada. It was made possible by the help and advice of Dr. Mitiku Haile, Dean, and field assistants Dereje Asefa, Alemtsehay Tsegay and Zelealem Tesfay, of Mekelle University College. I thank Sheila Boardman and Niall Finneran for their kind permission to use their unpublished excavation data, and Gordon Hillman for his valuable comments on reading this text. My deepest gratitude goes to the farmers of Adi Ainawalid and neighboring villages in Tigray for their patience, kindness, generosity and help.

REFERENCES

Abate Tedla, and Mohammed Saleem, M.A. (1992). Cropping systems for vertisols of the Ethiopian highlands. In *Reports and Papers on the Management of Vertisols (IBSRAM/AFRICALAND) Network Document No.1*, ILCA, Addis Ababa, pp. 55-66.

Adugna Negere, and Shelemew Mariam (1994). An overview of grass pea (*Lathyrus sativus*) production in Ethiopia. In Berhanu Abegaz, Redda Tekle-Haimanot, Palmer, V. S., and Spencer, P. S. (eds.), *Nutrition, Neurotoxins and Lathyrism: the ODAP Challenge*, Third World Medical Research Foundation, New York, pp. 67-72.

Alemayehu Konde (1993). *Report of Diagnostic Survey of Debre Medhanit Tabia in Dedebana Derga-Agen Wereda*, FARMAfrica, Addis Ababa.

Arkell, A .J. (1954). Four occupation sites at Agordat. *Kush* 2: 33-62.

Bahru Zewde (1991). *A History of Modern Ethiopia 1855-1974*, James Currey, London.

Bard, K. A., and Fattovich, R. (1997). The IUO/BU excavation at Beta Giyorgis (Aksum): an interim report. *Nyame Akuma* 48: 22-28.

Berhanu Gebremedin, and Mitiku Haile (1997). *Food Security and Dryland Agriculture: the Case of Tigray*, Utvikingsfundet (the Development Fund), http://www.u-fondet.no/engelsk/tema/konf/ 1-3.html.

Blench, R., and Marriage, Z. (1998). *The Social and Technical Construction of Weather: el Niño and Other Climatic Events in Sub-Saharan Africa*, Unpublished paper read at Southern Africa Regional Climate Outlook Forum, Pilanesberg, South Africa, May 12-15, 1998.

Boardman, S. (1996). Palaeoethnobotany. In Phillipson, D. W., BIEA Excavations at Aksum, Northern Ethiopia, 1995, pp. 126-129. *Azania* 31: 99-147.

Boardman, S. (1999). The agricultural foundation of the Aksumite empire, Ethiopia. In van der Veen, M. (ed.), *The Exploitation of Plant Resources in Ancient Africa*, Kluwer Academic/Plenum Publishers, New York, pp. 137-147.

Butler, E. A., Zilealem Tesfay, D'Andrea, A. C., and Lyons, D. E. (1999). The ethnobotany of *Lathyrus sativus* L. in Highland Ethiopia. In van der Veen, M. (ed.), *The Exploitation of Plant Resources in Ancient Africa*, Kluwer Academic/Plenum Publishers, New York, pp. 123-136.

Clark, J. D. (1962). Spread of food production in sub-Saharan Africa. *Journal of African History* 3: 211-228.

Clark, J. D. (1970). The spread of food production in sub-Saharan Africa. In Fage, J. D., and Oliver, R.A. (eds.), *Papers in African Prehistory*, Cambridge University Press, Cambridge, pp. 25-42.

Clark, J. D. (1976). The domestication process in sub-Saharan Africa with special reference to Ethiopia. In Harlan, J. R., de Wet, J. M. J., and Stemler, A. B. L. (eds.), *Origins of African Plant Domestication*, Mouton, The Hague, pp. 67-105.

D'Andrea, A. C., Mitiku Haile, Butler, E. A., and Lyons, D. E. (1997). Ethnoarchaeological research in the Ethiopian highlands. *Nyame Akuma* 47: 19-26.

D'Andrea, A. C., Lyons, D., Mitiku Haile, and Butler, A. (1999). Ethnoarchaeological approaches to the study of prehistoric agriculture in the Ethiopian highlands. In van der Veen, M. (ed.), *The Exploitation of Plant Resources in Ancient Africa*, Kluwer Academic/Plenum Publishers, New York, pp. 101-122.

Dessalegn Rahmato (1991a). Rural women in Ethiopia; problems and prospects. In Tsehai Berhane-Selassis (ed.), *Gender Issues in Ethiopia*, Institute of Ethiopian Studies, Addis Ababa, pp. 31-45.

Dessalegn Rahmato (1991b). *Famine and Social Strategies; a Case Study from Northeast Ethiopia*, Scandinavian Institute of African Studies, Uppsala.

Doggett, H. (1965). The development of the cultivated sorghums. In Hutchinson, J. (ed.), *Essays on Crop Plant Evolution*, Cambridge University Press, Cambridge, pp. 50-69.

Doggett, H. (1991). Sorghum history in relation to Ethiopia. In Engels, J. M. M., Hawkes, J. G., and Melaku Worede (eds.), *Plant Genetic Resources of Ethiopia*, Cambridge University Press, Cambridge, pp.140-159.

Dombrowski, J. (1970). Preliminary report on excavations in Lalibela and Natchabiet Caves, Begemder. *Annales d'Ethiopie* **8**: 21-29.

Ehret, C. (1980). On the antiquity of agriculture in Ethiopia. *Journal of African History* **20**: 161-177.

Elizabeth Wuhib, Redda Tekle-Haimanot, Angelina Kassina, Yemane Kidane, and Tadesse Alemu (1994). Survey in grasspea preparation and general dietary intake in rural communities. In Berhanu Mabegaz, Redda Tekle-Haimanot, Palmer, V., and Spencer, P. S. (eds.), *Nutrition, Neurotoxins and Lathyrism*, Third World Medical Foundation, New York, pp. 119-126.

Engels, J. M. M., Hawkes, J. G., and Melaku Worede (1991). *Plant Genetic Resources of Ethiopia*, Cambridge University Press, Cambridge.

FAO (1986). *Ethiopia: Economic Analysis of Land Use. Technical Report 8*, FAO, Rome.

Fattovich, R. (1997). The peopling of the Tigrean plateau in ancient and medieval times (ca. 4000 B.C. - A.D. 1500): evidence and synthesis. In Bard, K. A. (ed.), *The Environmental History and Human Ecology of Northern Ethiopia in the Late Holocene*, Istituto Universitario Orientale, Napoli, pp. 81-105.

Fekadu Bekele (1997). Ethiopian use of ENSO information in its seasonal forecasts. *The Internet Journal for African Studies* **2**: March, 1997.

Griffiths, J. F. (1972). Ethiopian Highlands. In Griffiths, J. F. (ed.), *World Survey of Climatology: Climates of Africa*, Vol. 10, Elsevier, Amsterdam, pp. 369-388.

Harlan, J.R. (1992). Indigenous African agriculture. In Cowan, C. S., and Watson, P. J. (eds.), *The Origins of Agriculture*, The Smithsonian Institution Press, Washington, pp. 59-70.

Harlan, J. R. (1993). The tropical African cereals. In Shaw, T., Sinclair, P., Bassey Andah, and Okopo, A. (eds.), *The Archaeology of Africa: Food, Metals and Towns*, Routledge, London, pp. 53-60.

Harlan, J. R., de Wet, J. M. J., and Stemler, A. (1976). Plant domestication and indigenous African agriculture. In Harlan, J. R., de Wet, J. M. J., and Stemler, A. B. L. (eds.), *Origins of African Plant Domestication*, Mouton, The Hague, pp. 3-19.

Hassan, F. (1988). The Predynastic of Egypt. *Journal of World Prehistory* **2**:135-185.

Hassan, F. (1998). Holocene climatic change and riverine dynamics in the Nile Valley. In di Lernia, S., and Manzi, G. (eds.), *Before Food Production in North Africa. Proceedings of the Homonymous Workshop held in Forli, September 1996, within the XIII World Congress of the International Union of the Prehistoric and Protohistoric Sciences*, ABACO Edizioni, Rome, pp. 43-51.

Holt, J., and Lawrence, M. (1993). *Making Ends Meet: a Survey of the Food Economy of the Ethiopian North-East Highlands*, Save the Children, UK, London.

Hunting Technical Services Ltd. (1973/4). *Tigray Rural Development Studies: Map of Landforms in Mekelle District: Gradients, Soil Depth and Soil Types (1-6)*, Ministry of Overseas Development, London.

Kiros, F. G. (1993). *The Subsistence Crisis in Africa: the Case of Ethiopia: Root Causes and Challenges of the New Century*, ICIPE Science Press, Nairobi.

Machado, M. J., Pérez-González, A., and Benito, G. (1998). Paleoenvironmental changes during the last 4000 yr in the Tigray, Northern Ethiopia. *Quaternary Research* **49**: 312-321.

McCann, J. C. (1995). *People of the Plow*, University of Wisconsin Press, Madison.

Melaku Worede (1996). Diversity News. *African Adversity: Seeds of Survival* **13**: 1-4.

Melaku Worede and Hailu Mekbib (1993). Linking genetic resource conservation to farmers in Ethiopia. In de Boef, W., Kojo Amanor, and Wellard, K. (eds.), *Cultivating Knowledge*, Intermediate Technology Publications, London, pp. 78-84.

Mesfin Wolde Mariam (1986). *Rural Vulnerability to Famine in Ethiopia, 1958-1977*, Intermediate Technology Publications, London.

Phillipson, D. W. (1977). The excavation of Gobedra rock-shelter, Axum. *Azania* **12**: 53-82.

Phillipson, D. W. (1993a). *African Archaeology*, 2nd edition, Cambridge University Press, Cambridge.

Phillipson, D. W. (1993b). The antiquity of cultivation and herding in Ethiopia. In Shaw, T., Sinclair, P., Bassey Andah, and Okopo, A. (eds) *The Archaeology of Africa: Food, Metals and Towns*, Routledge, London, pp. 344-357.

Phillipson, D. W., (1996). The BIEA Aksum excavation, 1995. *Nyame Akuma* **46**: 24-33.

Phillipson, D. W. (1998). *Ancient Ethiopia: Aksum: its Antecedents and Successors*, British Museum Publications, London.

Phillipson, D. W., and Reynolds, A. (1996). BIEA Excavations at Aksum, Northern Ethiopia, 1995. *Azania* **31**: 99-147.

Redda Tekle-Haimanot, Yemane Kidane, Elizabeth Wuhib, Angelina Kassina, Yohannes Endeshaw, Taddesse Alemu, and Spencer, P.S. (1994). The epidemiology of lathyrism in Ethiopia. In Berhanu Abegaz, Redda Tekle-Haimanot, Palmer, V.S., and Spencer, P.S. (eds.), *Nutrition, Neurotoxins and Lathyrism: the ODAP Challenge*, Third World Medical Research Foundation, New York, pp.1-9.

Riebsame,W.E. (1989). *Assessing the Social Implications of Climatic Fluctuations*, United Nations Environment Programme, Nairobi.

Simon Adebo (1993). *Report of Diagnostic Survey of Debri Tabia in Enderta Wereda*, FARMAfrica, Addis Ababa.

Simoons, F. (1960). *Northwest Ethiopia: Peoples and Economy*, University of Wisconsin Press, Madison.

Simoons, F. (1965). Some questions on the economic prehistory of Ethiopia. *Journal of African History* **6**:1-13.

Simoons, F. (1970). Some questions on the economic prehistory of Ethiopia. In Fage, J. D., and Oliver, R.A. (eds.), *Papers in African Prehistory*, Cambridge University Press, Cambridge, pp. 117-129.

Tesfaye Tesemma (1991). Improvement of indigenous durum wheat landraces in Ethiopia. In Engels, J. M. M., Hawkes, J. G., and Melaku Worede (eds.), *Plant Genetic Resources of Ethiopia*, Cambridge University Press, Cambridge, pp. 288-295.

Tsegay Fithanegast (1996). *Soil Classification and Management by Farmers in Central Tigray*, FARMAfrica, Addis Ababa.

Tsegay Wolde-Georgis. (1997). El Niño and early drought warning in Ethiopia. *The Internet Journal for African Studies* **2**: March 1997.

USAID (1998). *Famine early warning system, September 1997-1998*, http://gaia.info.usaid.gov/ fews/imagery/sat_ea.html.

Vavilov, N. I. (1992). *Origin and Geography of Cultivated Plants*, compilation of papers from 1926 onwards, translated by Löve, D., Cambridge University Press, Cambridge.

Westphal, E. (1974). *Pulses in Ethiopia, their Taxonomy and Agricultural Significance*, Centre for Agricultural Publishing and Documentation (PUDOC), Wageningen.

Westphal, E. (1975). *Agricultural Systems in Ethiopia*, Centre for Agricultural Publishing and Documentation (PUDOC), Wageningen.

Wetterstrom, W. (1993). Foraging and farming in Egypt: the transition from hunting and gathering to horticulture in the Nile Valley. In Shaw, T., Sinclair, P., Bassey Andah, and Okopo, A. (eds.), *The Archaeology of Africa: Food, Metals and Towns*, Routledge, London, pp.165-226.

SECTION III

PASTORALISM

What, however, is already certain is that the herdsmen occupied the central Sahara at a time of considerable humidity, since in the frescoes are to be seen also elephants, rhinoceros, hippopotamus and giraffe. All such animals, as well, of course, as the oxen, needed plenty of water to drink and plenty of grass to eat. We made also another discovery of importance. In a small shelter some two thousand feet above sea level, in the Aouanrhet massif (the highest in all the Tassili) we found on a rock-wall the picture of three reed canoes that seemed to be circling around three hippopotamuses.

What happened to these herdsmen and their cattle? Did they disappear without leaving a trace? It would surely be most improbable that such large numbers of people faded out completely. Rather we may think that, goaded on by increasing drought, they sought out new pastures in the Sudan steppes ['Sudan' refers here, as it used to, to Sahel countries of West Africa, *editor's note*] and to the south of the Sahara.

Henri Lhote, *The Search for the Tassili Frescoes: The Rock Paintings of the Sahara*, Dutton, New York, 1959.

The subject of animal domestication is introduced in this section by **Gautier** (Chapter 12), who defines animal domestication as a rapid microevolutionary process essentially due to considerable changes in selection to which the animals are exposed in the human econiche; as a result they acquire domestic traits. **Gautier** also re-examines the evidence for an early independent center of cattle domestication and pastoralism in the southern Western Desert of Egypt. He assesses the evidence in the light of the earliest data for the introduction of sheep and goats from Southwest Asia, probably accompanied by the dog, and most likely the pig. He remarks that the limitations of the data pose several questions, particularly concerning the nature of the assumed early domestication experiments in the Western Desert, the separate or conjunctive introduction of the mentioned domesticates in Africa and the possible routes of their introduction.

The cultural events predating the introduction of sheep/goats into Northeast Africa are examined in this volume by **Barich** (Chapter 13), who notes that the earliest moist phase of the Holocene (10-7.8 kyr bp) witnessed an extreme stylistic differentiation in the stone industry related to a broad spectrum of resource exploitation (fishing associated with wild plant and animal exploitation). She also remarks that the main event linked with climatic change that proved to be fundamental for subsequent cultural developments in Africa was the domestication of cattle dating back to the tenth millennium bp in the Eastern Sahara. The intensification of cattle keeping and pastoralism occurred in conjunction with the mid-Holocene arid phase (see also **Hassan**, Chapter 2).

In the Egyptian Western Desert (Farafra Oasis, Dakhla Oasis, Nabta Playa) the climatic oscillations following the major early Holocene warm episodes witnessed the beginning of sedentary settlements in favorable habitats in conjunction with an intensification in the use of wild strands of millet and sorghum together with herding sheep and goats ca. 7-6.9 kyr bp. The first bifacial implements appear at that time. Their prototypes have been recognized in the Western Desert Oases. Intensification of pastoral movements and more contact between the Sahara and the Nile Valley occurred during the late Holocene.

Di Lernia (Chapter 14) reports in this section the results of a recent systematic survey and archaeological excavations which shed new light on the climatic changes during the late Pleistocene and Holocene in the Western Fezzan (Libya, Central Sahara). **Di Lernia** recognizes that the 7.2-6.9 cal kyr BP event (see **Cremaschi** , Chapter 5) had a pronounced effect on human movements and interregional contacts. However, the main crisis leading to the onset of present full desert conditions happened around 5000-4800 bp (5.7 cal kyr BP), when lakes and swamps turned into sebkhas, shelter roofs often collapsed and many shelters were emptied by wind erosion. At that time, the advent of severe arid conditions had serious effects on mobility and organization of pastoral activity. Following this event, the Acacus mountain range became depopulated, and human groups either became concentrated in the proximity of localized oases or became engaged in high mobility, long-range pastoral economy.

In a comprehensive review, backed with a critical examination of radiocarbon age determinations, of the early spread of domestic cattle and ovicaprines during the late Holocene in western and central Africa, **Van Neer** (Chapter 15) places his emphasis on recent archaeozoological data from the savanna and forest areas of the region. He also discusses the role of hunting and fishing in subsistence. After examining the evidence he concludes that domestic stock keeping was vital to food provisioning but that hunting and fishing always remained important, and that exclusive reliance on livestock is a rather recent phenomenon. The combined use of different strategies for the acquisition of animal protein persisted for several millennia, and can be seen as an adaptive response to periods of environmental stress. The drawbacks in trying to quantify the relative importance of stock keeping, hunting, and fishing are briefly reviewed. **Van Neer** also demonstrates that proper sampling of faunal remains is a vital prerequisite to any meaningful analysis of data.

Responses to climatic fluctuations to gain more food security included, in addition to indigenous innovations in the range of food resources exploited or produced, significant technological innovations for food procurement and processing, such as the use of harpoons and hooks, the production of pottery vessels, the manufacture of grinding stones, sickle blades, adzes, gouges, construction of storage facilities, and the digging of deep wells. Moreover, certain social and ideological transformations provided a means of increasing the sustainability of human life in an unpredictable environment. These developments included means to maintain social networks among different communities and enhance social cohesion within families engaging in similar or complementary activities within a given territory, as well as the protection and defense of territorial ranges. Visible archaeological traces of such social and ideological modes range from monumental funerary architecture to personal adornments.

In the Lower Nile Valley, the introduction of farming ca. 5000 BC led, by 3200 BC, to the emergence of a unified nation-state following a series of political transformations including the emergence of provincial chiefdoms and petty states. The Egyptian state was sustained by an ideology that supported provincial nobles and a class of bureaucrats (scribes) and priests, with the manifest display of power in monumental temples, palaces, and tombs.

In the final chapter of this section, **Hendrickx** (Chapter 16) reveals that the curious association between cattle and ideology in Ancient Egypt provided a basis for the legitimization of kingship during late Predynastic and early Dynastic times, when the first steps toward a nation-state were forged. Egypt, an agrarian state, was thus indebted to its pastoral legacy for the ideology of kingship. Once in place, kingship provided a new element in the dynamic interaction between people and their changing climate.

12
THE EVIDENCE FOR THE EARLIEST LIVESTOCK IN NORTH AFRICA: OR ADVENTURES WITH LARGE BOVIDS, OVICAPRIDS, DOGS AND PIGS

A. Gautier

INTRODUCTION

One of the aims of this volume is to confront the supposedly earliest domestic animals in North Africa with evidence provided by other disciplines. Confrontations often benefit from explicitly formulated definitions and positions, and this introduction therefore focuses on the concept of animal domestication and related general topics, which I have already discussed repeatedly (Gautier, 1990, 1992, 1993, 1998).

Animal domestication involves the removal of animals of particular species—not all animal species can be domesticated—from their community in nature to reproduce under human control and for some benefit of the interfering humans. As a result, the animals acquire domestic traits, i.e. traits not or rarely found in natural populations. Domestication thus has a cultural dimension and a biological one, but the biological dimension has in the past been neglected and is still sometimes neglected today, especially by English speaking scholars. In fact, domestication is a microevolutionary process, a combination of genetic changes within biological species, or so-called intraspecific genetic changes, leading towards the origination of new species. The domestication process is caused mainly, if not completely, by changes in selection, or the differential survival and reproductive success of particular animals, with respect to selection in nature, and one can oppose artificial selection in a broad sense, i.e. selection as occurring in the human econiche, to selection in the wild or natural selection. Implicit in the foregoing circumscription of animal domestication is the fact that all domestic animals have a monophyletic origin. They all have only one wild ancestor, but in the domestication process domesticates may acquire markedly different traits, a phenomenon well illustrated by the extreme diversity of dogs.

The available evidence suggests that domestic traits appear after a limited number of generations or at most a few centuries. The distinction between a domestic animal *in statu nascendi* without domestic traits and a 'real' domestic animal, i.e. a domesticated animal with domestic traits is, in my opinion, not applicable in prehistory, because the chronologies are not refined enough. The mentioned neglect of the biological dimension of domestication processes and the

195

idea that, simply put, domestic traits would materialize but very slowly, led researchers to postulate a whole series of forgotten and episodic domesticates: animals of which the domestication was discontinued or which were domesticated, forgotten and then domesticated again; a good example of the latter is the Palaeolithic horse (Gautier, 1996). For all of these putative domesticates evidence is lacking, or at least extremely equivocal. Unfortunately, several of these ghost domesticates still haunt the literature and the minds of some archaeologists. An African example is the Barbary sheep, *Ammotragus lervia*, which also causes other problems discussed later.

The fact that domestication is a microevolutionary process confronts us with the question of whether domestic animals have evolved enough to merit Latin names of their own. Have they become separate biological species in the accepted sense of potentially interbreeding populations, separated from their wild ancestors by reproductive barriers? Unfortunately only incomplete reproductive barriers impede gene flow between wild animals and their domestic relatives and no consensus has yet been reached as to the taxonomic status of domestic animals and the Latin labels to be applied to them. As a result various labeling systems are in use. The nomenclatorial tangle is further complicated by the fact that in the past separate Latin labels were put on domesticates or remains of domesticates, essentially on purely typological grounds without consideration of genetic relations etc. Some workers still use such labels and add thus to the confusion of the non-initiated. The present author favors the Latin labeling system used by various European workers, stressing the conspecificity of domesticates and their wild relatives. Thus, all domestic cattle descend from the aurochs, *Bos primigenius*, and receives therefore the technical label *Bos primigenius* forma taurus and not *Bos taurus*, *Bos indicus* (zebu cattle), *Bos brachyceros* (shorthorn cattle) or some other name listed in Table 12.1. This same table lists also some Latin names of other domesticates dealt with in this paper and the nomenclature accepted by the author. If one does not want to stress the conspecificity of domestic animals with their ancestors, one might go back to the oldest original separate labels; i.e., *Bos taurus*, *Ovis aries*, *Capra hircus* etc., if necessary adding a qualification, for example *Bos taurus* (zebu type). Archaeologists confronted with the nomenclatorial tangle, briefly and incompletely referred to in the foregoing, should seek expert advice. To end this introduction, attention can be drawn to such terms as 'semi-domesticate' and others, which in the author's view should be avoided. They are vague and better replaced by more precise descriptions of the human control (*sensu* Hecker, 1982) one is trying to grasp.

IDENTIFYING DOMESTICATES

Various criteria are used to identify bone remains of early domesticates. In the domestication process animals, especially larger ones, become smaller and most researchers consider size decline a reliable signal of domestication. Morphological changes are another intrinsic diagnostic criterium, but these cannot be used often, because in incipient domestication only certain body parts are affected (e.g., skull, horncores and dentition) and archaeozoological remains

Table 12.1. Some Latin labels applied to animal domesticates

Older names[a]	Names adopted in this paper
Cattle	
Bos taurus	
Bos indicus (zebu)	
Bos brachyceros (short horn)	*Bos primigenius* f. taurus
Bos longifrons (idem)	
Bos africanus (Ancient Egyptian)	
Sheep	
Ovis aries	
Ovis longifers (*palaeoaegypticus*)	*Ovis ammon* f. aries
(long-legged, ancient Egypt)	
Goat	
Capra hircus	
Capra reversa (African dwarf goat)	*Capra aegagrus* f. hircus
Hircus reversus (idem)	
Dog	
Canis familiaris	
Canis (*familiaris*) *palustris*	*Canis lupus* f. familiaris
(Swiss lake dwellings)	
Pig	
Sus domesticus	*Sus scrofa* f. domestica

[a] for other older labels see, for example, Epstein (1971). The first names of each series
in this column can be used if one does not accept the nomenclature proposed in the
second column.

are generally very fragmented. Extrinsic criteria concern the former distribution of the ancestors of our domesticates and their ecology. The presence at a site of a species of which domesticated forms are known, and which does not occur as a wild animal in the region of the site, suggests introduction by people, most likely as a domesticate. The palaeoecological criterium is more refined: if the site lies in a region in which the ancestors of a species which has been domesticated could not live in the wild, introduction under human control is again a plausible explanation. Also used are age pyramids, or rather, marked changes in the age pyramids of possible domesticates killed by people. These changes may indicate that people had easier access to the animals involved because they controlled them and made particular culling choices. The abrupt appearance of a species known as a domesticate in a faunal sequence may also provide an indication that people had acquired access to that species, most likely as a domesticate. Preferably, various avenues using the archaeozoological evidence from several sites should be applied in identifying domestication processes or the arrival of domestic animals. In principle, archaeological evidence, in the form of artifacts, structures or iconography, can corroborate the archaeozoological evidence. However, if reliable, such evidence generally dates from periods when domesticates are already well established. I am personally loath to participate in the discussion on Saharan rock art featuring early domesticates or putative domestication scenes (see Lutz and Lutz, 1997), or on the so-called trapping or tethering stones in prehistoric North Africa (see Pachur, 1991; Lutz and Lutz, 1993), because of the problems concerning their interpretation and chronology.

EARLY DOMESTIC CATTLE IN NORTH AFRICA

The possible presence of early domestic cattle and ovicaprids in the archaeological sequence of (el) Nabta (Figure 12.1) was first announced in 1976 (Wendorf *et al.*, 1976). A detailed report came several years later (Gautier, 1980), followed by the analysis of remains collected from the Bir Kiseiba sequence, which also includes putative domestic cattle (Gautier, 1984). The early hypothetical presence of livestock, especially cattle, in the southern Western Desert and the origin of this livestock caused much debate (Wendorf *et al.*, 1987; Wendorf and Schild, 1994). Meanwhile, excavations between 1990 and 1998, mainly at Nabta, have produced more faunal remains; this material is still under study. It would however seem that the published arguments in favor of the proposed identifications are still acceptable.

The large bovid remains can in principle derive from wild cattle or aurochs (*Bos primigenius*), domestic cattle (*B. primigenius* f. taurus), the extant African buffalo (*Syncerus caffer*) or a fossil relative of this buffalo, the so-called African giant buffalo, until recently labeled *Pelorovis antiquus*. In my view and that of collaborators, this buffalo should not be put in a separate genus, as it is a close relative of the extant African buffalo, *Syncerus caffer*, to be called *Syncerus antiquus*, or even *S. caffer antiquus*, as an extinct large form of the extant buffalo (Gautier and Muzzolini, 1991; Peters *et al.*, 1994; but see Klein, 1994). Peters (1986b, 1988) studied the osteological differences between *Syncerus* and *Bos*. As to the small aurochs, which would have dwelt in North Africa under the name *Bos ibericus*, it has been put to rest as a composite of *Bos* material of various origins in the Maghreb (Gautier, 1988). One can most likely assign North African finds recorded in the literature as *Bos ibericus* to domestic cattle. Recently Wyrwoll (1997a,b) has tried to reintroduce the water buffalo (*Bubalis arnee*) in the Holocene wild fauna of North Africa. Parietal art of the Fezzan would depict this larger bovid, while the water buffaloes from the marshes around Djebel Ichkeul in Tunisia would show traits suggesting that they belong to a subspecies of wild water buffalo, endemic in North Africa. I have already expressed my distrust vis-à-vis rock art evidence. As to the Ichkeul buffaloes, they are most likely descendants of domestic water buffaloes introduced in the Maghreb during the Phoenician or Roman period; in the feral state they re-acquired 'wild' traits, as is known from other domestic animals escaping from the human econiche (Gautier, 1990; Gautier and Muzzolini, 1991).

Few specimens from Nabta or Bir Kiseiba can be tested for osteological characters distinguishing African buffaloes and *Bos* because the material is very poor. Nevertheless, a few small remains from the early Neolithic can be assigned to *Bos*; the same applies for the younger sequence. The osteometric data for the earlier sequence suggest animals within the lower size range of wild cattle as

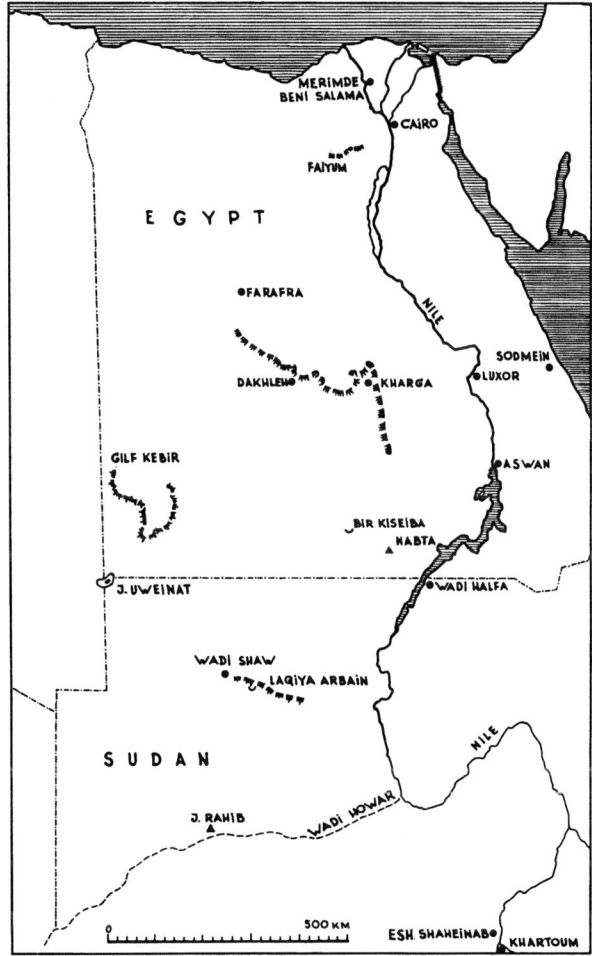

Figure 12.1. Location map of sites in Egypt and Sudan referred to in the text.

known from the Nile Valley. The osteometric data on African giant buffalo (Peters *et al.*, 1994; Peters, personal communication) and the finds I have seen (Gautier and Muzzolini, 1991, Appendix) confirm its vernacular name: this bovid is generally more robust and larger than wild cattle. The extrinsic arguments are palaeobiogeographical and palaeoecological in nature. The Holocene archaeo-zoological record of North Africa proves clearly that aurochs was well represented along the Mediterranean coast and along the Nile Valley as far as Aswan, but absent for the Holocene Sahara. We should however not be too categorical about this absence. For example, rock art from the Fezzan in Southeastern Libya suggests the presence there, sometime during the Holocene, of aurochs (Lutz and Lutz, 1995). Ecological corridors may have permitted the aurochs to colonize privileged regions in the Sahara, but I do not think such a corridor linked Nabta or Bir Kiseiba with the Nile Valley. Also, the faunal spectra in both regions are dominated by dorcas gazelle (*Gazella dorcas*) and hare (*Lepus capensis*) and no bovids intermediate in size between gazelle and large bovid have been collected. The low carrying capacity of both regions allowed but a poor and non-diversified mammalian fauna adapted to dry conditions. Moreover, along the Nile Valley and elsewhere, aurochs occurs always or very often together with hartebeest, *Alcelaphus buselaphus*; these herbivores would form a kind of a synecological couple. No trace of hartebeest was found in any of the Nabta and Bir Kiseiba sites. On the basis of the foregoing limited intrinsic and circumstantial evidence, I think the proposed identification of the earlier large bovids be maintained, but the poorly preserved and limited material continues to haunt the archaeozoological dreams (nightmares?) of their custodian, the writer of this paper.

As already said, the putative early domestic cattle of the southern Western Desert met with disbelief and incited their intrepid proposer to evaluate the evidence for North Africa, stressing the poor archaeological and archaeo-zoological record and how it can be manipulated in several ways (Gautier, 1987). As the literature is not easily accessible, the author missed several records and for additional, most, if not all, not easily decodable records, the reader is referred to Garcea (1993). Recently Chenal-Vélardé (1997) re-evaluated some of the older finds as well as some new ones, for example from the Fayum (von den Driesch, 1986), the Neolithic site of Kobadi (Raimbault *et al.*, 1987), and from Wadi Shaw in Northern Sudan (Van Neer and Uerpmann, 1989). To the latter, a few finds identified by the late Boessneck and recorded by Kröpelin (1993, p. 210) can be added. One find consists of some molars from the Middle Wadi Howar south of Jebel Rahib (K154-413/0) and is dated about 5.4 kyr bp. Corridi (1997) confirms the presence of both cattle and domestic ovicaprines in the seventh millennium bp in the Acacus. Van Neer (2000, this volume) has reviewed the earliest sites with domestic livestock in Western and Central Africa. Similarly, Hassan (2000, this volume) provides a calibrated chronology of the spread of pastoralism in Africa.

The revised and additional data in the papers cited in the previous paragraph do not alter the picture: the large bovids from Nabta and Bir Kiseiba remain the earliest putative domestic cattle. On the basis of the radiometric data known to me (Haas and Haynes, 1980; Connor, 1984; Close, 1984), and considering that the chances of survival of bone are inversely proportional to the time passed since their incorporation in the often shallow archaeological deposits, the Nabta and Bir Kiseiba bovids may date back to about 9.2 kyr bp. In the Asian Levant domestic cattle date back to 8.5 kyr bp (Gautier, 1990; Helmer, 1994; Benecke, 1994). We are hence forced to postulate an independent domestication of cattle in North Africa. Data concerning the Northern Sudanese Nilo-Saharan language group have been adduced to corroborate the hypothesis of very early cattle pastoralism (Wendorf and Schild, 1994). According to Grigson (1991) the so-called Sanga cattle from Africa may not result from cross breeding of older introduced cattle in Africa with zebu cattle, but it could descend from the original, autochthonous African cattle as exemplified by '*Bos africanus*' of ancient Egypt, which represents a mosaic of characteristics due to mixing with introduced cattle with and without hump in only the last few hundred years.

Recent analyses of DNA also seem to corroborate a separate origin of prehistoric cattle in Africa (Bradley *et al.*, 1996).

Wendorf and Schild (1994) suggest that domestication of cattle may have begun along the Nile, but good evidence is lacking. However, I am willing to accept that the late Palaeolithic hunters along the Nile Valley knew 'their' wild cattle well, having lived with these beasts in a close predator-prey relationship for many millennia. A Binfordian scenario seems more appropriate to me. People moving into the Sahara took with them small numbers of young wild cattle and tried to make them breed under their control. These early experiments may have had to be repeated quite often, especially since the would-be pastoralists had no previous experience of the domestication of small livestock, which preceded that of cattle in Asia (Gautier, 1990; Benecke, 1994). The foregoing might explain the large size of the early finds, but osteometric analysis is still underway and may prove that this large size is not exceptional but within the normal range of very early domestic bovids. The foregoing scenario does not preclude that in later prehistoric times domestic cattle from Asia, or elsewhere, was introduced in North Africa.

OVICAPRIDS IN NORTH AFRICA

In the case of 'true' sheep (*Ovis ammon*) and 'true' goat (*Capra aegagrus*), the palaeobiogeographical argument can be applied. Remains of these taxa in Africa are necessarily domestic, because the fossil record and modern biogeography indicate clearly that true sheep and goat never lived in Africa. Barbary sheep (*Ammotragus lervia*) is the typical wild ovicaprid of North Africa, originally ranging from the Atlantic coast to the Red Sea, but now much decimated by hunting and other forms of human interference. The Nubian ibex (*Capra ibex nubiana*) is, or was, found east of the Nile Valley (Kock, 1971; Haltenorth and Diller, 1979). Both prefer stony mountainous habitats, Barbary sheep occupying the lower stretches of such environments, ibex the higher ones in the regions where these ovicaprids occur sympatrically. Barbary sheep are large animals but sexual dimorphism is marked and small females may overlap in size with large sheep or goat. Osteologically they present a mixture of ovine and caprine characteristics (Gabler, 1985).

Ovicaprid remains excavated in the middle and late Neolithic site E-75-8 at Nabta were identified as those of sheep and goat (Gautier, 1980). The identification was based on the size of the finds and the presence of some specimens attributable to sheep and to goat according to the diagnostic criteria in use (Boessneck *et al.*, 1964; Boessneck, 1969). Furthermore, the rather flat region of Nabta did not seem to fit the description of the habitats preferred by Barbary sheep. However, the present day distribution of Barbary sheep reflects no doubt its retreat and survival in less accessible regions or refuges.

The presence of small livestock already in middle Neolithic contexts, between 7.1 and 6.2 kyr bp or somewhere around 6.7 kyr bp (Haas and Haynes, 1980), also met with resistance (see for example Muzzolini, 1990). In the light of more recent finds of domestic sheep and goat elsewhere in the Egyptian deserts, the early presence of sheep and goat in the southern Western Desert appears now to be quite acceptable. Indeed, domestic ovicaprids appear to be present quite early in Dakhla Oasis, where a mid-Holocene cultural unit, baptized Bashendi, subdivided into Bashendi A and B, spans more than two millennia (McDonald, 1991, 1996). Bashendi A sites consist of extensive scatters of hearths and artifacts eroding out of basin floor silts and are assumed to represent the remains of aggregation camps of pastoralists dated between 7.6 and 6.9 kyr bp. Faunal remains, collected mostly *in situ*, comprise gazelle, presumably dorcas gazelle, hartebeest, hare, fox and birds including ostrich, but no firm evidence of livestock (Churcher, personal communication and in McDonald, 1991, p. 45). Bashendi B sites comprise mainly hearth mounds on the basin edges above the level of the playa silts, dated between 6.5 and 5.5 kyr bp. A faunal collection excavated at one of the Bashendi B sites (site 271) "consists almost entirely of bones of cattle and goat, both most probably domesticated" (Churcher, in McDonald, 1991, p. 47). Small surface collections from Bashendi B sites include "a few scraps of gazelle and perhaps hartebeest, and, in several cases, tooth fragments of *Bos*" (Churcher, in McDonald, 1991, p. 47). In general Bashendi A would correspond to the middle Neolithic of the southern reaches of the Egyptian Western Desert as defined by Wendorf and associates (1984), Bashendi B to the late Neolithic. Until more is known about the faunal record, we can assume that the Bashendi A people were pastoralists still heavily reliant on hunting, while the Bashendi B people exploited their herds much more intensively; the herds of Bashendi B people apparently included cattle and ovicaprids, but I find it strange that only goats would be present. Anyhow, if goat is present from about 6.5 kyr bp on, sheep were no doubt also present.

Still from the Western Desert, faunal remains from the Hidden Valley Village site in the Farafra Oasis have been submitted to the author. Excavated since 1995, this site comprises permanent or semi-permanent stone-lined hut structures in playa deposits, dated from the seventh millennium bp (Barich and Hassan, 2000, and personal communication). This small, very fragmentary and poorly preserved collection contains some bone remains of birds, including ostrich, dorcas gazelle and small livestock. A fragment of a horncore of a goat with twisted horns was collected in the upper stratum II (6.9 kyr bp), while in stratum III a distal anterior canonbone or metacarpus smaller than those of female Barbary sheep measured by Gabler (1985) can be attributed to a large (male?) sheep.

The foregoing records corroborate the early presence in the southern Western Desert of small livestock, but meanwhile new problems have arisen at Nabta. Material collected since 1990 provides limited but undeniable evidence that Barbary sheep is present among the ovicaprid remains of site E-75-8, making it difficult to identify less diagnostic specimens. Moreover, how should we explain the absence of Barbary sheep in early Neolithic sites both at Nabta and Bir Kiseiba, although some of these produced substantial samples? An explanation could be that Barbary sheep were attracted by the domestic ovicaprids and became the victims of their own behavior and thus visible in the archaeozoological record. Various ruminant species living in herds are known to associate and Barbary sheep have been seen together with other wild ruminants

(Kowalski and Kowalska, 1991, p. 201). Nievergelt (1966) records the association of Alpine ibex (*Capra* ibex) with domestic goats. It will probably be impossible to document comparable behavior by Barbary sheep, since this ovicaprid has become rather rare.

Nearer to the Nile, well identified material of domestic ovicaprids, including skull fragments of sheep, is known from the Fayum A Neolithic, for which the oldest date is about 6.5 kyr bp (Gautier, 1976; von den Driesch, 1986), while small livestock, including ovicaprids and pig, at Merimde Beni Salama in the Nile Delta, may go back to 6.1 kyr bp (von den Driesch and Boessneck, 1985; Hawass *et al.*, 1988). These same animals, as well as cattle, are found in Predynastic sites along the Nile (see for example McArdle, 1992), but these represent younger developments. In the Central Sudan, livestock, including cattle and ovicaprids, is well represented in the so-called Khartoum Neolithic, with dates going back to about 6 kyr bp (el Mahi, 1982, 1988; Hassan, 1986).

The recorded finds suggest that goat and sheep were adopted earlier in the desert than along the Nile Valley. Small livestock finds from the Sodmein Cave site in the Eastern Desert corroborate the foregoing (Vermeersch *et al.*, 1994). The younger backfill of this cave contains mainly small livestock, but also dorcas gazelle and rock dassie (*Procavia capensis*). One bone element was identified as goat, others are either goat or sheep. Moreover, the upper younger backfill consists mainly of dung, especially from goat and sheep. The cave was probably used intermittently by pastoralists in the period between 7.1 kyr and 6.3 kyr bp and the oldest sheep and goats would date from shortly after 7 kyr bp.

DOGS IN NORTH AFRICA

The history of the dog in Africa may help to understand that of livestock. Dogs, however variable, all descend from the wolf. Morey (1996) recently brought together some of the oldest records, which clearly indicate that the origin of the dog goes back to the end of the Palaeolithic. Wolves do not occur in Africa, except in Northern Libya and Egypt, where small desert wolves, related to those from Arabia and adjacent regions, would live sympatrically with the golden jackal (*Canis aureus*); these small wolves were until now labeled *Canis aureus lupaster* (Ferguson, 1981). Dogs were identified in the late Neolithic of Nabta and Bir Kiseiba (scarp survey sites; Gautier, 1980, 1984), but not yet published material from site E-75-8 at Nabta suggests that dogs were already present in the middle Neolithic. The earliest dogs along the Nile were identified at Merimde Beni Salama (von den Driesch and Boessneck, 1985) and in the Khartoum Neolithic at Esh Shaheinab (Peters, 1986a). The foregoing may indicate that domestic canids came to Africa together with small livestock, most likely as herding dogs. In a recent paper, Cesarino (1997) lists some more dog finds as well as representations of dogs in the parietal art of North Africa. He suggests that dogs may have been domesticated in North Africa, whereas this author would use parietal dogs to test the hypothesis of their arrival concurrent with that of small livestock and the chronology of the rock art scenes in which one recognizes dogs.

DOMESTIC PIG IN NORTH AFRICA

Domestic pig has already been mentioned in passing as part of the small livestock in Neolithic and Predynastic sites of the Nile Valley and we should try to bring this suid in the picture. One still reads that wild boar (*Sus* scrofa) was part of the game fauna along the Nile Valley and that it is the ancestor of the original domestic swine in Egypt (see for example Houlihan, 1995, p. 25). Archaeo-zoological and biogeographical data indicate that wild boar was probably restricted to coastal Egypt and the Delta. In theory, pig domestication may have occurred in the Delta, from where the results spread along the Nile. The domestic status of the pigs from Neolithic Merimde Beni Salama is unquestionable (von den Driesch and Boessneck, 1985) and the domestication of the animal predates no doubt the earliest occupation of the site. The most cautious scenario brings sheep, goats, dogs and pigs to Merimde Beni Salama from Asia, but then why not also cattle? Among the cattle remains of Merimde Beni Salama, some horncores occur, attributed to longhorns (von den Driesch and Boessneck, 1985.). Knowledge of the kind of Asian cattle, which could have been introduced in Neolithic Egypt, may help to solve the problem posed.

CONCLUSIONS

The conclusions of this rapid survey of the oldest livestock in North Africa can be brief. First of all, we should remain aware of the fact that the archaeozoological record is still limited, especially when considering the regions and millennia to be covered. There exists also an annoying gap in the archaeological and archaeo-zoological record of the Nile Valley and its extensions between the latest post-Palaeolithic industries and the earliest Neolithic manifestations, to which Hassan (1988) drew attention. It may span more than a millennium (7.2-6.3 kyr bp). Anyhow, it would appear that the eastern Western Desert witnessed very early (9.2 kyr bp) experiments in cattle pastoralism combined with hunting. These experiments do not exclude the introduction of domestic cattle from Asia, after domestication there (8.5 kyr bp). Sheep and goat are beyond doubt domesticates introduced from Asia, very likely together with the dog. We perceive them first in the deserts of Egypt around 7.0 kyr bp. Most probably they came via Sinai along the Red Sea coast and along the Mediterranean coast. Cattle, sheep, goat, dog and pig become visible in Nilotic Egypt only by about 6.1 kyr bp or somewhat later (at Merimde Beni Salama). As with sheep, goat and dog, domestic pig was probably introduced from Asia. If the foregoing domesticates are immigrants, it is not unreasonable to assume that cattle may also have been introduced into northern Egypt. Since pig cannot have reached the Nile from the adjacent deserts, it traveled south along the Nile to be found in Predynastic and later sites along the Nile. Again one can ask whether it traveled alone or accompanied by other livestock. No doubt the attentive reader recognizes that these conclusions and questions imply research projects which may not be realizable.

REFERENCES

Barich, B. E., and Hassan, F. A. (2000). A stratified sequence from Wadi El Obeyid, Farafra: new data on subsistence and chronology of the Egyptian Western Desert. In Krzyzaniak, L., Kroeper, K., and Kobusiewicz, M. (eds.), *Recent Research Into the Stone Age of Northeastern Africa*, Studies in African Archaeology 7, Poznan Archaeological Museum, Poznan, pp. 11-20.

Benecke, N. (1994). *Der Mensch und seine Haustiere. Die Geschichte einer jahrtausendealten Beziehung*, Theis, Stuttgart.

Boessneck, J. (1969). Osteological differences between sheep (*Ovis aries* Linné) and goat (*Capra hircus* Linné). In Brothwell, D., and Higgs, E. (eds.), *Science in Archaeology*, Thames and Hudson, London, pp. 331-358.

Boessneck, J., Müller, H.-H., and Teichert, M. (1964). Osteologische Unterscheidungsmerkmale zwischen Schaf (*Ovis aries* LINNÉ) und Ziege (*Capra hircus* LINNÉ). *Kühn-Archiv* 78(1-2): 1-129.

Bradley, D. G., Machugh, D. E., Cunningham, P., and Loftus, R. T. (1996). Mitochondrial diversity and the origin of African and European cattle. *Proceedings of the National Academy of Science USA* 93: 5131-5135.

Cesarino, F. (1997). I cani del Sahara. *Sahara* 9: 93-113.

Chenal-Vélardé, I. (1997). Les premières traces de boeuf domestique en Afrique du Nord: état de la recherche centré sur les données archéozoologiques. In Gautier, A. (ed.), Animals and People in the Holocene of North Africa. *ArchaeoZoologia* 9(1/2): 11-40.

Close, A.E. (1984). Report on site E-80-4. In Wendorf, F., and Schild, R. (ass.), Close, A.E. (ed.), *Cattle-Keepers of the Eastern Sahara: The Neolithic of Bir Kiseiba*. Southern Methodist University Press, Dallas, pp. 325-349.

Connor, D.R. (1984). Report on Site E-79-8. In Wendorf, F., and Schild, R. (ass.), Close, A.E. (ed.), *Cattle-Keepers of the Eastern Sahara: The Neolithic of Bir Kiseiba*, Southern Methodist University Press, Dallas, pp. 217-250.

Corridi, C. (1997). Some new archaeozoological data from the Tadrart Acacus, Libya (9th to 5th millennium B.P.). In Gautier, A. (ed.), Animals and People in the Holocene of North Africa. *ArchaeoZoologia* 9(1/2): 41-48.

El Mahi, A.T. (1982). *Fauna, Ecology and Socio-Economic Conditions in the Khartoum Nile Environment*, Ph.D. thesis, University of Bergen, Bergen. [unpublished]

El Mahi, A.T. (1988). *Zooarchaeology in the Middle Nile Valley. A Study of Four Neolithic Sites Near Khartoum*, Cambridge Monographs in African Archaeology 27, British Archaeological Reports International Series 418, Oxford.

Epstein, H. (1971). *The Origin of the Domestic Animals of Africa*, Vol. I & II, Africana Publishing Corporation, New York/London/Munich.

Ferguson, W.W. (1981). The systematic position of *Canis aureus lupaster* (Canivora: Canidae) and the occurrence of *Canis lupus* in North Africa, Egypt and Sinai. *Mammalia* 45(4): 459-465.

Gabler, K.O. (1985). *Osteologische Unterscheidungsmerkmale am postkranialen Skelett zwischen Mähnenspringer (Ammotragus lervia), Hausschaf (Ovis aries) und Hausziege (Capra hircus)*, Inaugural Dissertation, Ludwig-Maximilians-Universität, Munich.

Garcea, E.A.A. (1993). *Cultural Dynamics in the Saharo-Sudanese Prehistory*, Gruppo Editoriale Internazionale, Rome.

Gautier, A. (1976). Animal remains from archaeological sites of terminal Paleolithic to Old Kingdom age in the Fayum. In Wendorf, F., and Schild, R. (eds.), *Prehistory of the Nile Valley*, Appendix I, Academic Press, New York, pp. 369-381.

Gautier, A. (1980). Contributions to the archaeozoology of Egypt. In Wendorf, F., and Schild, R. (eds.), *Prehistory of the Eastern Sahara*, Appendix 4, Academic Press, New York, pp. 317-344.

Gautier, A. (1984). Archaeozoology of the Bir Kiseiba region, Eastern Sahara. In Wendorf, F., and Schild, R. (ass.), Close, A.E. (ed.), *Cattle-Keepers of the Eastern Sahara: The Neolithic of Bir Kiseiba*, Southern Methodist University Press, Dallas, pp. 49-72.

Gautier, A. (1987). Prehistoric men and cattle in North Africa: a dearth of data and a surfeit of models. In Close, A. E. (ed.), *Prehistory of Arid North Africa, Essays in Honor of Fred Wendorf*, Southern Methodist University Press, Dallas, pp. 163-187.

Gautier, A. (1988). The final demise of *Bos ibericus? Sahara* 1: 37-48.

Gautier, A. (1990). *La Domestication. Et l'Homme créa ses Animaux*, Errance, Paris.

Gautier, A. (1992). Domestication animale et animaux domestiques prétendument oubliés. In Bodson, L. (ed.), *Contributions à l'Histoire de la Domestication 3*, Université de Liège, Colloques d'histoire des connaissances zoologiques, Journée d'étude 02.03.1991, Liège, pp. 31-36.

Gautier, A. (1993). 'What's in a name?'. A short history of the Latin and other labels proposed for domestic animals. In Close, A., Payne, S., and Uerpmann, J.-P. (eds.), *Skeletons in her Cupboard, Festschrift for Juliet Clutton-Brock*, Oxbow Monograph 34, Oxbow Books, Oxford, pp. 91-98.

Gautier, A. (1996). The 'unacceptable face' of the Western European Palaeolithic revisited: the evidence for the presumed domestication of the horse during that period. *The Workshops and the Posters of the XIII International Congress of Prehistoric and Proto-historic Sciences (Forlì, Italia, 8-14 September 1996)*, 2. Abstracts, ABACO Edizioni, Forlì, pp. 13-14.

Gautier, A. (1998). Once more: the names of domestic animals. In Kokabi, M., and Washl, J. (eds.), Proceedings of the 7th International Conference for Archaeozoology, ICAZ, Constance, September 1994. *Anthropozoologica* **25-26**: 113-118.

Gautier, A., and Muzzolini, A. (1991). The life and times of the giant buffalo alias *Bubalus/Homoioceras/Pelorovis antiquus* in North Africa. *ArchaeoZoologia* 4(1): 39-92.

Grigson, C. (1991). An African origin for African cattle? - some archaeological evidence. *African Archaeological Review* 9: 119-144.

Haas, H., and Haynes, C.V. (1980). Discussion of radiocarbon dates from the Western Sahara. In Wendorf, F., and Schild, R. (eds.), *Prehistory of the Nile Valley*, Academic Press, New York, pp. 373-378.

Haltenorth, T., and Diller, H. (1979). *Elseviers Gids van de Afrikaanse Zoogdieren*, Elsevier, Amsterdam.

Hassan, F.A. (1986). Chronology of the Khartoum 'Mesolithic' and 'Neolithic' and related sites in the Sudan: statistical analysis and comparisons with Egypt. *African Archaeological Review* 4: 83-102.

Hassan, F.A. (1988). The Predynastic of Egypt. *Journal of World Prehistory* 2(2): 135-185.

Hassan, F. A. (2000). Climate and cattle in north Africa. In Blench, R., and MacDonald, K. C. (eds.), *The Origins and Development of African Livestock: Archaeology, Genetics, Linguistics, and Ethnography*, University College London Press, London, pp. 61-86.

Hawass, Z., Hassan, F.A., and Gautier, A. (1988). Chronology, sediments, and subsistence at Merimde Beni Salama. *Journal of Egyptian Archaeology* **74**: 31-38.

Hecker, H.M. (1982). Domestication revisited: its implications for faunal analysis. *Journal of Field Archaeology* 9: 217-236.

Helmer, D. (1994). La domestication des animaux d'embouche dans le Levant Nord (Syrie du Nord et Sinjar) du milieu du IXe millénaire BP à la fin du VIIIe millénaire BP. Nouvelles données d'après les fouilles récentes. *Anthropozoologica* 20: 41-54.

Houlihan, P.F. (1995). *The Animal World of the Pharaohs*, American University in Cairo Press, Cairo.

Klein, R.G. (1994). The long-horned African buffalo (*Pelorovis antiquus*) is an extinct species. *Journal of Archaeological Science* 21: 725-733.

Kock, D. (1971). Zur Verbreitung von Mähnenschaf und Steinbock im Nilgebiet. *Säugetierkundliche Mitteilungen* **19**: 28-39.

Kowalski, K., and Rzebik-Kowalska, B. (1991). *Mammals of Algeria*, Zaklad Narodowy Imienia Ossolinskich Wydawnictwo Polskiej Akademii Nauk, Warsaw.

Kröpelin, S. (1993). Zur Rekonstruktion der spätquartären Umwelt am unteren Wadi Howar (Südöstliche Sahara / NW-Sudan). *Berliner Geographische Abhandlungen* 54.

Lutz, R., and Lutz, G. (1993). From picture to hieroglyphic inscription. The trapping stone and its function in the Messak Sattafet (Fezzan, Libya). *Sahara* **5**: 71-78.

Lutz, R., and Lutz, G. (1995). *Das Geheimnis der Wüste. Die Felskunst im Messak Sattafet und Messak Mellet - Libyen*, Golf Verlag, Innsbruck.

Lutz, R., and Lutz, G. (1997). The domestic cattle in prehistoric Sahara. *Sahara* 9: 137-140.

McArdle, J.E. (1992). Preliminary observations on the mammalian fauna from Predynastic localities at Hierakonpolis. In Friedman, R., and Adams, B. (eds.), *The Followers of Horus. Studies dedicated to Michael Allan Hoffman*, Egyptian Studies Association 2, Oxbow, Oxford, pp. 54-56.

McDonald, M.M.A. (1991). Origins of the Neolithic in the Nile Valley as seen from Dakhleh Oasis in the Egyptian Western Desert. *Sahara* 4: 41-52.

McDonald, M.M.A. (1996). Relations between Dakhleh Oasis and the Nile Valley in the Mid-Holocene: a discussion. In Krzyzaniak, L., Kroeper, K., and Kobusiewicz, M. (eds.), *Interregional Contacts in the Later Prehistory of Northeastern Africa*, Studies in African Archaeology 5, Poznan Archaeological Museum, Poznan, pp. 93-99.

Morey, D.F. (1996). L'origine de plus vieil ami de l'homme. Le chien domestique, un loup préhistorique resté louveteau. *La Recherche* **288**: 72-77.

Muzzolini, F. (1990). The sheep in Saharan rock art. *Rock Art Research* **7**(2): 93-109.

Nievergelt, B. (1966). *Der Alpensteinbock (Capra ibex L.) in seinem Lebensraum. Ein ökologischer Vergleich*, Parey, Hamburg.

Pachur, H.-J. (1991). Tethering stones as palaeoenvironmental indicators. *Sahara* **4**: 13-32.

Peters, J. (1986a). A revision of the faunal remains from two Central Sudanese sites: Khartoum Hospital and Esh Shaheinab. *ArchaeoZoologia*, Mélanges publiés à l'occasion du 5e Congrès International d'Archéozoologie (Bordeaux Août 1986): 11-35.

Peters, J. (1986b). *Osteomorphology and osteometry of the appendicular skeleton of African Buffalo, Syncerus caffer (SPARRMAN, 1779) and Cattle Bos* primigenius *f. taurus BOJANUS, 1827*, Occasional Papers, Lab. voor Paleontologie, Rijksuniversiteit Gent, No. 1, p. 83

Peters, J. (1988). Osteomorphological features of the appendicular skeleton of African buffalo, *Syncerus caffer* (Sparrman, 1779) and of domestic cattle, *Bos primigenius* f. taurus Bojanus, 1827. *Zeitschrift für Säugetierkunde* **53**: 108-123.

Peters, J., Gautier, A., Brink, J.S., and Haenen, W. (1994). Late Quaternary extinction of ungulates in Sub-Saharan Africa: a reductionist's approach. *Journal of Archaeological Science* **21**: 17-28.

Raimbault, M., Guérin, C., and Faure, M. (1987). Les vertébrés du gisement néolithique de Kobadi (Mali). *ArchaeoZoologia* I(2): 219-238.

Van Neer, W. (2000). Domestic animals from archaeological sites in Central and West-Central Africa. In Blench, R., and MacDonald, K. C. (eds.), *The Origins and Development of African Livestock: Archaeology, Genetics, Linguistics, and Ethnography*, University College London Press, London, pp. 163-190.

Van Neer, W., and Uerpmann, H.-P. (1989). Palaeoecological significance of the Holocene faunal remains of the B.O.S.-Missions. In Kuper, R., (ed.), Forschungen zur Umweltgeschichte der Ostsahara. *Africa Praehistorica* **2**: 307-341.

Vermeersch, P. M., Van Peer, P., Moeyersons, J., and Van Neer, W. (1994). Sodmein Cave site, Red Sea Mountains (Egypt). *Sahara* **6**: 31-40.

von den Driesch, A. (1986). Tierknochenfunde aus Qasr el-Sagha / Fayum (Neolithikum und Mittleres Reich). *Mitteilungen des Deutschen Archäologischen Instituts, Abteilung Kairo* **42**: 1-8.

von den Driesch, A., and Boessneck, J. (1985). *Die Tierknochenfunde aus der Neolithischen Siedlung von Merimde-Benisalâme am Westlichen Nildelta*, Deutsches Archäologisches Institut, Abteilung Kairo, Munich.

Wendorf, F., and Schild, R. (1994). Are the early Holocene cattle in the Eastern Sahara domestic or wild? *Evolutionary Anthropology* **3**(4): 118-128.

Wendorf, F., and Schild, R. (ass.), Close, A.E. (ed.) (1984). *Cattle-Keepers of the Eastern Sahara: The Neolithic of Bir Kiseiba*, Southern Methodist University Press, Dallas.

Wendorf, F., Schild, R., Said, R., Haynes, C.V., Gautier, A., and Kobusiewicz, M. (1976). The prehistory of the Egyptian Sahara. *Science* **139**: 103-104.

Wyrwoll, T.W. (1997a). Zur Verbreitung des wilden Wasserbüffels (*Bubalus arnee*) auf der Arabischen Halbinsel. *Zeitschrift der Deutschen Morgenländischen Gesellschaft* **147**(2): 480-485.

Wyrwoll, T.W. (1997b). Ein archäologischer Beitrag zum ehemaliger Vorkommen des wilden Wasserbüffels (*Bubalus arnee* (Kerr, 1792)) in Nordafrika und auf der Arabischen Halbinsel. *Säugetierkundliche Mitteilungen* **39**(3): 103-114.

13
CULTURAL RESPONSES TO CLIMATIC CHANGES IN NORTH AFRICA: BEGINNING AND SPREAD OF PASTORALISM IN THE SAHARA

B. E. Barich

INTRODUCTION

Since the mid-1960s, research programmes in Northeastern Africa have provided data and promoted debates which have helped progressively to put into focus the social dimension of cultures. At the same time, because of the new trends in archaeology, the interaction between environment and culture has begun to be interpreted from a rather different perspective (Barich, 1997, pp. 28-29). The 'abrupt climatic change' concept (Hassan, 1996, 1997) has encouraged the investigation of changes that may have had a direct impact on the cultural system. Currently, it is believed that to understand how the external changes are perceived and internally represented by society is the first priority. In light of this, I will try to outline a history of the interaction between climate and culture in the Sahara during the Holocene, referring to the most meaningful and complete information. From this point of view, Libya and Niger on the one hand, and the Egyptian Western Desert on the other, provide more material than other regions for investigating human adaptation and illustrating the interrelation among different territories.

EARLY HOLOCENE: THE BROAD SPECTRUM OF RESOURCE EXPLOITATION

In the earliest moist phase of the Holocene, water was the main factor influencing social behavior (from settlement patterns to aquatic resource exploitation and, obviously, to the implements which were, then, mainly microlithic). Subsequently the climatic trend towards increasing aridity, recorded throughout the Sahara between 8500 and 8200 radiocarbon years bp, seems to have induced a more permanent exploitation of sites where water was accessible all year round, as observed at Uan Afuda (unit 1) and Ti-n-Torha East (layers II-III) in the Libyan Tadrart Acacus. In the same time range, semi-permanent settlements featuring wells for storing water appear in the Nabta Playa region in the contexts of the El

Ghorab and El Nabta early Neolithic. These societies, based on intensive resource exploitation, tended towards logistic mobility, seasonal food storage and probably had a more complex internal structure. This broad spectrum resource exploitation model is notably the same as the one attested in the middle Holocene contexts of the Central Sudan. According to Ehret (1993), such contexts can be related to the development of Nilo-Saharan languages.

Increasing sedentism seems to have favored the growth of ceramic production. Its rarity in the tenth millennium bp layers at Tagalagal and Adrar Bous 10 in Niger, the oldest ceramic contexts known up to now, would indicate that ceramic vessels were precious.[1] At that time, pottery must have had more of a symbolic meaning than an economic role. However, starting from 8500 radiocarbon years bp, its growth proceeded in parallel with the intensification of wild cereal collection. Diagnostic ceramics of the broad spectrum resource adaptation are decorated with broad-spanning side combs, or spatulae, used with rocker technique. The dotted wavy line is one of the most typical motifs. Based on its presence in the ceramic repertoires, intergroup exchanges between the Sahara and the Nile have been inferred.

In this same wide context, the Egyptian Western Desert has an important, fundamental, position for the study of developmental social processes related to cattle husbandry. The reoccupation of a territory where meat was obtained primarily by hunting hares and gazelles would have induced the immigrants to bring some specimens of *Bos* into an incipient domestic status with them, probably from the Nile Valley (Qadan and Arkinian contexts). Cattle would have facilitated the use of the desert by providing food resources like milk and blood. This thesis, propounded by Wendorf and Schild (1980, 1994), still meets opposition (e.g., Clutton-Brock, 1989; Robertshaw, 1989; Smith, 1992; Muzzolini, 1993). However, in the light of all evidence, I think that the thesis of an 'incipiently domestic *Bos*' at Kiseiba and Nabta is the only one which goes along with the entire set of archaeological, faunal and chronological data coming from the other Saharan contexts. Moreover, this would explain the 'explosion' of pastoralism in western areas during the seventh millennium bp, and also a few older examples (which go back to mid-eighth millennium bp) in Libya and Chad.[2] As a matter of fact, the stratigraphic data show that the dry oscillation between 7900 and 7700 radiocarbon years bp (Wendorf and Schild, 1984, p. 405; Close, 1992, p. 169; Hassan, 1997, p.216) put the broad spectrum exploitation strategy into a real crisis, and, in contrast, promoted the intensification of the first experiments in cattle herding among the western communities. Holl has recently observed that "... the appearance of a large bovid species at Ti-n-Torha East in a context of constantly broadening faunal spectrum... may signal a solution to an enduring subsistence crisis" (Holl, 1998, p. 84). Moreover, di Lernia (1998) has highlighted a quite unsuspected mastery in wild animal management from the early Holocene.

MIDDLE AND LATE HOLOCENE—THE SPREAD OF PASTORALISM

The Uan Muhuggiag rock shelter in the Libyan Tadrart Acacus (Figure 13.1), with its sequence covering the whole range of the middle and late Holocene, is of the utmost importance for the study of pastoralism. It was excavated on two different occasions, by Pasa during the sixties (Pasa and Pasa Durante, 1962), and by Barich during the eighties (Barich, 1978, 1987). Recently Cremaschi and collaborators have carried out further tests for micromorphology and radiometric age determinations at the site (Cremaschi *et al.*, 1996, pp. 92-93). On the whole, the middle Holocene sequence shows a climatic trend going from moderately wet to dry. Moister conditions are recorded between 7 and 6.5 kyr bp and are followed by worsening climate conditions throughout the sixth millennium bp. Around 5 kyr bp, an arid crisis occurred which caused the vault to collapse, and the subsequent abandonment of the shelter. The most recent occupation phase, which continued up to the second millennium bp (Table 13.1) represented a completely different occupation episode, perhaps related to the 'horse' peoples.

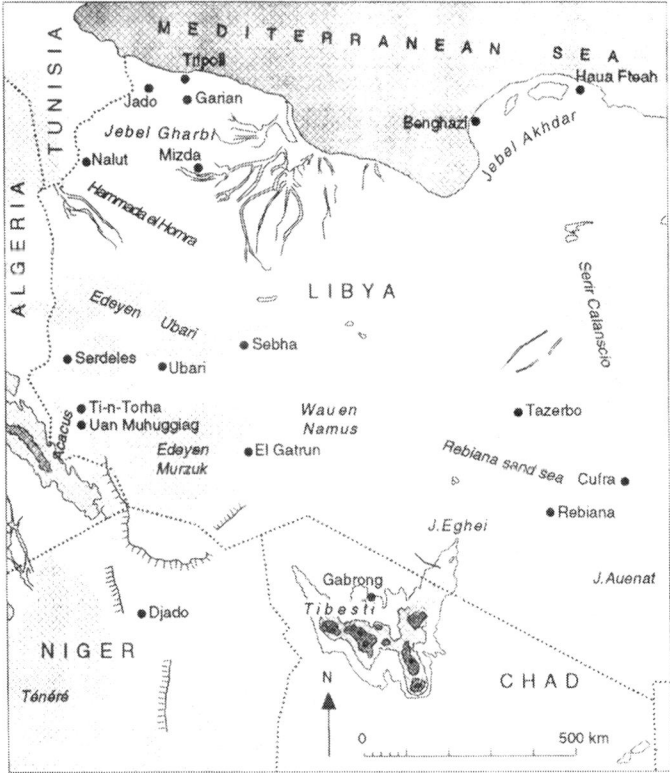

Figure 13.1. Map of the Libyan Sahara.

Table 13.1. Libya: synthetic radiocarbon chronology of deposits at Uan Muhuggiag (modified after Pazdur, 1993)

Sector A[a]				Unit	Sector B[a]			
Layer	Age ^{14}C bp	Ref.	cal BC		Layer	Age ^{14}C bp	Ref.	cal BC
1a	3770 ± 200	Gd-224	2410-2040	Unit 1	1	2220 ± 220	Gd-4290	490-110
1a	3800 ± 140	Gd-4363	2390-2130		1	2770 ± 80	Gd-4288	1000-880
					1	3810 ± 80	Gd-2854	2360-2190
2	5290 ± 110	Gd-4362	4220-4050	Unit 2	1a	4980 ± 110	Gd-4357	3890-3710
2	6035 ± 110	Ud-225	?-4840		2	5340 ± 120	Gd-2959	4270-4080
2a	6030 ± 80	Gd-2853	?-4840		2a	5480 ± 120	Gd-4361	4430-4230
2c	5780 ± 80	Gd-4358	4720-4590		2b	5420 ± 50	Gd-5337	4320-4250
					2b	5350 ± 200	Ud-226	4330-4040
	6690 ± 130[b]	NA-59	-	Unit 3	-	-	-	-
	6900 ± 220[b]	NA-60	-			-	-	-
	7438 ± 220[c]	PI	-			-	-	-

[a] excavation by Barich, 1982
[b] from Cremaschi, 1998
[c] from Pasa and Pasa Durante, 1962

Among the archaeological materials, pottery is one of the most typical examples of pastoral (or Bovidian) production (Figure 13.2). This is a black, thin-walled, quartz-tempered pottery, decorated with double pronged implements or spatulae, in place of combs. The decorations are quite monotonous, consisting of zigzags or punctuations, either forming paired-line patterns or distributed evenly on the body, providing a smocking effect (Caneva, 1987). In the lithic industry backed blades and geometric microliths are replaced by tools of much lower technological sophistication (apart from rather refined quartzite arrow-heads).

Palynological and faunal data also testify to the presence of an arid trend at the site. In fact, Gautier (1987) has emphasized that cattle remains have a much higher incidence in the lowest levels, dated to 7 kyr bp, while in the middle layers sheep/goats become increasingly more important. Dry conditions allowed for a good preservation of floral remains, particularly in the mid-upper layers of the sequence. They form a large sample of grasses primarily belonging to the subfamily *Panicoideae (Cenchrus, Digitaria, Echinochloa, Panicum, Setaria, Urochloa)* and attest to the importance of wild grass exploitation in the human diet (Barich, 1992; Wasylikowa, 1992, 1993).

The results of the most recent research in the Acacus by Cremaschi and di Lernia (1996) provide a useful contribution for a better understanding of the settlement pattern at a regional scale. Numerous open air sites, real villages, were detected in the Erg Uan Kasa and in the Edeyen Murzuq along the shores of ancient lakes which reached their maximum expansion at about 6.6 kyr bp. Cattle remains are very abundant at these sites, and therefore such sites could represent the principal areas of occupation from which task groups seasonally moved, probably with small livestock, to reach pastures which were then available in the mountains.

The late Pastoral Neolithic in the Acacus can be dated between 5 and 3.8 kyr bp. Except for ephemeral camps and hearthplaces, the plain territories were abandoned.[3] The occupation then moved, instead, into the mountains, where numerous deposits belonging to this phase are recorded. The economic pattern is mainly based on ovicaprines and on the use of secondary products. The pottery is either undecorated, brown or reddish, or is decorated with impressed motifs all around the neck. Among the artifacts, exotic items are numerous, such as axes in volcanic rocks and bifacial Predynastic-type products. All this indicates the groups' circulation across distances of hundreds of kilometers.

In Niger the very beginning of the Pastoral Neolithic occurred around the middle of the seventh millennium bp. It is testified by a number of sites (Dogomboulo and Rocher Toubeau) and, above all, by some cattle burials in the Adrar Bous area. Because of the lack of information, the relationship of this phase with the Tenerian occupation is not clear. The latter was particularly marked around 5 kyr bp, paradoxically at the point of a return to dry conditions. Findings pertaining to the Tenerian are particularly widespread along the eastern edges of the Aïr, between the Adrar Bous and the Areschima latitude. Roset suggested that the arid conditions pushed the groups to gather in the mountains where much humidity survived (Roset, 1987).

The Tenerian—defined by Tixier in the 1960s—is characterized by the presence of large bifacial tools for woodworking and for tilling the soil, such as

Figure 13.2. Pottery with rocker impressions from Uan Muhuggiag, Tadrart Acacus, Libya (from Barich, 1987).

gouges, disc-knives, sickle-knives, arrowheads and axes (Tixier, 1962). On the other hand, the association of microlithic-backed types with these implements is dubious (Roset, 1987, pp. 207-208). Ceramics have globular shapes and very refined fabrics. Among the decorations, which are always impressed in the paste, spatula or plain comb zig-zag patterns, and punctuations with a 'smocking' effect prevail. A notable affinity can be observed with the Uan Muhuggiag repertoire.

The Tenerian pastoralists' economy featured herding, hunting and, occasionally, fishing. Plant collection is attested by grain impressions in the pottery, which prove the presence of *Brachiaria* and *Sorghum*. Furthermore, Paris' studies of Neolithic and post-Neolithic burials in the Adrar Bous region provide new data relating to the population make-up (Paris, 1996, 1997). The latest Neolithic phase (between the mid-fifth and mid-fourth millennia bp) seems to record the immigration of a people who bring funerary monuments (platform cairns) to the region, and differ from the common Sudanese type in terms of their anthropological traits.[4] The few examples of ceramics (reddish with conical bases) found in these tombs are different from the Tenerian ceramics. Paris ponders the relationship between the local Tenerians and the monument builders. Were they immigrants? Where were they from?[5] Since funerary monuments, on the whole, are few and reserved for males, the author suggests that they were designated for prestigious figures such as chiefs, within an aristocratic, chiefdom-type organization. A further worsening of climatic conditions put an end to the Tenerian occupation in the Adrar Bous around 3800 radiocarbon years bp.

During the mid-Holocene, the Western Desert (Figure 13.3) witnessed the growing importance of the oases region, which experienced a relevant occupational phenomenon between 7.7 and 6.6 kyr bp. It is well-illustrated, above all, by field research programmes carried out since the eighties in the Farafra and Dakhla Oases. In both oases the frequent moist oscillations in the early Holocene favored the first hunter-gatherer aggregations. Subsequently, the occurrence of the short arid interval between 7.1 and 6.9 kyr bp (Hassan, 1986, 1996, p. 84), could have induced the groups to exploit more intensively the places where water was available all year round, leading to residential lifestyles and to some settled nuclei. This general trend can be inferred from the artifact density, the use of local raw materials, and the number of assemblages. Radiocarbon age determinations obtained from both the Farafra and Dakhla regions (McDonald, 1998), bear witness to a more intensive occupation during the seventh millennium bp which, on the whole, records a gradual environmental deterioration. At that time these areas, real water reservoirs, must have exerted a special attraction for the groups which then lived scattered throughout the desert. The Hidden Valley settlement at Farafra was continually re-occupied between 7200 and 6200 radiocarbon years bp. Inhabitants extensively used the cereals which grew on the edges of playas including *Sorghum* and *Panicum* (Barich and Hassan, 2000; Barich, 1998; Barakat and Fahmy, 1999). The bifacial tools used as sickles and axes for tilling the soil, which hint at Predynastic models, particularly from the Badari and the Armant-Qurna region, can be dated to between 7 and 6.7 kyr bp. In addition, sheep/goat herding and hunting form part of the economic model of these groups (Gautier, in litt.). A clue to the societal organization, myths and beliefs is provided by a number of examples of rock art and 'art mobilier'; a clay

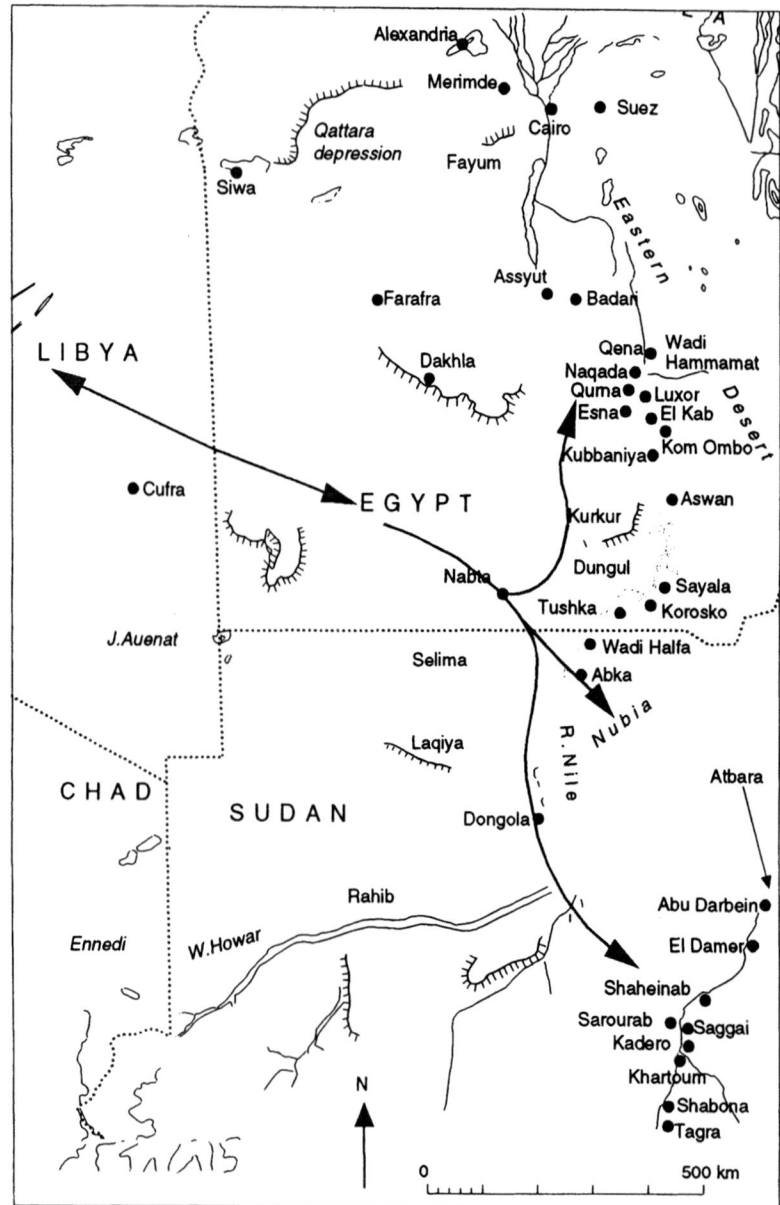

Figure 13.3. Map of the Egyptian Western Desert, showing directions of the dispersal of domestic cattle.

figurine representing a woman anticipates analogous later examples from Badarian contexts. The palaeoclimatic results from Farafra show a climatic crisis around 6000 radiocarbon years bp (Hassan, 1997, p. 217). Afterwards, the area was probably abandoned. More recent radiocarbon dates in the sixth and fifth millennia bp indicate that the oasis was reoccupied during the late Holocene. At Dakhla the occupation continued until the fifth millennium bp with the Sheikh Muftah phase in the innermost area of the oasis (McDonald, 1993).

In contrast, at Nabta Playa, a short arid interval (ca. 6.6 kyr bp) separates the middle from the late Neolithic, whose beginning is conventionally placed around 6.5 kyr bp, the date of the last influential moist phase in the Western Desert.[6] The middle Neolithic experiences a great density of villages. The very large site E-75-8 shows signs of having been continually reoccupied betweeen 7120 ± 150 bp (SMU-242) and 6700 ± 50 bp (SMU-261). Moreover, a greater density of villages together with campsites and hearthplaces are indicative of movements over only short distances (Close, 1990). In contrast, in the late Neolithic, none of the hearths contained more than a few scraps of bone and only a few edible, charred plants. The occurrence of raw materials not locally available (e.g., turquoise, perhaps from Sinai, Red Sea shells, and Nile chert and shells) suggests a very mobile pattern, including frequent contacts with the Nile Valley and beyond. While the Saharan tradition in ceramics comes to an end then, the late Neolithic ceramics, red or black burnished and smudged, could be better related to the Abkan and Badarian examples (Wendorf and Schild, 1994, p.121).

The discovery of a ceremonial center, not far from site E-75-8, is extremely important in that it provides a useful insight into the societal structure and the set of values which inspired late Neolithic communities. According to Wendorf and his team, during the mid-Neolithic this site played a particular role as a place of aggregation where Neolithic groups gathered for ceremonial purposes, after the summer rains. The recent unearthing of two calendar circles and three megalithic alignments (datable to 6 kyr bp) might indicate the persistence of a ceremonial tradition tied to this area, even during the late Neolithic (Wendorf, et al., 1997). Two tumuli, covering cattle burials, are significant. One of them is dated to 6470 ± 270 radiocarbon years bp (Wendorf, et al., 1997, p. 96), approximately the same date as the cattle burials found in Niger. This clear focus on cattle probably anticipates the complex of religious beliefs tied to 'cows' in Ancient Egypt.

OVERVIEW

The integrated reading of the sequences examined separately until now highlights significant links and interrelations for Holocene settlement between the Sahara and the Nile (Figure 13.4). In the Egyptian Western Desert, the Nabta Playa and Bir Kiseiba area has a strategic position in such exchanges. It is a focal point which receives and redistributes Central Saharan influences towards the Nile. A good example is the continuation at Nabta of the Saharan ceramic tradition up to the mid-Neolithic (Close, 1995). Moreover, there are stylistic similarities between the middle and upper layers at Ti-n-Torha East and the El Ghorab and El Nabta

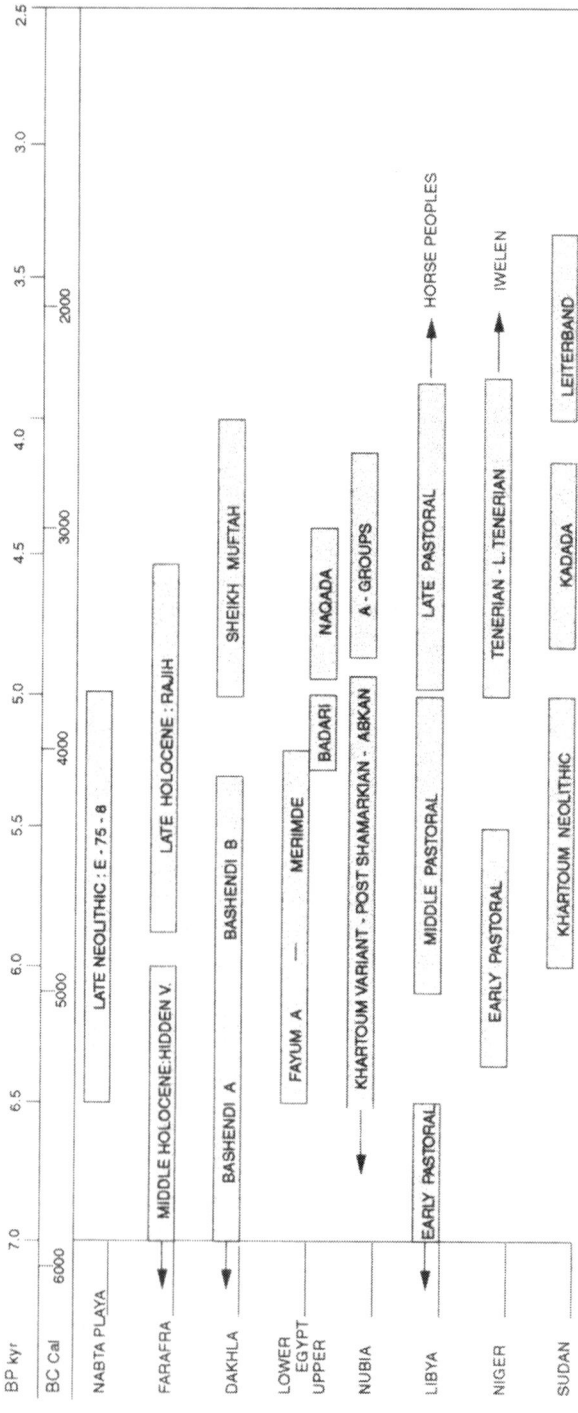

Figure 13.4. Mid-late Holocene cultural sequence in the Sahara and the Nile Valley.

early neolithic complexes (between 8500 and 7900 radiocarbon years bp; Close, 1987).

I would like to suggest that through prolonged exchanges, probably by task units which periodically left their base camps, on the one hand the Saharan hunter-gatherers acquired the first cattle specimens, and on the other Saharan decorations spread among the Western Desert inhabitants through the exchange of pots, or better, direct contact with potters.[7] In a similar way, the Saharan decorative techniques from the Western Desert spread gradually into the Upper Nile region, together with cattle and sheep/goats. Between 6 and 5 kyr bp Saharan ceramics in the dotted wavy line style, together with domestic livestock, appear in the Khartoum area in the early Khartoum Neolithic layers. The latter period is still characterized by a broad spectrum of resource economy, where herding and proto-cultivation are associated with local traditional activities, such as hunting and fishing.

The middle Holocene (7.5-6.6 kyr bp) witnessed a degree of climatic and cultural stability throughout the Sahara. In the Libyan region the pastoral occupation pattern is made up of both plain sites, inhabited all year round, and mountain sites, occupied only seasonally. The trend towards aridity, particularly severe around the mid-seventh millennium bp, could have intensified migratory movements towards other regions. We can hypothesize that at that time groups moving southwards may have originated the first pastoral facies in the Ténéré, anticipating the actual Tenerian occupation of the sixth-fifth millennia bp.

At the same time, in the Western Desert Oases (25°-27°N), a mixed economy prevailed, which included herding, hunting and, above all, proto-cultivation. At that time, the first bifacial implements designated for sorghum and millet gathering appeared. Both the settlement and the economic model presaged the characteristics of the Egyptian Neolithic societies flourishing between the fifth and fourth millennia in calibrated years BC: Fayum (6500 radiocarbon years bp), Merimde (5800 bp), and Badari (5300 bp).

In contrast, again in the Western Desert but at a more southern latitude (22°N), as early as 6.5 kyr bp, the area witnessed a more severe climatic change which gave way to a much more mobile pastoral occupation. It is reasonable to think that these Western Desert herders, pressured by unfavorable climatic conditions, tended towards the proto-agricultural spheres of the oases and the valley. In fact, during the Nabta late Neolithic, many artifacts (scrapers, sickles, side-blow flakes) typical of the Farafra repertoire appeared, which would later become a part of the Predynastic repertoires.[8] A type of ceramic also appeared, hinting at Tarifian and Badarian examples. The final abandonment of the area is dated to around 5000-4600 radiocarbon years bp. Some groups probably went towards the middle Nile Valley, while others took the route towards Nubia. Schön (1996) has proposed that the latter could represent the 'Saharan nomadic cattle-raising' component of the A-Groups.

This scenario is similar to the late pastoral conditions in the Acacus (5-4 kyr bp), when the transhumance model within the regional range failed and the groups initiated contacts at a wider range. I support Aumassip's (1996) suggestion that the Bovidian culture was a continuum, including the Acacus, Tassili n'Ajjer, Tefedest, Erg Admer and Niger. I would also not overlook eventual contacts with the sub-coastal region (Aurès), where other pastoral communities, belonging to the Neolithic of Capsian Tradition sphere (Roubet, 1979), were located. Seemingly it was through repeated contacts of this kind, and through small scale migratory movements, that local peopling was progressively modified.[9] Moreover, the appearance in the latest Tenerian phase of monumental tombs reserved for prestigious figures (males), could indicate a parallel modification within the social structure and the beginning of ranking. This fact is quite important and goes along with the appearance of the so-called 'white-skinned Mediterranean' individuals in the rock paintings attributed to the middle and late Pastoral phases (Mori, 1965, 1998).

These Saharan Bovidian communities between 5 and 4 kyr bp may have initiated more frequent itineraries towards the Nile Valley. The routes which were followed then passed at a more southern latitude (17°), along the edges of the Tibesti, Borkou, Auenat and Ennedi, and from there through the Wadi Howar (Kröpelin, 1993) towards the Nile. Specialized nomadic pastoralism, the use of secondary products, and, moreover, the material culture features are the common traits among these groups. Their arrival in the Khartoum area put an end to the broad spectrum adaptation (Shaheinab) and brought about a more strongly pastoral based economy (Kadada; Haaland, 1992). Proof of this is the spread of pottery in the Uan Muhuggiag and Ténéré style, characterized by zig zag patterns and punctuations on paired dotted lines. These ceramics appeared at Kadada around 4800/4600 radiocarbon years bp (ca. 3600 cal BC) and in the final layers at Shaqadud (4050 bp, 2650 cal BC; Caneva, 1996). This same pottery forms the basis of the so-called 'Leiterband' repertoire which Keding (1993) recognized at Wadi Howar, and which can be correlated to the C-Groups.

Haaland (1991, 1992) has emphasized a clear-cut contrast between the pastoral-nomadic adaptation and the broad spectrum of resources typical of the Khartoum Neolithic. While admitting the difficulties in associating archaeological aspects with languages, Haaland suggested identifying the groups with prevalent pastoral economies with the proto-Cushites. Their arrival in Central Sudan must have caused the displacement of groups, originating the subsequent division between Nilo-Saharan and Afro-Asiatic languages.

NOTES

(1) Two sherds of these units have been recently dated by thermoluminescence, obtaining dates that go back to around 10,000 years ago, which is consistent with the calibrated radiocarbon dates from charcoals at the same locations (Guiber, *et al.*,1994).

(2) The remains from Bardagué in the Tibesti mountains are dated to 7445 ± 180.

(3) In the Acacus range and in the Messak Settafet Plateau the desert varnish forms are indicative of a moderately dry climate (Cremaschi, 1996).

(4) According to Paris (1997, p. 60) they are characterized by lower prognatism and nasal index, slimmer cranium, post-cephalic skeleton less robust and lighter stature.

(5) It is tempting to think that the newcomers came from the northern regions where the climatic deterioration antedates the analogous phenomenon of the southern regions.

(6) Late Neolithic sites are: E-94-2, E-94-3 and E-77-1.

(7) We cannot even exclude that the diffusion of the decorative techniques was due to exogamic marriages, women being the earliest pottery makers.

(8)   The proto-gouge and the gouge, which Farafra and Dakhla give the most ancient examples of, appear also in the Tenerian repertoire.
(9)   Hassan has recently presented a model for the dispersal of pastoralism in North Africa indicating a rate of 12-30 km distance per generation (Hassan, 2000).

REFERENCES

Aumassip, G. (1996). Propos sur le Bovidien. In Aumassip, G., Clark, J. D., and Mori, F. (eds.), *The Concept of the "Neolithic" in Africa with Particular Reference to the Saharan Region,* XIII International Congress of the UISPP, Colloquium XXX, Forlì, pp. 209-218.

Barakat, H. N., and Fahmy, A.G. (1999). Wild grasses as 'Neolithic' food resources in the Eastern Sahara: A review of the evidence from Egypt. In van der Veen, M. (ed.), *The Exploitation of Plant Resources in Ancient Africa.* Kluwer Academic/ Plenum Publishers, New York, pp. 33-46

Barich, B.E. (1978). Lo scavo di Uan Muhuggiag (Teshuinat) e attività di survey nell'Acacus settentrionale. *Libya Antiqua* 15: 305-316.

Barich, B.E. (1987). The Uan Muhuggiag rock shelter. In Barich, B.E. (ed.), *Archaeology and Environment in the Libyan Sahara. The excavations in the Tadrart Acacus 1978-1983,* British Archaeological Reports International Series 368, Oxford, pp.123-219.

Barich, B.E. (1992). The botanical collections from Ti-n-Torha/Two Caves and Uan Muhuggiag (Tadrart Acacus, Libya). An archaeological commentary. *Origini* 16: 109-123.

Barich, B.E (1997). Dynamics of populations, movements and responses to climatic changes in North Africa and the Nile Valley. *Human Evolution* 12(1-2): 25-31.

Barich, B.E. (1998). Early to mid-Holocene occupation at Farafra (Western Desert, Egypt): A social approach. In Krzyzaniak, L. (ed.), *Later Prehistoric and Protohistoric Social Groups in Northeastern Africa,* Proceedings of the XIII Congress of the IUSPP, Forlì 1996, Vol.6, Workshop 8, ABACO Edizioni, Forlì.

Barich, B. E., and Hassan, F. A. (2000). A stratified sequence from Wadi El Obeyid, Farafra: new data on subsistence and chronology of the Egyptian Western Desert. In Krzyzaniak, L., Kroeper, K., and Kobusiewicz, M. (eds.), *Recent Research Into the Stone Age of Northeastern Africa,* Studies in African Archaeology 7, Poznan Archaeological Museum, Poznan, pp. 11-20.

Caneva, I. (1987). Pottery decoration in prehistoric Sahara and upper Nile: a new perspective. In Barich, B.E. (ed.), *Archaeology and Environment in the Libyan Sahara. The excavations in the Tadrart Acacus 1978-1983,* British Archaeological Reports International Series 368, Oxford, pp. 231-254.

Caneva, I. (1996). The influence of Saharan prehistoric cultures on the Nile Valley. In Aumassip, G., Clark, D. J., and Mori, F. (eds.), *The Concept of the "Neolithic" in Africa with Particular Reference to the Saharan Region,* XIII International Congress of the UISPP, Colloquium XXX, Forlì 1995, ABACO Edizioni, Forlì, pp. 231-239.

Close, A.E. (1987). The lithic sequence from the Wadi Ti-n-Torha (Tadrart Acacus). In Barich, B.E. (ed.), *Archaeology and Environment in the Libyan Sahara . The excavations in the Tadrart Acacus 1978-1983,* British Archaeological Reports International Series 368, Oxford, pp. 63-85.

Close, A.E. (1990). Living on the edge: neolithic herders in the eastern Sahara. *Antiquity* 64: 79-96.

Close, A. E. (1992). Holocene occupation of the Eastern Sahara. In Klees, F., and Kuper, R. (eds.), *New Light on the Northeast African Past,* Heinrich Barth Institut, Köln, pp. 155-183.

Close, A.E. (1995). Few and far between. Early ceramics in North Africa. In Barnett, W.K., and Hoopes, J.W. (eds.), *The Emergence of Pottery. Technology and Innovation in Ancient Societies,* Smithsonian Institution Press, Washington and London, pp. 23-37.

Clutton-Brock J. (1989). Cattle in ancient North Africa. In Clutton-Brock, J. (ed.), *The Walking Larder: Patterns of Domestication, Pastoralism and Predation,* Unwin Hyman, London, pp. 200-206.

Cremaschi, M. (1996). The rock varnish in the Messak Settafet (Fezzan, Libyan Sahara). Age, archaeological context, and paleoenvironmental implication. *Geoarchaeology* 11(5): 393-421.

Cremaschi, M. (1998). Geological evidence for late Pleistocene and Holocene environmental changes in south-western Fezzan (Central Sahara, Libya). In di Lernia, S., and Manzi, G. (eds.), *Before Food Production in North Africa. Proceedings of the Homonymous Workshop held in Forli, September 1996, within the XIII World Congress of the International Union of the Prehistoric and Protohistoric Sciences*, ABACO Edizioni, Rome, pp. 53-69.

Cremaschi, M., and di Lernia, S. (1996). Climatic changes and human adaptive strategies in the Central Saharan massifs. The Tadrart Acacus and Messak Settafet perspective (Fezzan, Libya). In Pwiti, G., and Soper, R. (eds.), *Aspects of African Archaeology. Papers from the 10th Congress of the Panafrican Association for Prehistory and Related Studies*, University of Zimbabwe Publications, Harare, pp. 39-51.

Cremaschi, M., di Lernia, S., and Trombino, L. (1996). From taming to pastoralism in a drying environment. Site formation processes in the shelters of the Tadrart Acacus massif (Libya, Central Sahara). In Castelletti, L., and Cremaschi, M. (eds.), *Micromorphology of Deposits of Anthropogenic origin*, XIII International Congress of the UISPP, Volume 3, Paleoecology, ColloquiumVI, Forlì 1995, ABACO Edizioni, Forlì, pp. 87-106.

di Lernia, S. (1998). Cultural control over wild animals during the early Holocene: The case of Barbary sheep in Central Sahara. In di Lernia, S., and Manzi, G. (eds.), *Before Food Production in North Africa. Proceedings of the Homonymous Workshop held in Forli, September 1996, within the XIII World Congress of the International Union of the Prehistoric and Protohistoric Sciences*, ABACO Edizioni, Rome, pp. 113-126.

Ehret, C. (1993). Nilo-Saharans and the Saharo-Sudanese Neolithic. In Shaw, T., Sinclair, P., Andah, B., and Okpoko, A. (eds.), *The Archaeology of Africa: Food, Metals and Towns*, Routledge, London and New York, pp.104-125.

Gautier, A. (1987). The archaeozoological sequence of the Acacus. In Barich, B.E. (ed.), *Archaeology and Environment in the Libyan Sahara. The Excavations in the Tadrart Acacus 1978-1983*, British Archaeological Reports International Series 368, Oxford, pp. 283-308.

Guiber, P., Schvoerer, M., Etcheverry, M.P., Szepertyski, B., and Ney, C. (1994). IXth millennium B.C. ceramics from Niger: detection of a U-Series disequilibrium and TL dating. *Quaternary Geochronology (Quaternary Science Reviews)* 13: 555-561.

Haaland, R. (1991). Specialized pastoralism and the use of secondary products in prehistoric central Sudan. *Archéologie du Nil Moyen* 5: 149-155.

Haaland, R. (1992). Fish, pots and grains: Early and Mid-Holocene adaptations in the Central Sudan. *African Archaeological Review* 10: 43-64.

Hassan, F.A. (1986). Desert environment and origins of agriculture in Egypt. *Norwegian Archaeological Review* 19(2): 63-76.

Hassan, F.A. (1996). Abrupt Holocene climatic events in Africa. In Pwiti, G., and Soper, R. (eds.), *Aspects of African Archaeology, Papers from the 10th Congress of the Panafrican Association for Prehistory and Related Studies*, University of Zimbabwe Publications, Harare, pp. 83-89.

Hassan, F.A. (1997). Holocene paleoclimates of Africa. *African Archaeological Review* 14(4): 213-230.

Hassan, F.A. (2000). Climate and cattle in North Africa. In Blench, R., and MacDonald, K. C. (eds.), *The Origins and Development of African Livestock: Archaeology, Genetics, Linguistics, and Ethnography*, University College London Press, London, pp. 61-86.

Holl, A.F.C. (1998). The dawn of African pastoralism: An introductory note. *Journal of Anthropological Archaeology* 17: 81-96.

Keding, B. (1993). Leiterband sites in the Wadi Howar, North Sudan. In Krzyzaniak, L., Kobusiewicz, M., and Alexander, J. (eds.), *Environmental Change and Human Culture in the Nile Basin and Northern Africa until the Second Millennium B.C.*, Poznan Archaeological Museum, Poznan, pp. 371-380.

Kröpelin, S. (1993). The Gilf Kebir and lower Wadi Howar: contrasting early and mid-Holocene environments in the Eastern Sahara. In Krzyzaniak, L., Kobusiewicz, M., and Alexander, J. (eds.), *Environmental Change and Human Culture in the Nile Basin and Northern Africa until the Second Millennium B.C.*, Poznan Archaeological Museum, Poznan, pp. 249-258.

McDonald, M.M.A. (1993). Cultural adaptations in Dakhleh Oasis, Egypt, in the early and mid-Holocene. In Krzyzaniak, L., Kobusiewicz, M., and Alexander, J. (eds.), *Environmental Change and Human Culture in the Nile Basin and Northern Africa until the Second Millennium B.C.*, Poznan Archaeological Museum, Poznan, pp. 199-209.

McDonald, M.M.A. (1998). Adaptive variability in the Eastern Sahara during the early Holocene. In di Lernia, S., and Manzi, G. (eds.), *Before Food Production in North Africa. Proceedings of the Homonymous Workshop held in Forli, September 1996, within the XIII World Congress of the International Union of the Prehistoric and Protohistoric Sciences*, ABACO Edizioni, Rome, pp. 127-136.

Mori, F. (1965). *Tadrart Acacus – Arte Rupestre del Sahara Preistorico*, Torino, Einaudi.

Mori, F. (1998). *The Great Civilisations of the Ancient Sahara*, 'L'Erma' di Bretschneider, Roma.

Muzzolini, A. (1993). The emergence of a food-producing economy in the Sahara. In Shaw, T., Sinclair, P., Andah, B., and Okpoko, A. (eds.), *The Archaeology of Africa: Food, Metals and Towns*, Routledge, London and New York, pp. 227-239.

Paris, F. (1996). Les sepultures monumentales d'Iwelen (Niger). *Journal de la Société des Africanistes* 60(1): 47-74.

Paris, F. (1997). Burials and the peopling of the Adrar Bous region. In Barich, B.E., and Gatto, M.C. (eds.), *Dynamics of Populations, Movements and Responses to Climatic Change in Africa*, Forum for African Archaeology and Cultural Heritage, Rome, pp. 49-61.

Pasa, A., and Pasa Durante, M.V. (1962). Analisi paleoclimatiche nel deposito di Uan Muhuggiag, nel massiccio dell'Acacus (Fezzan meridionale). *Memorie Museo Civico di Storia Naturale di Verona* 10: 251-255.

Pazdur, M.F. (1993). Evaluation of radiocarbon dates of organic samples from Uan Muhuggiag and Ti-n-Torha, Southwestern Libya. In Krzyzaniak, L., Kobusiewicz, M., and Alexander, J. (eds.), *Environmental Change and Human Culture in the Nile Basin and Northern Africa until the Second Millennium B.C.*, Poznan Archaeological Museum, Poznan, pp. 43-47.

Robertshaw, P. (1989). The development of pastoralism in East Africa. In Clutton-Brock, J. (ed.), *The Walking Larder: Patterns of Domestication, Pastoralism and Predation*, Unwin Hyman, London, pp. 207-214.

Roset, J.P. (1987). Néolithisation, Néolithique et post-Néolithique au Niger Nord-oriental. *Bulletin de l'Association Française pour l'Etude du Quaternaire* 24(32): 203-214.

Roubet, C. (1979). *Economie Pastorale Préagricole en Algerie Orientale: le Néolithique de Tradition Capsienne, Exemple: l'Aurès*, Editions du CNRS, Paris.

Schön, W. (1996). The late Neolithic of Gilf Kebir: evolution and relations. In Krzyzaniak, L., Kroeper, K., and Kobusiewicz, M. (eds.), *Interregional Contacts in the Later Prehistory of Northeastern Africa*, Poznan Archaeological Museum, Poznan, pp.115-123.

Smith, A.B. (1992). *Pastoralism in Africa. Origins and Development Ecology*, Hurst and Company, London.

Tixier, J. (1962). Le "Ténéréen" de l'Adrar Bous III. In Hugot, H.J. (ed.), *Mission Berliet Ténéré-Tchad*, AMG, Paris, pp. 333-348.

Wasylikowa, K. (1992). Holocene flora of the Tadrart Acacus area, SW Libya, based on plant macrofossils from Uan Muhuggiag and Ti-n-Torha/Two Caves archaeological sites. *Origini* 16: 125-159.

Wasylikowa, K. (1993). Plant macrofossils from the archaeological sites of Uan Muhuggiag and Ti-n-Torha, Southwestern Libya. In Krzyzaniak, L., Kobusiewicz, M., and Alexander, J. (eds.), *Environmental Change and Human Culture in the Nile Basin and Northern Africa until the Second Millennium B.C.*, Poznan Archaeological Museum, Poznan, pp. 25-41.

Wendorf, F., and Schild, R. (1980). *Prehistory of the Eastern Sahara*, Academic Press, New York.

Wendorf, F., and Schild, R. (1984). Conclusions. In Wendorf, F., and Schild, R. (ass.), Close, A. E. (ed.), *Cattle Keepers of the Eastern Sahara: The Neolithic of Bir Kiseiba*, Southern Methodist University Press, Dallas, pp. 404-428.

Wendorf, F., and Schild, R. (1994). Are the early Holocene cattle in the Eastern Sahara domestic or wild? *Evolutionary Anthropology* 3: 118-128.

Wendorf, F., Schild, R., Applegate, A., and Gautier, A. (1997). Tumuli, cattle burials and society in the Eastern Sahara. In Barich, B.E., and Gatto, M.C. (eds.), *Dynamics of Populations, Movements and Responses to Climatic Change in Africa*, Forum for African Archaeology and Cultural Heritage, Rome, pp. 90-104.

14
DRY CLIMATIC EVENTS AND CULTURAL TRAJECTORIES: ADJUSTING MIDDLE HOLOCENE PASTORAL ECONOMY OF THE LIBYAN SAHARA

S. di Lernia

INTRODUCTION

The effects of climate change are obvious to all of us. Summers are becoming increasingly hot, which has many consequences on the environment and human landscapes. The greenhouse effect, the destruction of ozone, and the heavy impact of humans on the environment are but a few causes of the disasters throughout the world.

Human activity has probably had some responsibility in such disasters, and even if the climate has been affected by human impact in recent times, surely climate was what shaped and motivated economic and social dynamics for the major part of human history. The crux of this process may be placed, as far as the African continent is concerned, at the turning point marked by the emergence of food producing activities. The search for the beginning of the imbalance in marginal environments appears to be a strategic way to analyze food security among humans in this part of the world.

As archaeologists, we should probably focus our attention on the interaction between human activity and the environment, in order to identify possible sustainable developments for the future. Among the immediate effects of climate affecting human activity, we must mention food security and the consequent higher or lower life expectancy dependent on the availability and predictability of resources.

Africa is among the most impressive examples of how dramatically climatic uncertainty, whether droughts or flood, can change the destiny of human groups. As noted by Hassan (1996, 1997), abrupt climatic episodes appear to be related to global climatic events in the zones within the Intertropical Convergence Zone monsoon belt. Such shifts heavily influenced food resources as much in the past as at present, and affected mostly cereal crops. This has been recorded also in the recent period from 1961 to 1997 (FAO estimates, 1998), but the yields from pastoral activities and their byproducts were much less influenced (Figure 14.1).

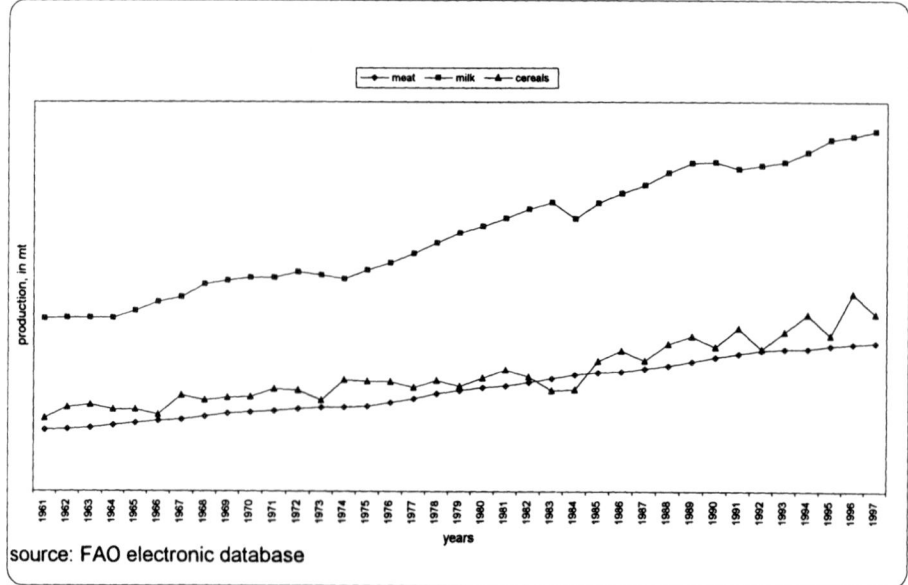

source: FAO electronic database

Figure 14.1. Recent production of the main alimentary sources in North Africa. Note the instability of cereal yields. Meat, in particular, shows a greater stability and minor sensibility to recent droughts.

Probably, the greater stability of pastoralism may be related to the long tradition of this activity (di Lernia, 1999a,b). Such marked changes in continental-scale climatic mechanisms are not new, and they were particularly effective in the late Quaternary.

Many regional and cross-regional studies have demonstrated how climatic changes were continental, if not global, and how regional variations are the result of specific features of the local environment, such as topography and latitudinal shifts. Another factor is the timing of these phenomena, which can vary from one area to another, involving different regions with different effects (Hassan, this volume). High-resolution data are becoming available to the scientific community (Bonnefille *et al.*, 1990; Bryson and Bryson, 1997; Hassan, 1997), and we are beginning to better understand the dynamics of change on a more refined scale.

Based on these observations, some general trends may be outlined for the climate during the Holocene in the Central Sahara, and in particular the Libyan Sahara. A long, wet period began somewhere around 15,000 uncalibrated years bp. The climate became progressively drier and several smaller variations punctuated these millennia, alternating wet periods with dry episodes (Maley, 1981; Vernet, 1994; Hassan, 1997; Cremaschi, 1998). The onset of present day arid conditions in the Libyan Sahara is placed at approximately 5100 uncalibrated years bp, that is ca. 5.9-5.6 kyr cal BP (Cremaschi and di Lernia, 1998; Cremaschi, 1998).

Recent approaches to palaeoclimate studies reveal a greater emphasis on short, abrupt dry intervals, believed to have particularly influenced human activity and more generally the cultural dynamics in the area. If we consider these dynamics as expressions of human groups in relation to a changing landscape, and not merely as assemblages of lithics and ceramics, we probably have an adequate perspective. As a matter of fact, on a generation scale, the environment was probably more static than dynamic; but short, drastic events may have had particular relevance in shaping decision-making strategies among groups.

This paper will consider short, or even abrupt, dry intervals which conditioned human dynamics in the Central Sahara, taken as examples of the interrelation of environmental changes and cultural adjustments in the past. The cases deal with mid-Holocene food-producing groups, whose economic basis relied on pastoral herding and intensive plant exploitation. The study area is the Acacus mountain range and its surroundings (Figure 14.2), an area already intensely studied in the past, and currently being investigated by the Italo-Libyan Joint Mission of the University of Rome, "*La Sapienza*", directed by Prof. Mario Liverani.

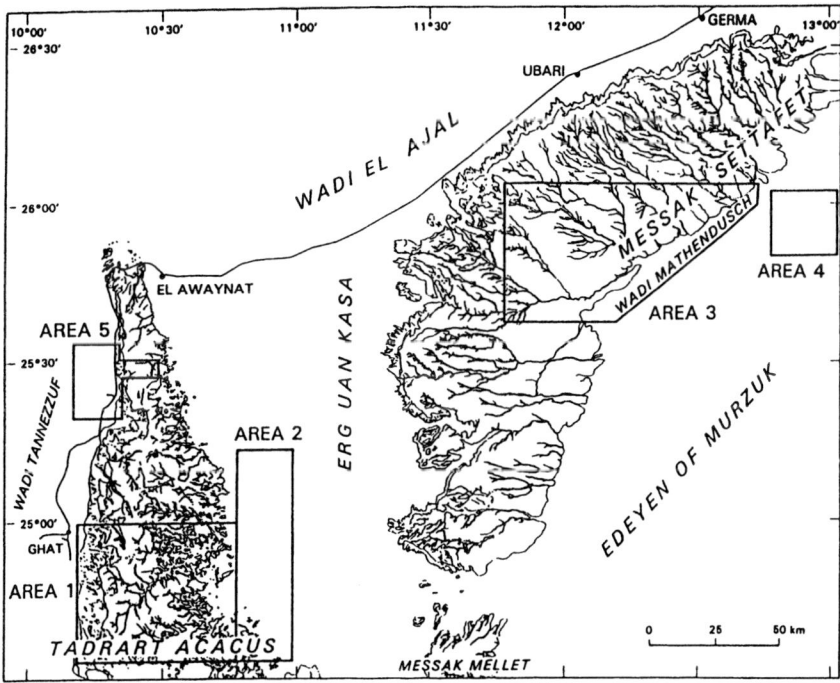

Figure 14.2. The area licensed to the Italo-Libyan Joint Mission. The surveyed areas (nos. 1-5) intercept different physiographic units: mountains, plateaus, sand seas, and fluvial valleys.

FOOD PRODUCTION DURING THE HOLOCENE: THE PASTORAL 'NEOLITHIC' OF THE ACACUS AND SURROUNDINGS

Interdisciplinary research started in the Acacus in the 1960s, with the first excavations of Angelo Pasa and Fabrizio Mori at Uan Muhuggiag (Pasa and Pasa Durante, 1962; Mori, 1965), which proved to be a key site for the analysis of the emergence of food production in North Africa. More recently, detailed work has been done by Barbara Barich on this site and other pastoral Neolithic sites of the Acacus, providing an in-depth view of the complex climatic, social and cultural features characterizing this area (Barich, 1987a). Archaeozoological studies were first undertaken by Pasa (Mori, 1965), and thereafter were more deeply dealt with by the work of Gautier (1987).

One of the more indicative pieces of information provided by this important site was the presence of cattle remains found in the lowest levels. This evidence suggested that there was a possible process of autonomous domestication in this region (Mori, 1961, 1965; Barich, 1987a).

Generally speaking, two major trends were identified in the economic basis of these Holocene food-producing groups of the Acacus: a first, earlier, phase with both cattle and sheep and goats; and a more recent phase with a predominance of ovicaprines, goat prevailing over sheep (Gautier, 1987; Corridi, 1998). The deterioration of the environment and the different water requirements of these animals served as an explanation of this change through time (Gautier, 1987). Even rock art seems to reflect this general pattern (Mori, 1965).

We are deeply indebted to these earlier investigations, which introduced a modern approach to the study of the region and sketched a general picture of the peoples of the past, which is still relatively valid today. However, if a limit may be indicated in such approaches, we probably should point to the absence of a true regional scale of analysis, since those studies favored investigation of the mountain range, ignoring the surrounding dune fields and fluvial systems. Pastoral economies and the related strategies of food systems need to be studied on a different scale to be fully understood. This can be accomplished only with a regional approach, as already indicated some twenty years ago by Hole (1978), among others.

Concerning the possible autonomous domestication of cattle, archaeozoological work undertaken in the Acacus area, first by Gautier (1987) and later by Corridi (1998), failed to find any traces of wild cattle in the pre-pastoral sites of the mountain range, such as Ti-n-Torha, Uan Tabu, Uan Afuda and others. Such discontinuity in faunal remains led recently to the rise of the hypothesis of the introduction of domesticated cattle from outside, probably already at that time together with sheep and goats, somewhere near the very end of the eighth millennium (di Lernia, 1997, 1998, 1999a). Consequently, the previous hypothesis of an autonomous, local, process of domestication of the aurochs (Mori, 1961, 1965; Barich, 1987b) should be discarded. Discrepancies between those animals present in the rock art, both paintings and engravings, and faunal remains found in archaeological deposits was noted and discussed at length (Gautier, 1993). Perhaps such a clear nonconformity should be considered a classical case of decision-making among human groups, and analysis should address non-anthropic sedimentary sequences. Recent surveys carried out in the region (Cremaschi and di Lernia, 1998, di Lernia *et al.*, 2000), have started to discover wild game fauna scattered in lacustrine deposits, or along dune slopes, not linked to human habitations: hippopotamus and elephant, to name but a few, probably of mid-Holocene age, that have never been found in the past in archaeological contexts (see also Pachur and Braun, 1980, for similar discoveries in other regions of Libya).

However, beginning in the lowest levels of Uan Muhuggiag, and at other sites in the Acacus, pastoral activity is in fact characterized by a multi-specific organization of herds, including both cattle and sheep/goat. The difference in herd composition appears to be related to site organization, mobility and seasonality (Cremaschi *et al.*, 1996; Cremaschi and di Lernia, 1998; di Lernia, 1999b). The oldest dating from Mori's excavation of the 1960s (7438 ± 220 yr bp), is no longer confirmed by recent analyses. However, age determinations from 7200 to 6900 uncalibrated years bp (8.0-7.65 kyr cal BP), obtained at the base of the sequence in other parts of the shelter (di Lernia and Manzi, 1998; Cremaschi, 1998), suggest an early phase (at least from 7000 bp, 7.85 kyr cal BP) of human occupation relying on cattle and ovicaprine herding.

The economic basis in the Acacus during the Holocene, whether cattle were domesticated in the area or introduced by way of the Eastern Sahara and Nile Valley, shows three important complexes so far. In a framework of cultural continuity which characterized primarily the interrelations among hunter-gatherers and early pastoralists, and subsequently the contact between different groups of pastoralists, an important role was played by short dry intervals. This seems particularly true for the mid-Holocene, when food procurement was based on the integration of the pastoral economy and wild plant exploitation. Evidence for these intervals has to be sought in the geoarchaeological record, environmental studies, radiocarbon age determinations, and material culture.

THE FIRST MID-HOLOCENE CRISISTHE DRY INTERVAL AT 6410-6082 YEARS bp (7.3-6.9 KYR CAL BP) AND THE SHIFT TOWARD AN ORGANIZED PASTORAL SOCIETY

Evidence from the Area for the Abrupt Dry Interval

A recent intensive palaeonvironmental study by Cremaschi (1998, this volume) in the Libyan Sahara, provides a more detailed picture of the climatic changes in this region during the late Quaternary. Generally speaking, much evidence, well known to be present throughout North Africa, has also been documented here and furnishes, at present, the basis for a correct understanding of the cultural dynamics in the Acacus from the early Holocene hunter-gatherers up to the mid-Holocene specialized nomadic pastoralists.

Leaving aside the dry interval, radiocarbon dated to approximately 8000-7500 years bp (8.8-8.3 kyr cal BP; Cremaschi, 1998; Cremaschi and di Lernia, 1998), which may have had some influence in accelerating the integration of cattle herding at the end of the eighth millennium bp (di Lernia, 1996, 1999a), another important interval can be recognized slightly before 6000 bp (6.9 kyr cal BP) that was first recognized on the basis of radiocarbon. More than 130 radiocarbon determinations have been taken into consideration, and a clearly visible hiatus is placed between 6410-6082 years bp (ca. 7.3-6.9 kyr cal BP; Figure 14.3). The gap may possibly be due to sampling but, considering the high number of analyses and the existing difference also in the standard deviation, I believe this gap is related to a drop in human occupation. Significantly, this gap is not at present coherent with the evidence of the sedimentary sequence studied in several exposed archaeological deposits. Conversely, the dry period dated to 8000-7500 years bp (8.8-8.3 kyr cal BP) is particularly recognizable in inconsistencies and stratigraphic lacuna present in the sedimentary record, as indicated by both cave fills and fluctuations in lake levels (Cremaschi, 1998).

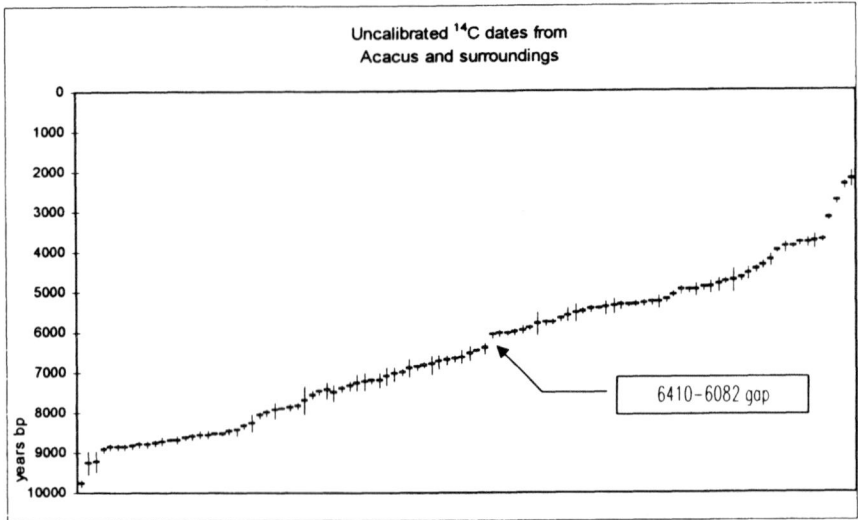

Figure 14.3. Dating of Holocene archaeological sites in the Acacus and surroundings, in radiocarbon uncalibrated years before present (after Mori, 1965; Barich, 1987b; Cremaschi and di Lernia, 1998).

Palynological data also confirm this earlier arid interval, although several sites show some continuity in human occupation, as indicated by, among other things, radiocarbon dates at 7900 bp (8.75 kyr cal BP), 7800 bp (8.6 kyr cal BP) and 7550 bp (8.35 kyr cal BP; di Lernia, 1997; Cremaschi and di Lernia, 1998).

It is interesting to note that Barich (1987b) pointed out the presence of a break in human occupation on the basis of a roof collapse in level III of the Ti-n-Torha North rock shelter. She placed this break at the end of the seventh millennium bp, and more probably between 6500 and 6200 years bp (ca. 7.35-6.9

kyr cal BP; Barich, 1987a, p. 255). Interestingly, Schulz (1987) did not report any evidence for a discontinuity in the pollen diagram for the stratigraphic sequences studied at that time.

Recent studies undertaken in the Acacus (Cremaschi and di Lernia, 1998) reconsidered the important sequence of Uan Muhuggiag (Cremaschi, 1998; di Lernia and Manzi, 1998). Palynological study (Mercuri *et al.*, 1998) brought to light a possible discontinuity in the pollen diagram. At the end of the seventh millennium bp, radiocarbon dated between approximately 6400 and 6100 years bp (ca. 7.25-6.9 kyr cal BP), the vegetation in the Central Acacus was still fairly open, but hygro-hydrophylous pollen assemblages strongly decreased, indicating also the existence of strong seasonal fluctuations. Shrub-land species, such as *Artemisia herba-alba* and Cheno-Amaranthaceae began to spread, together with *Tamarix* and *Acacia* communities. *Panicum* sp. peaks in frequency at this point. According to Mercuri and associates, the initial diffusion of the *Acacia tortilis— Panicum turgidum* association, which characterizes the present vegetation of the wadi valley of the Acacus, is probably to be attributed to this period.

Other evidence of environmental deterioration comes from the malacological analyses of mollusc shells collected from the lacustrine sequences located in the Edeyen of Murzuq and Erg Uan Kasa (Girod, 1998). Disappearance of some molluscs, such as *V. nilotica* and *H. aponensis duveyrieri*, may be considered as indicative of stagnant water with poor alimentation from influent streams, and of a marginal environment (sample 106b). Such phenomenon is recorded immediately after 6500 years bp (7.35 kyr cal BP), and probably indicates fluctuations in lake levels (or a progressive recession of lakes), which have been rarely documented until now (Cremaschi, 1998).

At any rate, apart from some differences between northern and central-southern Acacus, probably related to micro-climatic variations, there is evidence for a brief arid interval in the area at ca. 6400-6100 uncalibrated years bp (7.3-6.9 kyr cal BP). The calibration of this interval, by means of the Oxford Calibrating Program (OxCal 2.18, Stuiver *et al.*, 1993), places the interval between 5600-4950 cal BC (upper limit) and 5220-4800 cal BC (lower limit; both at 95.4% confidence).

The arid interval isolated at ca. 6410-6082 years bp (7.3-6.9 kyr cal BP), although not completely matched by the sedimentary record, should be considered evidence from the Acacus of the so-called 'Bougdouma-Oyo' event (Hassan, 1996, 1997), which is recorded in different regions of central-northern Africa.

Outside the Tadrart Acacus, an arid phase was recognized by Alimen (1987) in the Algerian Sahara in the interval 6500-6150 years bp (7.3-6.95 kyr cal BP). This arid event occurs roughly at the same time in Chad (Maley, 1981), Niger (Maley, 1981) and Mali (Petit-Maire *et al.*, 1983). A sharp decrease of *Typha* at ca. 6300 uncalibrated years bp has been interpreted as evidence of a dry episode by Ritchie (1987) at Bir Atrun, Northwestern Sudan. Neumann (1991) recognizes a dry period shortly before 6000 years bp (6.85 kyr cal BP) in different areas of North Africa. Needless to say, a short dry interval at ca. 6000 bp (6.8 kyr cal BP) is recorded at Oyo and Wadi Shaw, and also at Bougdouma (Gassc and van Campo, 1994; Vernet, 1994; Hassan, 1997).

Therefore, we are faced with two main forms of evidence: the first concerning the geographical expanse of the phenomenon, with inconsistent information concerning these arid spells, but with several common elements showing a patchy distribution; the second concerning the duration—where recorded, it seems to last for around 300 years.

So, in an interval of a few centuries (only some ten generations), we should probably consider the possibility of contractions of populations, short and long-range movements, contact and waves of migration in a territory stretching from the Central Sahara to the Nile Valley along the Sahelian belt. What was the influence of this dry interval in the Acacus region? And did it have a continental scale effect?

Cultural Responses and Implications: Reorganization of the Economic Basis and the Emergence of Seasonal Pastoralism

The dry interval from 6410-6082 uncalibrated years bp (7.3-6.9 kyr cal BP) appears to have particular relevance in the cultural dynamics of the pastoral groups of the Acacus and surroundings. Comparing the organization of pastoral activity in the Early Pastoral and Middle Pastoral periods, a different pattern is evident: although both phases developed in good environmental conditions, the earlier pastoralists appear to be more sedentarized in the mountain range, whereas a flourishing in the exploitation of the lowlands is recorded in the later phase, with the emergence of a strongly seasonal oriented vertical pastoralism (*sensu* Khazanov, 1984). The hinge between these two phases may be represented by the dry interval discussed above.

More specifically, early pastoralism, based on cattle and sheep herding, is radiocarbon dated from the very end of the eighth millennium bp to the seventh millennium bp in the mountain range, but much later in the surrounding dune fields. Plant resources, namely wild cereals, were largely integrated in the diet. The material culture also shows common features, as extensively indicated by Barich (1974, 1987a), Garcea (1993) and di Lernia (Cremaschi and di Lernia, 1998). Among the main features, attention should be drawn to the presence of impressed pottery decorated by rocker technique and the important innovation of the 'alternately pivoting stamp' technique (Caneva, 1987). Vessels appear to be mostly of globular shapes. The lithic industry features macro-size artifacts of silicified sandstone: among tools, end-scrapers, scrapers and arrowheads were of particular importance, as was grinding equipment. As far as settlement pattern is concerned, sites seem to be distributed mainly in the mountain areas; lake areas in the lowlands were also occupied, but much less than the Acacus range. One explanation of this difference relates to the time necessary for the water table to recharge, and subsequently favor the formation of ponds and shallow lakes. A gap between the mountain area and lowlands probably existed, resulting in a scaling of the effect of the general climatic amelioration recorded in that period; in the mountain first and lowlands later.

The Middle Pastoral groups which inhabited the region after the dry interval discussed, radiocarbon dated to the sixth millennium bp, show deep differences in economic organization, settlement system, material culture, and probably also in ritual customs.

What is interesting is that environmental conditions in both periods were very similar, but cultural responses and organization were different. The Middle Pastoral period probably enjoyed better conditions, since lakes reached high levels, but a picture of open savanna conditions may be inferred for both the early and middle phases of pastoralism. Hydrophilous plants, such as *Typha* and Cyperaceae, are common, and indicate habitats with perennial or at least seasonal water reserves. Semi-permanent settlements in the dune fields were probably located along the lakeshores, showing large sized, articulate structures and well arranged intrasite organization (Figure 14.4).

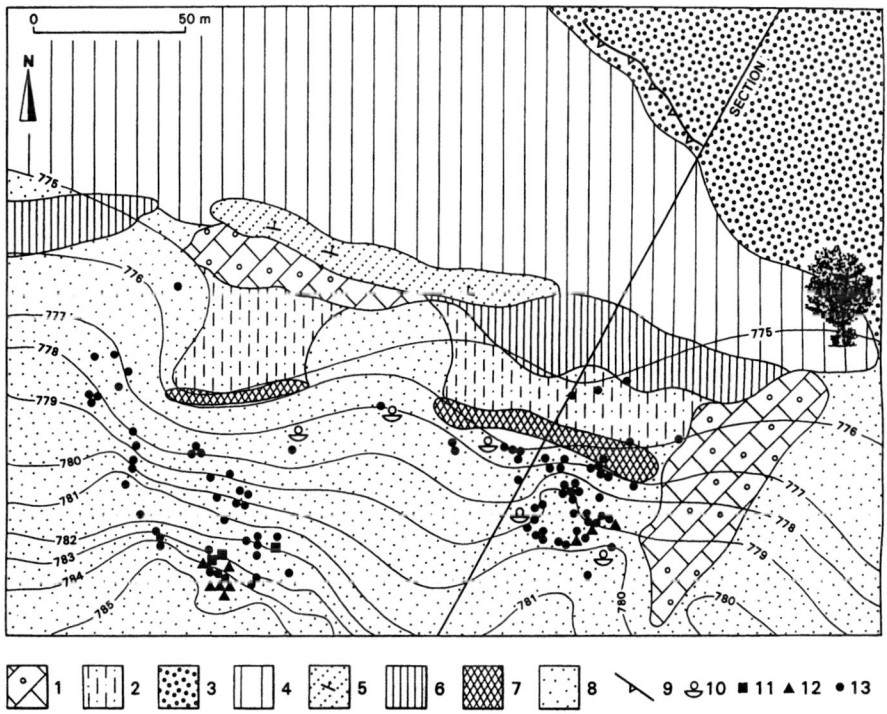

Figure 14.4. A Middle Pastoral site in the erg Uan Kasa. The site 94/63 shows an articulated intrasite organisation and evidence of re-occupations. Key: 1. bedrock; 2. lacustrine deposits; 3. gravel terrace; 4. eroded corridor bottom; 5. gypsum crust; 6. lacustrine organic sand; 7. shore organic sand; 8. dune sand; 9. scarp; 10. grinding stones; 11. concentration of sherds; 12. pit including faunal remains; 13. fireplaces (after Cremaschi and di Lernia, 1998).

For this period we have the first substantial evidence of the seasonal use of shelters, which were mainly frequented during the dry season, that is, from late winter to early spring (Cremaschi *et al.*, 1996; Trevisan Grandi *et al.*, 1998). They were small in size, and more often used as sheep and goat pens. Site deposits are characterized by alternating lenses of ash and organic sand, but spherulites and coprolites in the deposits indicate the use of the shelters as animal enclosures, as well as demonstrating an increase in pastoral activities with ovicaprines.

This type of site organization was due to a vertical, seasonal-based transhumance, with an east to west orientation: during the dry season, probably only sheep and goat moved toward the Acacus mountains, whereas most of the groups with some cattle remained in the vicinity of the lake areas (Figure 14.5). We cannot discount the possibility of an exclusive occupation of these lowland areas during the rainy season, but a much longer occupation probably better fits the available data of marked intrasite organization. More data are needed to resolve this question.

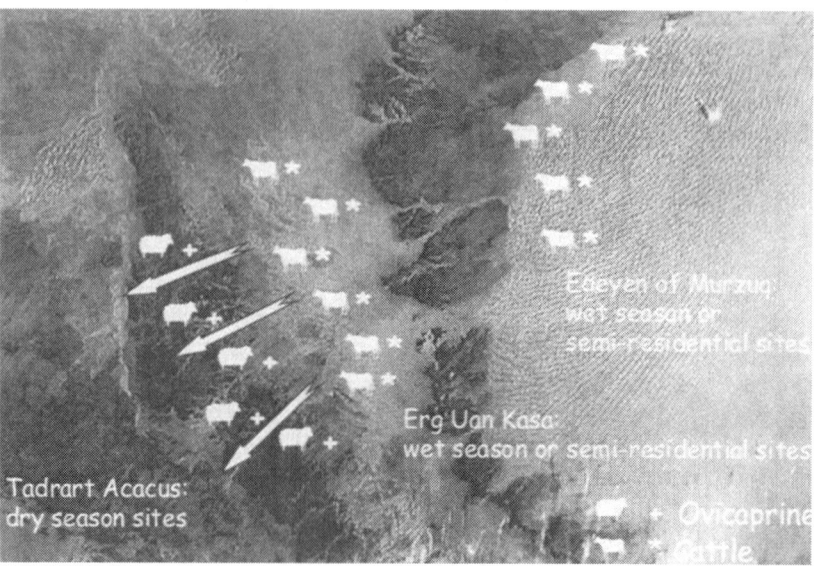

Figure 14.5. A model of the Middle Pastoral settlement pattern in the Acacus and surroundings, based on a seasonal vertical transhumance between lowlands and mountains.

However, such organization could also have been adopted to avoid the over-exploitation of pastures during the harshest period of the year. This seems to better explain the differences in the archaeozoological record existing between mountain sites, with ovicaprine remains, hunting of wild animals and only very few, if any, traces of cattle, and lake sites, characterized by the systematic presence of cattle bones (Table 14.1; di Lernia, 1999b; Cremaschi and di Lernia, 1998). The dispersal of groups and the avoidance of cattle-slaughter in favor of exploiting secondary products are typical responses of African pastoralists (Close and Wendorf, 1992; Wendorf and Schild, 1984; Gautier, 1987; Holl, 1990; Hassan, 1996).

Table 14.1. Domestic animals from Holocene archaeological sites in the Acacus and surroundings

Sites	Economic Basis		Environmental Location	
	Cattle	Sheep/Goat	Mountains	Lowlands
Ti-n-Torha North I	+	-	x	
Uan Muhuggiag upper	+	++	x	
Uan Muhuggiag wadi	+	++	x	
Uan Muhuggiag lower	++	++	x	
Uan Telocat II	-	++	x	
Uan Telocat III	+	++	x	
TH 118	+	++	x	
TH 124	+	++	x	
THA 14	++	+		x
TH 103	++	+		x
TH 10	++	-		x
TH 105	+	+		x
94/15	++	+		x
94/63	++	+		x
TH 13	+	+		x
94/108	++	-		x
94/75	++	+		x
MT 13	++	-		x
MT 134b	++	-		x
MT 136	++	+		x

After Gautier and Van Neer, 1977-1982; Gautier, 1987; Corridi, 1998
- absent
+ present
++ frequent

As indicated elsewhere, the Middle Pastoral period looks like a new event for the Acacus and the surrounding area. For the first time in the region there appears to be evidence for a capillary human presence, a heavy exploitation of the environment, extensive use of resources, and planned forms of raw material procurement (di Lernia and Cremaschi, 1997; di Lernia et al., 1997; Cremaschi and di Lernia, 1998).

It could be suggested that the cultural adjustments and discontinuities evident in the archaeological record of the Middle Pastoral period in the Acacus and surroundings are clearly a result of contact, integration and replacements with the Early Pastoral groups which occurred during and immediately after the dry interval recorded here.

As is to be expected, the effects of the climate on humans are not clearly traceable, and we cannot see the region as a precisely delimited area with clear and sharp decline in population, movements, departures and arrivals. Instead they appear to indicate an intermittent reality. In this way, it is interesting to observe how economic organization documented in the area radically changed after the dry interval in respect of the previous Early Pastoral period. This change was probably conditioned by the mixture of human groups that took place in the neighboring regions, all affected by this arid spell, but with different times and modalities. Therefore, it seems useful to consider the new cultural organization of the middle pastoral groups as an integration of the former socio-cultural reality with new innovations and partial additions reflecting both inner changes and external influences.

Variability of human types, and consequent intermixing, are evident in the few human remains of the Acacus, as well as in the different modalities of inhumation and treatment of corpses. Some non-conformities may be cautiously claimed in cranial and post-cranial remains of the Acacus people, but sharp differences are mainly recorded in the trend of dental reduction beginning in the early Holocene. Particularly interesting is the case of the mummy of Uan Muhuggiag (Figure 14.6), radiocarbon dated to 5400 uncalibrated years bp (6.25 kyr cal BP), which shows different physical traits if compared to other contemporary burials, such as Imenennaden and more so at Fozzigiaren (di Lernia and Manzi, 1998).

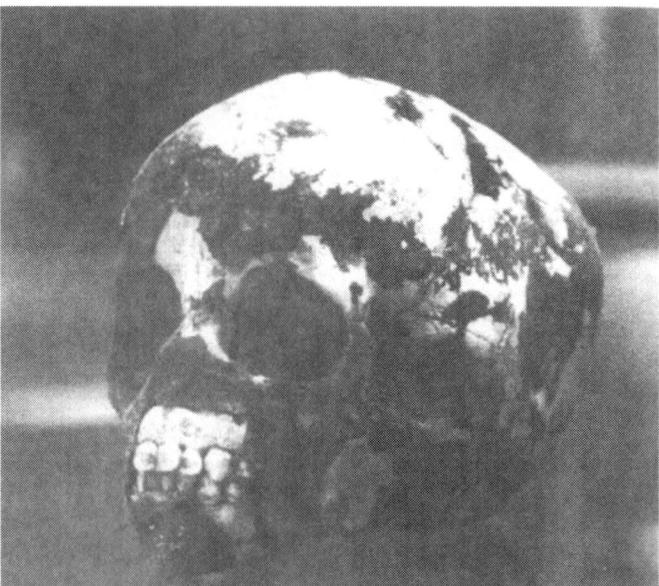

Figure 14.6. The cranium of the mummified infant from Uan Muhuggiag (UMG.H1). Courtesy of the Museum of the Great Jamahiriya (Tripoli).

Besides the differences in human types, we should outline how, even in a small sample, strong variations are present in funerary practices. Inhumation in deep burials, bodies just a few centimeters below the sand surface, single, double and multiple burials, inhumation and desiccation of corpses, with or without grave goods, and so on. We are facing a typical case of multi-faceted socio-cultural organization, probably reflecting the mixture of races and cultures following the dry period at the end of the seventh millennium bp.

Rock art also seems to confirm this scenario, with different peoples represented in the paintings and engravings in the regions, depicted in daily activities, initiation ceremonies and rituals of various kinds. Such a variety of human types was already indicated by Mori (1965), mostly on the basis of face profiles, physique and types of hairstyle. More recently, Smith has recognized the difficulties in this kind of exercise. He nevertheless notes the contemporary presence of 'Sub-Saharan' and 'Proto-Mediterranean' people in the central massifs of the Sahara during the middle Holocene (Smith, 1993, p. 475).

The dry interval has important implications also for the interregional contacts outside of central massifs, and for the diffusion of some forms of food. The hypothesis of a sort of migration southward is not new, and the spreading of cattle herding has often been related to this deterioration of environmental conditions in the Central Sahara. Cattle keepers moved southward, perhaps indicating a retreat of the savanna vegetational zone from north to south that would have followed the shift of the monsoonal belt. Therefore, it is interesting to observe that northern territories were almost ignored by Saharan pastoralists, showing some kind of cultural preference, or pointing to worse environmental conditions that were diffused northward. We can follow the movements of these cattle keepers, in Niger, at Adrar Bous (Clark et al., 1973; Smith, 1984), and then along the 'Saharan waterway' of the Tilemsi Valley (Smith, 1979), up to the Nile Valley, where some decorative motifs in the pottery clearly indicate a Saharan influence in the area of Khartoum, Central Sudan (Caneva and Marks, 1990, p. 19; Caneva, 1996).

Food security may have been guaranteed by a new organization of pastoral activity, restructuring the settlement system and, for the first time in this region, instituting a form of transhumance from the lowlands to the mountains. This kind of vertical pastoralism (sensu Khazanov, 1984; but see also Bar-Yosef and Khazanov, 1992) was based on cattle and ovicaprines: probably most of the diet was assured by plant resources, ovicaprine meat, and secondary cattle products. The shift from cattle to ovicaprines seems a clear response in searching for increased security in food procurement. Among the innovations and discontinuities present in the archaeological record should be emphasized a different mobility pattern; a diversified ceramic assemblage and new funerary and ritual practices.

It is then possible to affirm that the 6410-6082 bp (7.3-6.9 kyr cal BP) dry spell had impacts on a continental scale, and that the Saharan cattle-keepers had a relevant role in drawing specific trajectories, which shaped the subsequent cultural dynamics over an enormous territory.

THE SECOND MID-HOLOCENE CRISIS: THE DRY INTERVAL OF
5100-4800 YEARS BP (5.85-5.6 KYR CAL BP) AND THE ONSET OF
PRESENT DESERT CONDITIONS IN THE CENTRAL SAHARA

The Ultimate Landscape: A Definitive Transition?

The deterioration of climate at the end of the sixth millennium bp was even more
dramatic for humans and the environment in general than the earlier dry event.
Geological studies by Cremaschi indicate great alterations in the landscape (it
would be interesting to evaluate what kind of impact such modifications in the
past had on human perception, and their subsequent decision making), affecting
both the mountain range and dune fields. Landslides, the collapse of cave and
shelter roofs, the accumulation of aeolian sand, the erosion of surfaces and the
desiccation of lakes are only some of the more visible effects of this arid spell
(Cremaschi, 1998, this volume).

   This scenario formed only a part of a global change which surely had a
continental effect. It also affected the Nile Valley, as testified by a dramatic drop
in water discharge of the Nile, recorded in the Fayum during the fifth millennium
bp (Hassan, 1986, 1996). As Hassan recently remarked (1996, 1997), this is
probably related to the so-called Malha event, roughly radiocarbon dated at 4.5
kyr bp. Evidence for this dry spell is spread throughout northern and equatorial
Africa, from Grotte Capelletti (Vernet, 1994) and Tigalmamine C86 Core in
Morocco, at Bougdouma in the Sahel, at Kashiru, Nudurumu, Kuruyange in
Burundi, Lake Rudolf in Turkana, and Lake Malha in northern Darfur (Gasse and
van Campo, 1994; Lamb et al., 1995; Bonnefille et al., 1990; Mees et al., 1991;
see review in Hassan, 1997, 1998).

   Two concepts are of interest here: the continental relevance and the
exact time range. If the first is accepted, apparent small variations in the time
scale may be due to specific regional variations. However, a more refined time-
resolution of this event is urgently needed.

   Evidence in the Acacus, which also needs to be better dated, seems to
place the event at the very beginning of the fifth millennium bp, probably
between 5100 and 4800 years bp (5.85-5.6 kyr cal BP; Cremaschi and di Lernia,
1998; Cremaschi, 1998, this volume). In any case, these conditions seem to
present themselves rather suddenly. As late as 5500 years bp (6.35 kyr cal BP)
lakes were still at a high level. Cattle remains in the pastoral sites located along
the lakeshores in the erg Uan Kasa and Edeyen of Murzuq were also still
abundant at 5200 years bp (6.05 kyr cal BP; Corridi, 1998; Cremaschi and di
Lernia, 1998). At 5100 bp (5.8 kyr cal BP) and immediately afterward, we have
evidence for a vegetal cover of plants requiring much less water, or other true
indicators of aridity, such as Cheno-Amaranthaceae, Compositae (and particularly
Pulicaria and *Artemisia*), Caryophyllaceae such as *Paronnychia*, Boraginaceae,
such as *Echium, Calligonum, Moltkiopsis* cf. *ciliata*. It marks the definitive, and
unchanging, establishment of the present day *Acacia* and *Tamarix* community
(Mercuri et al., 1998).

Evidence of an arid crisis is also given by the perfect state of preservation of organic matter, due to the almost total absence of bacterial activity. In the archaeological deposits of the mountain range, dozens of caves are characterized by perfectly preserved dung layers, with age determinations concentrating around the interval from ca. 5000 up to 3900 uncalibrated years bp (5.7-4.3 kyr cal BP; Figure 14.7). In the lowlands outside the Acacus, shallow ponds and lakes turned into sebkhas, and it is likely that the region became inhospitable, and was progressively abandoned.

Figure 14.7. A rock shelter in the Acacus used as sheep/goat dwelling during the Late Pastoral. The dung of site TH 125 dates to 4960 ± 175 bp.

Even if this arid crisis was punctuated by a few short periods of relief, it seems plausible that the first dramatic changes immediately forced human groups to abruptly adjust their organization to face the new environmental conditions.

It is interesting to note, in fact, changes in material culture and settlement organization that appear, if possible, even more rapidly in the archaeological record. It seems a crucial phase in the history of past societies, and it is tempting to suggest a causal relationship between this crisis and the emergence of the first social stratification in the human groups of the area.

Cultural Responses and Implications: Monospecific Nomadic Pastoralism Versus Increasing Sedentism in the Palaeo-Oasis

Present day conditions, whose onset can be placed at ca. 5100 bp (5.8 kyr cal BP), have been subsequently punctuated by other minor fluctuations which surely affected human activity in the area. It seems promising to analyze the effects of its very beginning on the dynamics of the pastoral groups of the Acacus and surroundings. Droughts at 6400-6000 bp (ca. 7.3-6.9 kyr cal BP) definitely forced Saharan pastoralists to rely on other, more resistant herds for food security, and the increasing numbers of sheep and goat, alongside cattle herding, may be explained in this perspective (see also Hassan, 1997). Actually, the archaeozoological record of the Acacus and surroundings during the sixth millennium bp is characterized by a particular form of vertical pastoralism, as explained above, with cattle herding in the lowlands around the shallow lakes and ponds, and mostly goat and sheep management in the mountain. Movements of shepherds with more resistant animals toward the mountain range may be explained by the necessity to avoid over-exploitation of the pastures, while the cattle needed to remain near the lakes, even at low-levels, or the marshes, for their daily water requirements. Mountain sites, such as Uan Muhuggiag, Uan Telocat, Tadrak and others appear to be characterized by human occupation during the dry season (Cremaschi *et al.*, 1996; Trevisan Grandi *et al.*, 1998), whereas sites in the lowlands and dune fields are larger in size and more articulated, featuring the abundant presence of cattle.

Immediately after the dry spell at ca. 5000 bp (5.7 kyr cal BP), probably within a few generations, we record profound differences in several segments of these societies. Rather than describing each specific difference in material culture, settlement system and economic basis, it is preferable to discuss the general features of adaptive strategies recorded in the area immediately after this abrupt event.

Even if the environmental and climatic conditions at ca. 5000-4500 bp (ca. 5.7-5.2 kyr cal BP) were substantially homogenous over a large territory covering the Acacus Mountains, the dune fields and the fluvial valleys, the cultural answers were not. Two different cultural models appear to emerge.

The first system, in some ways already identified, concerns the Acacus Mountain and an area extending eastward, that is at least as far as the erg Uan Kasa, the Edeyen of Murzuq and probably the Messak Settafet Plateau. In this area, a socio-cultural organization, based on a monospecific, nomadic pastoralism, existed (Figure 14.8). A year round, large scale movement, similar to the present and sub-recent specialized desert pastoralists, has been detected thanks to the presence of small, specialized sites in the Acacus, mostly represented by thick sheep/goat dung layers. Other, transient camps are located along the dunes: they consist of a few fireplaces and very rare artifacts, such as undecorated pottery and some lithics. Since lakes were completely dry, these transient camps have to be interpreted as brief temporary stops during crossings of the dune fields. The location of these sites, both on dune fringes and lake bottoms, supports this model. Pasture and water, even if dramatically decreased, were probably still available over large areas, and therefore required long and repeated movements.

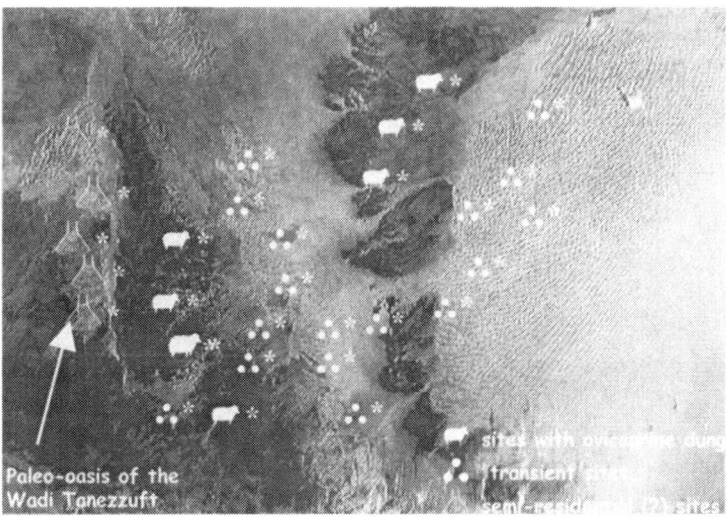

Figure 14.8. A model of the Late Pastoral settlement pattern of nomadic groups in the Acacus and surroundings, characterized by a high mobility and large-scale movements. Note the semi-residential sites located in the palaeo-oasis of the Wadi Tanezzuft.

In these sites, cattle had almost completely, if not totally, disappeared. Radiocarbon dates are consistent with this pattern, and palynological study of several samples of dung testifies to occupation in the Acacus from early winter to late spring, that is during the dry season (Trevisan Grandi *et al.*, 1998).

Therefore, we are dealing in this case with a specialized pastoralism which produced large scale movements. Interregional contact is evident in some elements of the material culture, such as Predynastic tools (Figure 14.9), exotic raw materials and new decorative motifs on the pottery (di Lernia *et al.*, 1997). Petrographic determinations show source areas hundreds of kilometers apart, such as Serir Tibesti and eventually the Nile Valley. It is still hard to decide if such contacts reflect either direct access or some form of boundary reciprocal exchange, but we catch here glimpses of a form of pastoralism perfectly adapted to arid conditions, which will be completely affirmed during the first millennium BC, and will be fully stabilized into the present form of nomadism of the pastoral groups of the Sahara (Smith, 1980). It is interesting to note how this settlement pattern and economic organization identifiable in the archaeological record are so impressively similar to the pastoralism of the Tuaregs and other groups which inhabited the Acacus Mountain and the surrounding areas at the beginning of this century (Figure 14.10).

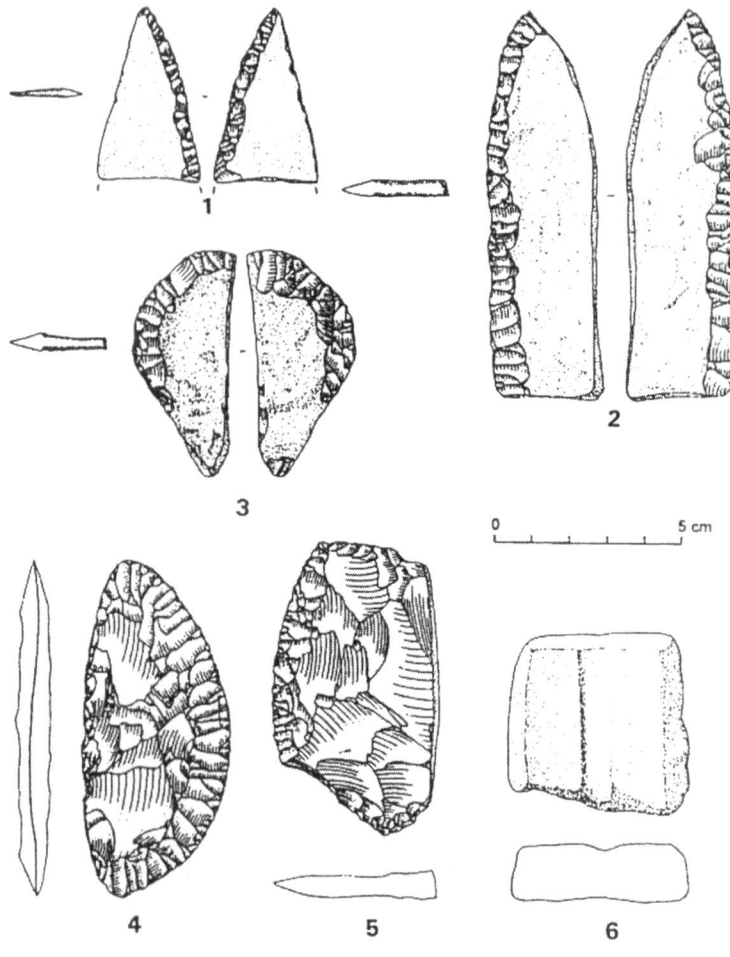

Figure 14.9. Examples of 'exotic' tools found in Late Pastoral sites.

Conversely, in the same period, or perhaps slightly later, some pastoral groups decided to abandon areas occupied for millennia and tried an alternative path: in the Wadi Tanezzuft, adjacent to the eastern scarp of the Acacus, they attempted to settle in proximity to the oases, such as in the Ghat-Tahala-Serdeles system of the Wadi Tanezzuft area, and also in the other great fluvial systems of the region (Wadi El Ajal).

These populations are characterized by an increasing site density, and a major reorganization of the settlement system. In an area of ca. 2500 square km

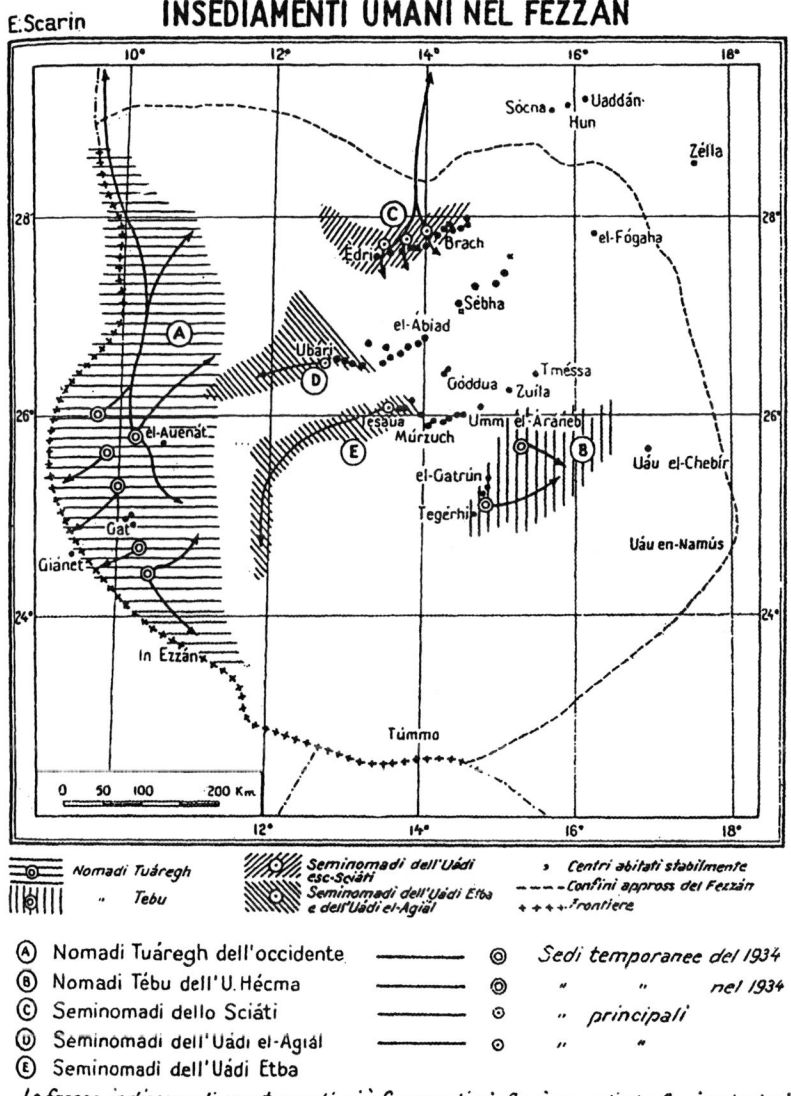

Figure 14.10. The settlement pattern of nomadic and semi-nomadic pastoral groups in the area of the Acacus Mountains at the beginning of the century (after Scarin, 1937).

more than 300 sites have been found during the recent survey performed by the Joint Mission (di Lernia *et al.*, 2000), and such a high site density was not (and never will be) reached in the Acacus region (Figure 14.11). Among the sharp differences observed are the strong increase in the numbers of plant-related tools, possibly cultivation-related tools, such as gouges and hoes, which may signal a

AREA 5

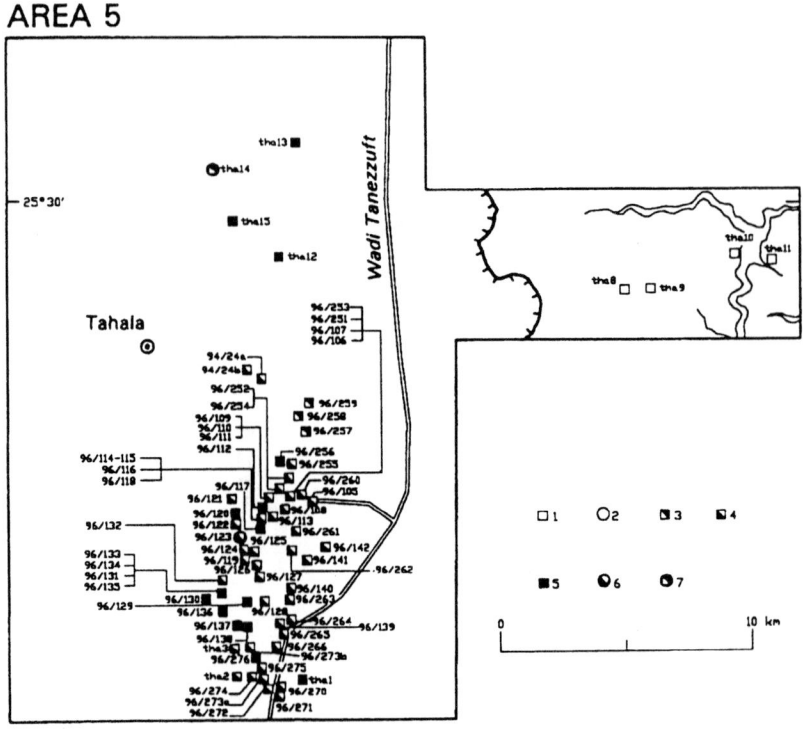

Figure 14.11. The semi-sedentary groups of the Late Pastoral in the Wadi Tanezzuft Valley. Sites feature a high density and are concentrated near the palaeo-oasis of Tahala.

first step toward a systematic and sedentary system of land exploitation. It is tempting to suggest, with such a high site density, some form of internal stress related to the emergence of some type of land rights with, among the consequences, the emergence of social stratification.

In this area domesticated cattle seem to be still present in a few cases, testifying to good water availability, as indicated by the existence of some fluvial activity at 3900 bp (ca. 4.3 kyr cal BP, Cremaschi and di Lernia, 1996; Cremaschi, 1998). However, economic emphasis seems to be on ovicaprine herding and extensive plant exploitation. Needless to say, the increased ratio of sheep to goat surely had heavy consequences on a changing landscape, where land is restricted and pasture decreasing. Therefore, over-exploitation of the environment may be another co-causal factor of landscape deterioration.

Work is still in progress (di Lernia *et al.*, 2000), but some competition and conflict might have happened in the region, leading to the birth of some form of social stratification. The most important indication of this is the emergence of megalithic architecture (Figure 14.12). Some of these structures were excavated

Figure 14.12. Circular tumuli in the eastern slopes of the Acacus Mountain, facing the erg Uan Kasa. Megalithic monuments are diffused in the neighboring regions of the Acacus and impressive concentrations dated from 5000 to 3000 years bp are located in the Wadi Tanezzuft and in the Messak Settafet.

or tested (di Lernia and Cremaschi, 1997; Cremaschi and di Lernia, 1998), and they seem, so far, to invariably belong to the very end of the sixth and fifth millennia bp. We have indications of stone tumuli, megalith alignments, and other more complex megalithic structures, much more recent, known in the region as 'pre-Islamic' monuments (see, for example Barich, 1974). Older stone tumuli in our area appear to be single burials demonstrating a particular effort in the building of the monument, implying high social organization and probably the presence of important individuals within human groups. Worthy of note is the presence at In Habeter III of a votive stele with the ritual deposition of two vessels and the post-cranial part of domesticated cattle, radiocarbon dated to 5200 years bp (5.9 kyr cal BP, Cremaschi and di Lernia, 1998). This find could be another step toward the discovery of the existence of a 'cattle cult' (probably linked to the progressive climatic deterioration which forced cattle to move out of the area), of which we have important evidence in the Eastern Sahara (Wendorf *et al.*, 1997, but see also Hassan, 1994, there quoted). It might improve our understanding of rock art images, particularly engravings, which cover almost all rock walls of the nearby plateaus of the Messak Settafet, and where evidence of important individuals seems recurrent. Finally, I think the evidence of emergence of fighting represented in the late paintings of the Acacus should not be overlooked. If this late pastoral style is really datable to approximately after 5000 bp (ca. 5.8 kyr cal BP), as indicated by Mori (1965), then evidence of increasing stress and competition may have led, for the first time in the Acacus, to the emergence of warfare among groups as a final, dramatic effect.

SHORT, OR ABRUPT, DRY INTERVALS AND CULTURAL DYNAMICS: MANY MOVERS, NO MOVERS

The search for prime movers in the study of change in past societies was a leading approach until the 60s, after which processual and systemic approaches prevailed, pointing out the importance of multi-factor causes in shaping change among human societies. Archaeological and palaeonvironmental studies in the Central Sahara over the last decades show a particular blend of geographic determinism and processual theory, unexpectedly forming an original 'Africanist' school of thought which only partially shares the effects of post-processual influences. Probably, the particularly extreme environmental conditions of these regions forced scholars to find an equilibrium, both theoretical and operational, to bridge the gap between the archaeological data and the interpretation of the cultural dynamics of the past, with the climatic factor as a fundamental background to the debate.

Needless to say, changes in human societies are due to both internal and external causes, but North African, and particularly Saharan, regions indicate an unusual situation, where high stresses over short periods are likely to be considered important factors of cultural adjustments, rather than low stresses over longer periods. In the latter case, a classic cause of a strong break in the system may be due to demographic growth. As far as the former is concerned, we have in the Acacus evidence for sharp, rapid changes, with the intermixing of human groups, social reorganization and the formulation of new social realities. Among mid-Holocene groups, food requirement and its security was a major factor in shaping cultural trajectories. Changes in cultural dynamics were related to the reorganization of food procurement, and it is interesting to observe how greater changes mostly affected the organization of pastoral activity. The intensive use of plants, which were not domestic until historical periods, is another important factor. The relationships between herding and forms of cultivation are still to be fully explored, but surely they affected the organization of human groups, at least so far as settlement patterns and site mobility were concerned. Interrelations between pastoral societies and 'agricultural' societies are fundamental in defining what kind of pastoralism is practiced (see Khazanov, 1984). It does not seem to be chance that the heavy transformation of the last two factors seems to be recorded in the Wadi Tanezzuft, where the first forms of cultivation may be seen.

There are two important lessons to be learnt from the Libyan Sahara: abrupt dry periods forced humans to reorganize food security and related arrangements, to keep pace with the climatic effects which were taking place. After the abrupt changes, cultural trajectories seem to be directed toward stabilization. After the dry interval of 6.4-6.1 uncalibrated kyr bp (7.3-6.9 kyr cal BP), we have evidence for an immediate integration of human types, leading to a replacement of site organization with the institution of a vertical form of pastoralism, seasonally organized: mountain sites occupied in the dry seasons, and lacustrine areas probably inhabited for much longer periods. This organization, formed after the above mentioned dry interval, did not change for almost a millennium. Even other, minor, fluctuations did not affect human organization as much as the dry interval did, and we do not find traces of these environmental changes in the archaeological record. Only one other abrupt, dry event, at 5000 bp (ca. 5.7 kyr cal BP), forced human groups to radically change again.

The second lesson is that human answers may differ under similar environmental conditions in a restricted area, as in the Acacus and surroundings, which had two different answers from human groups at 5000 bp (ca. 5.7 kyr cal BP). Mobility is probably the best archaeological indicator of these different cultural dynamics, which are differentiated with regard to settlement pattern, intra-site organization, economic basis, and material culture. If these two pastoral systems have to be considered culturally different, it is not possible to define what kind of relationship existed between nomadic pastoralists and sedentary groups in the palaeo-oases. They may have originated from the same socio-cultural entity, that is from the Middle Pastoral groups. But they also could reflect some type of arrival or integration from other regions of the southern Sahara, and possibly even the Eastern Sahara and the Nile Valley (see an analogous question in Paris, 1997). It is tempting to suggest a radical shift in this period toward ethnic differentiation. The fragmentation recorded in historical sources (e.g., Herodotus and others) may have its origins in the river valleys of the Central Sahara.

REFERENCES

Alimen, M.H. (1987). Evolution du climat et des civilisations depuis 40000 ans du Nord au Sud du Sahara occidental (Prémières conceptiones confrontées aux donnees recentes). *Bulletin de l'Association Français pour l'Étude Quaternaire* 4: 215-227.

Barich, B.E. (1974). La serie stratigrafica dell'Uadi Ti-n-Torha (Acacus, Libia). *Origini* VIII: 7-157.

Barich, B.E. (1987a). Uan Muhuggiag and the pastoralism in central Sahara. In Barich, B.E. (ed.), *Archaeology and Environment in the Libyan Sahara. The Excavations in the Tadrart Acacus, 1978-1983*, British Archaeological Reports International Series 368, Oxford, pp. 255-266.

Barich, B.E. (ed.), (1987b). *Archaeology and Environment in the Libyan Sahara. The excavations in the Tadrart Acacus, 1978-1983*, British Archaeological Reports International Series 368, Oxford.

Bar-Yosef, O., and Khazanov, A. (eds.) (1992). *Pastoralism in the Levant. Archaeological Materials in Anthropological Perspectives*, Monographs in World Archaeology 10, Prehistory Press, Madison.

Bonnefille, R., Roeland, J.C., and Guiot, J. (1990). Temperature and rainfall estimates for the past 40,000 years in equatorial Africa. *Palaeogeography, Palaeoclimatology, Palaeoecology* 109: 331-343.

Bryson, R.A., and Bryson, R.U. (1997). Macrophysical climatic modeling of Africa's late quaternary climate: Site-specific, high-resolution applications for archaeology. *African Archaeological Review* 14: 143-160.

Caneva, I. (1987). Pottery decoration in prehistoric Sahara and Upper Nile: a new perspective. In Barich, B.E. (ed.), *Archaeology and Environment in the Libyan Sahara. The Excavations in the Tadrart Acacus, 1978-1983*, British Archaeological Reports International Series 368, Oxford, pp. 231-254.

Caneva, I. (1996). The influence of Saharan prehistoric cultures on the Nile Valley. In Aumassip, G., Clark, J.D., and Mori, F. (eds.), *The Prehistory of Africa*, XIII Congress UISPP Vol. 15, ABACO, Forlì, pp. 231-239.

Caneva, I., and Marks, A. (1990). More on the Shaqadud pottery: evidence for Saharo-Nilotic connections during the 6th-4th Millennium B.C. *Archéologie du Nil Moyen* 4: 11-35

Clark, J.D., Williams, M.A.J., and Smith, A.B. (1973). The geomorphology and archaeology of Adrar Bous, central Sahara: a preliminary report. *Quaternaria* 17: 245-297.

Close, A.E., and Wendorf, F. (1992). The beginning of food production in the Eastern Sahara. In Gebauer, A.B., and Price T.D. (eds.), *Transitions to Agriculture in Prehistory*, Monographs in World Archaeology 4, Prehistory Press, Madison, pp. 63-72.

Corridi, C. (1998). Faunal remains from Holocene archaeological sites of the Tadrart Acacus and surroundings (Libyan Sahara). In Cremaschi, M., and di Lernia, S. (eds.), *Wadi Teshuinat. Palaeoenvironment and Prehistory in South-western Fezzan (Libyan Sahara)*, Quaderni di Geodinamica Alpina e Quaternaria 7, C.N.R, Milan, pp. 89-94.

Cremaschi, M. (1998). Late Quaternary geological evidence for environmental changes in Western Fezzan (Libyan Sahara). In Cremaschi, M., and di Lernia, S. (eds.), *Wadi Teshuinat. Palaeoenvironment and Prehistory in South-western Fezzan (Libyan Sahara)*, Quaderni di Geodinamica Alpina e Quaternaria 7, C.N.R, Milan, pp. 13-48.

Cremaschi, M., and di Lernia, S. (1996). Current research on the prehistory of the Tadrart Acacus (Libyan Sahara). Survey and excavations 1991-1995. *Nyame Akuma* 45: 50-59.

Cremaschi, M., and di Lernia, S. (1998). The geoarchaeological survey in central Tadrart Acacus and surroundings (Libyan Sahara). Environment and cultures. In Cremaschi, M., and di Lernia, S. (eds.), *Wadi Teshuinat. Palaeoenvironment and Prehistory in South-western Fezzan (Libyan Sahara)*, Quaderni di Geodinamica Alpina e Quaternaria 7, C.N.R, Milan, pp. 245-298.

Cremaschi, M., di Lernia, S., and Trombino, L. (1996). From taming to pastoralism in a drying environment. Site formation processes in the shelters of the Tadrart Acacus massif (Libya, central Sahara). In Castelletti, L., and Cremaschi, M. (eds.), *Micromorphology of Deposits of Anthropogenic Origin*. XIII International Congress of the UISPP, Volume 3, Paleoecology, Colloquium VI, Forlì 1995, ABACO Edizioni, Forlì, pp. 87-116.

di Lernia, S. (1996). Changing adaptive strategies: a long-term process in the central Saharan massifs from Late Pleistocene to Early Holocene. The Tadrart Acacus perspective (Libyan Sahara). In Aumassip, G., Clark, J. D., and Mori, F. (eds.), *The Prehistory of Africa*, XIII Congress UISPP, Colloquium XXX, Vol. 15, ABACO, Forlì, pp. 195-208.

di Lernia, S. (1997). *Condizioni Culturali e Forme di Adattamento Prima della Produzione del Cibo: i Massicci Centrali Sahariani Nell'antico Olocene (10000-7500 bp)*, Tesi di Dottorato, Università "La Sapienza" di Roma, Rome. [Unpublished]

di Lernia, S. (1998). Cultural control over wild animals during the early Holocene: the case of Barbary sheep in central Sahara. In di Lernia, S., and Manzi, G. (eds.), *Before Food Production in North Africa. Proceedings of the Homonymous Workshop held in Forli, September 1996, within the XIII World Congress of the International Union of the Prehistoric and Protohistoric Sciences*, ABACO Edizioni, Rome, pp. 113-126.

di Lernia, S. (1999a). Alle origini del pastoralismo africano: riflessioni su alcune forme di gestione animale nell'antico Olocene. *La Ricerca Folclorica* 40: 13-24.

di Lernia, S. (1999b). Discussing pastoralism. The case of the Acacus and surroundings (Libyan Sahara). *Sahara* 11: 7-20.

di Lernia, S., and Cremaschi, M. (1997). Processing quartzite in Central Sahara: a case-study from In Habeter IIIa - Wadi Mathendusc (Messak Settafet, Libya). In Schild, R., and Sulgostowska, Z. (eds.), *Man and Flint*, Institute of Archaeology and Ethnology, Polish Academy of Sciences, Warsaw, pp. 225-232.

di Lernia, S., and Manzi, G. (1998). Funerary practices and anthropological features at 8000-5000 BP. Some evidence from central-southern Acacus (Libyan Sahara). In Cremaschi, M., and di Lernia, S. (eds.), *Wadi Teshuinat. Palaeoenvironment and Prehistory in South-western Fezzan (Libyan Sahara)*, Quaderni di Geodinamica Alpina e Quaternaria 7, C.N.R, Milan, pp. 219-244.

di Lernia, S., Cremaschi, M., and Notarpietro, A. (1997). Procurement, exploitation and circulation of raw material: analysis of the early and middle Holocene lithic complexes from south-western Libya (Tadrart Acacus and Messak Settafet). In Schild, R., and Sulgostowska, Z. (eds.), *Man and Flint*, Institute of Archaeology and Ethnology, Polish Academy of Sciences, Warsaw, pp. 233-242.

di Lernia, S., Cremaschi, M., Castelli, R., Grassi, G., Merighi, F., and Trombino, L. (2000). Environmental changes and settlement systems in the mid-Holocene palaeo-oasis of the wadi Tanezzuft (Libyan Sahara). Preliminary results of an intensive survey. *Abstracts of the SafA Biennial Conference, Cambridge, 12-15th July*, 2000.

Garcea, E.A.A. (1993). *Cultural Dynamics in the Saharo-Sudanese Prehistory*, GEI, Rome.

Gasse, F., and van Campo, E. (1994). Abrupt post-glacial climate events in West Asia and African monsoon domains. *Earth and Planetary Science Letters* **1256**: 435-456.

Gautier, A. (1987). The archaeozoological sequence in the Acacus. In Barich, B.E. (ed.), *Archaeology and Environment in the Libyan Sahara. The Excavations in the Tadrart Acacus, 1978-1983*, British Archaeological Reports International Series 368, Oxford, pp. 283-312.

Gautier, A. (1993). Mammifères holocènes du Sahara d'après l'art rupestre et l'archéozoologie. In Calegari, G. (ed.), *L'arte e l'Ambiente Preistorico: Dati e Interpretazioni*, Memorie del Società Italiana delle Scienze Naturali e Museo Civico di Storia Naturali di Verona, Milan, XXVI.II, pp. 261-268.

Gautier, A., and Van Neer, W. (1977-1982). Prehistoric fauna from Ti-n-Torha (Tadrart Acacus, Libya). *Origini* **XI**: 87-127.

Girod, A. (1998). Molluscs and Palaeoenvironment of Holocene lacustrine deposits in the erg Uan Kasa and in the edeyen of Murzuq (Libyan Sahara). In Cremaschi, M., and di Lernia, S. (eds.), *Wadi Teshuinat. Palaeoenvironment and Prehistory in South-western Fezzan (Libyan Sahara)*, Quaderni di Geodinamica Alpina e Quaternaria 7, C.N.R, Milan, pp. 73-88.

Hassan, F. (1986). Holocene lakes and prehistoric settlements of the Western Faiyum. *Journal of Archaeological Science* **13**: 483-501.

Hassan, F. (1996). Abrupt Holocene climatic events in Africa. In Pwiti, G., and Soper, R. (eds.), *Aspects of African Archaeology. Papers from the 10th Congress Pan-African Association for Prehistory and Related Studies*, University of Zimbabwe Publications, Harare, pp. 83-89.

Hassan, F. (1997). Holocene Palaeoclimates of Africa. *African Archaeological Review* **14**(4): 213-230.

Hassan, F. (1998). Holocene climatic change and riverine dynamics. In di Lernia, S., and Manzi, G. (eds.), *Before Food Production in North Africa. Proceedings of the Homonymous Workshop held in Forli, September 1996, within the XIII World Congress of the International Union of the Prehistoric and Protohistoric Sciences*, ABACO Edizioni, Rome, pp. 43-51.

Hole, F. (1978). Pastoral nomadism in Western Iran. In Gould, R.A. (ed.), *Explorations in Ethnoarchaeology*, Cambridge University Press, Cambridge, pp. 126-167.

Holl, A. (1990). Les formes du pastoralisme au Sahara néolithique (9000-3000 B.P.). In Francfort, H.-P. (ed.), *Nomades et Sèdentaries en Asia Centrale*, Editions du CNRS, Paris, pp. 141-155.

Khazanov, A.M. (1984). *Nomads and the Outside World*, Cambridge University Press, Cambridge.

Lamb, H.F., Gasse, F., Benkaddour, A., el-Hamouti, N., van der Kaars, S., Perkins, W.T., Pearce, N.J., and Roberts, C.N. (1995). Relation between century-scale Holocene arid intervals in tropical and temperate zones. *Nature* **373**: 134-137.

Maley, J. (1981). Etudes palynologiques dans le bassin du Tchad et paléoclimatologie de l'Afrique nord-tropicale de 30000 and à l'époque actuelle. *Palaeoecology of Africa* **13**: 45-52.

Mees, F., Verschuren, D., Nijs, R., and Dumont, H.J. (1991). Holocene evolution of the crater lake at the Malha, Northwest Sudan. *Journal of Palaeolimnology* **5**: 227-253.

Mercuri, A.M., Trevisan Grandi, G., Mariotti Lippi, M., and Cremaschi, M. (1998). New pollen data from the Uan Muhuggiag rockshelter (Libyan Sahara). In Cremaschi, M., and di Lernia, S. (eds.), *Wadi Teshuinat. Palaeoenvironment and Prehistory in South-western Fezzan (Libyan Sahara)*, Quaderni di Geodinamica Alpina e Quaternaria 7, C.N.R, Milan, pp. 107-122.

Mori, F. (1961). Aspetti di cronologia sahariana alla luce dei ritrovamenti della V Missione Paletnologica nell'Acacus (1960-1961). *Ricerca Scientifica* **31**: 204-215.

Mori, F. (1965). *Tadrart Acacus. Arte Rupestre e Culture del Sahara Preistorico*, Einaudi, Turin.

Neumann, K. (1991). In search of the green Sahara: palinology and botanical macro-remains. *Palaeoecology of Africa* **22**: 203-212.

Pachur, H.J., and Braun, G. (1980). The palaeoclimate of the central Sahara, Libya and the Libyan Desert. *Palaeoecology of Africa* **12**: 351-363.

Paris, F. (1997). Burials and the peopling of the Adrar Bous region. In Barich, B.E., and Gatto, M.C. (eds.), *Dynamics of Populations, Movements and Responses to Climatic Change in Africa*, Bonsignori, Rome, pp. 49-61.

Pasa, A., and Pasa Durante, M.V. (1962). Analisi paleoclimatiche nel deposito di Uan Muhuggiag, nel massiccio dell'Acacus (Fezzan Meridionale). *Memorie del Museo Civico di Storia Naturali di Verona* **X**: 251-255.

Petit-Maire, N., Celles, J.C., Commelin, D., and Raimbault, M. (1983). The Sahara in northern Mali:
    man and his environment between 10000 and 3500 years bp (Preliminary results). *African
    Archaeological Review* 1: 105-125.
Ritchie, J. (1987). A Holocene pollen record from Bir Atrun, Northwestern Sudan. *Pollen et Spores*
    29: 391-410.
Scarin, E. (1937). Insediamenti e tipi di dimore. In *Fezzán e Oasi di Gat*, Reale Società Geograifca,
    Rome, pp. 515-560.
Schulz, E. (1987). Holocene vegetation in the Tadrart Acacus: the pollen record of two early ceramic
    sites. In Barich, B.E. (ed.), *Archaeology and Environment in the Libyan Sahara. The
    excavations in the Tadrart Acacus 1978-1983*, British Archaeological Reports International
    Series 368, Oxford, pp. 313-326.
Smith, A.B. (1979). Biogeographical considerations of the colonization in the lower Tilemsi valley in
    the second millennium BC. *Journal of Arid Environments* 2: 355-361.
Smith, A.B. (1984). Origins of the Neolithic in the Sahara. In Clark, J.D., and Brandt, S.A. (eds.),
    From Hunters to Farmers: The Causes and Consequences of Food Production in Africa,
    University of California Press, Berkeley, pp. 84-92.
Smith, A.B. (1993). New approaches to Saharan rock art. In Calegari, G. (ed.), *L'Arte e l'Ambiente
    Preistorico: Dati e Interpretazioni*, Memorie del Società Italiana delle Scienze Naturali e
    Museo Civico di Storia Naturali, Milano XXVI. II, pp. 467-478.
Smith, S. (1980). The environmental adaptation of nomads in the West African Sahel: a key to
    understanding prehistoric pastoralists. In Williams, M.A.J., and Faure, H. (eds.), *The
    Sahara and the Nile*, Balkema, Rotterdam, pp. 454-487.
Stuiver, M., Long, A., and Kra, K. R. S. (1993). Calibration Issue. *Radiocarbon* 35(1).
Trevisan Grandi, G., Mariotti Lippi, M., and Mercuri, A.M. (1998). Pollen in dung layers from rock-
    shelters and caves of wadi Teshuinat (Libyan Sahara). In Cremaschi, M., and di Lernia, S.
    (eds.), *Wadi Teshuinat. Palaeoenvironment and Prehistory in South-western Fezzan
    (Libyan Sahara)*, Quaderni di Geodinamica Alpina e Quaternaria 7, C.N.R, Milan, pp. 95-
    106.
Vernet, R. (1994). Les Paléoenvironnements du Nord de l'Afrique Depuis 600.000 Ans. *Dossiers et
    Recerches sur l'Afrique* 3, CNRS, Muedon.
Wendorf, F., and Schild, R. (ass.), Close, A. E. (ed.) (1984). *Cattle Keepers of the Eastern Sahara: the
    Neolithic of Bir Kisieba*, Southern Methodist University Press, Dallas.
Wendorf, F., Schild, R., Applegate, A., and Gautier, A. (1997). Tumuli, cattle burials and society in
    the Eastern Sahara. In Barich, B.E., and Gatto, M.C. (eds.), *Dynamics of Populations,
    Movements and Responses to Climatic Change in Africa*, Bonsignori, Rome, pp. 90-104.

15

FOOD SECURITY IN WESTERN AND CENTRAL AFRICA DURING THE LATE HOLOCENE: THE ROLE OF DOMESTIC STOCK KEEPING, HUNTING AND FISHING

W. Van Neer

INTRODUCTION

The first critical review of early cattle finds in northern and western Africa was made by Gautier (1987), who demonstrated that the number of sites with well preserved, securely identified, and well dated remains is low. An update of the North African data has been given by Garcea (1993), Chenal-Vélardé (1997) and Hassan (2000, this volume). West African data on early cattle are also dealt with in an overview by MacDonald and MacDonald (2000), whereas the Central African early finds for domestic cattle and ovicaprines have been reviewed by Van Neer (2000). The dispersal in the Sahara still needs to be refined, but there is a trend showing that the dates for cattle become gradually younger as one moves from the Eastern Sahara westward. In addition, a north-south diachronic trend can be observed. Despite the low number of data points, several models of the early spread of cattle have been published (e.g., Shaw, 1981; Roche, 1991; Krzyzaniak, 1992; Breunig and Neumann, 1996; Hassan, 2000, this volume). The poor number of finds and the lack of sufficient chronological resolution hamper the establishment of a precise relation between the colonization of new areas by pastoralists and short term climatic events. It is likely, however, that the introduction of cattle into the Central Sahara (Tibesti, Acacus and Aïr) during the seventh and sixth millennia bp results from possible population movements and long-range transhumance following pressure that arid spells exerted on food security (Hassan, 1996). New data from Uan Muhuggiag and Uan Telocat indicate also that small livestock were already present in the Acacus region by the beginning of the seventh millennium bp (Corridi, 1997). The southward migration of pastoralists from the Sahara into the West African Sahel is linked to the late Holocene arid phase which apparently started in the north, at latitude 21°20', at ca. 5000 bp and shifted southward thereafter (Hassan, 1996, p. 85). The spread of pastoralism in the southern part of the Sahel and the northern savannas from the fourth millennium bp onwards has been poorly documented thus far, but new data have become available recently that will be reviewed below. This will be followed by a summary of the archaeozoological data for the forested areas of West and Central Africa and the savannas just south of the equatorial forest. Finally, the degree of reliance on domestic livestock, hunting and fishing will be considered for the different time periods and regions, wherever quantitative data allow this. The West and Central African sites discussed below are indicated on Figure 15.1. All the dates mentioned in the text are approximate uncalibrated radiocarbon years: Table 15.1 includes the original, uncalibrated radiocarbon dates and the calibration into years BC/AD (minimum and maximum dates 1 sigma, calculated with the OxCal v3.0 program).

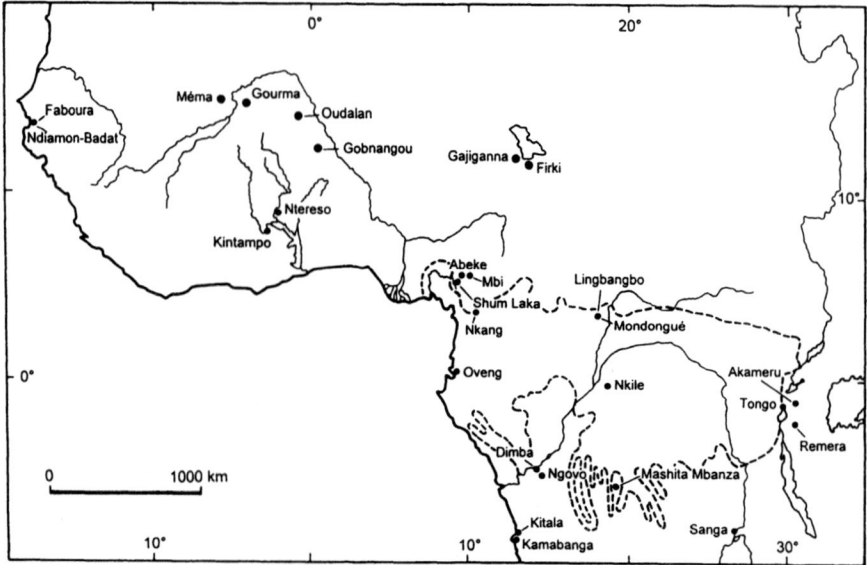

Figure 15.1. Map of the major localities mentioned for West and Central Africa. The limit of the equatorial forest is indicated.

THE ARRIVAL AND SPREAD OF PASTORALISM IN THE PRESENT-DAY SAHEL AND NORTHERN SAVANNA

Clear north-south diachronic trends illustrating the southward migration of pastoralists from the Sahara towards the Sahel can be seen in northern Niger and in Mali. In the Adrar Bous area, located at the eastern border of the Aïr Mountains, cattle skeletons have been dated between 6350-6200 bp (Paris, 1996, 1997) and 5800-4900 bp (Carter and Clark, 1976). Southwest of Adrar Bous, in the Talak-Timersoï region, cattle inhumations were dated to ca. 4000 bp at Arlit-Somaïr 4 and to ca. 5400 bp at Arlit (Paris, 1997). Still farther southwest, in the Eghazer Basin, cattle inhumations date between ca. 3900 and ca. 3300 bp at Chin

Tafidet, whereas at In Tuduf dates range from about 3700 to 3400 bp (Paris, 1997). A southward movement of Saharan pastoralists is also observed along the Azaouak and Tilemsi basins. Cattle dated to 4100 bp were found at Taferjit, in the upper reaches of the Azouak, and comparable sites further south along the basin have also yielded cattle (Joleaud, 1936; Gautier, 1987). Around 4000 bp, cattle and small livestock appear in the Tilemsi Valley at Karkarichinkat and at Menaka (Smith 1980a,b; Gautier, 1987). Data on animals other than cattle are rare for the aforementioned sites, either because they represent inhumations, or because the archaeozoological data from the living sites are restricted to a list of species with, sometimes, vague indications about the proportions in which they occur. Wherever data are available, however, it is clear that hunting and fishing were practiced.

New archaeozoological data from Mali include finds from the Méma and Gourma regions published by MacDonald (1994). Previous research at Kobadi had yielded cattle dated between ca. 3300 and 2900 bp (Raimbault et al., 1987). Both cattle and ovicaprines were found at the nearby site of Kolima Sud, from which a cattle tooth was AMS dated at ca. 3100 bp (MacDonald and Van Neer, 1994). A comparative study of the ceramics (MacDonald, 1996a) suggests that the appearance of livestock at the site reflects an influx of people from the Mauritanian Sahel (Tichitt-Oualata). This movement resulted from a desertification in the home area and was facilitated by a retreat of inundated areas in the Méma region. The recent archaeological exploration of the Gourma region, in the eastern Inner Niger Delta, allowed the distinction of three facies (MacDonald, 1996b). The Gourma facies has affinities with Kobadi, but yielded only wild animals. The Windé Koroji facies, dated between ca. 3600 and 3100 bp, as well as the Zampia facies, dated to 2600 bp, yielded both small and large livestock. At all the aforementioned sites conditions were more humid than today, as indicated by various mammal species presently found in more southerly regions. In addition, there is an abundant aquatic fauna in most instances, with species typical of permanent and well oxygenated waters (e.g., Nile perch Lates niloticus) no longer found in the region.

In Burkina Faso, new fieldwork has been carried out in the northeast, in the Oudalan Province, and in the southeast, in the Chaîne de Gobnangou (Breunig and Wotzka, 1991; Ballouche et al., 1993; Breunig and Neumann, 1996; Neumann and Vogelsang, 1996; Frank et al., in press). The landscape in the Oudalan province comprises east-west oriented fossil dunes several kilometers wide. Today the depressions between the dunes are filled with water during the rainy season and attract pastoralists. Comparable land use may have been practiced in the past as well. Despite extensive archaeological survey, the region seems to have been uninhabited prior to 4000 bp (Breunig and Neumann, 1999, this volume). A similar conclusion was reached for the Gourma region (MacDonald, 1996b). Faunal preservation is poor in the dune sites of Ti-n-Akof and remains usually consist only of tooth fragments. At the pottery site BF 94/133 at Ti-n-Akof, large bovid teeth and unidentifiable long bone fragments were found in levels dated to 3500 bp. During a first inspection of the faunal remains the teeth were provisionally attributed to domestic cattle. However, subsequent detailed comparison with reference material showed that the initial identification,

Table 15.1. Radiocarbon dates mentioned in the text

Site	Lab Code	Material	Reference	^{14}C date yr bp	range 1 sigma cal BC
Senegal					
Faboura	Mc-2073	shell	Descamps et al., 1977	1315 ± 80	AD 630-790
Faboura	Mc-1390	Anadara shell	Descamps et al., 1977	1940 ± 80	60 BC-AD 140
Ndiamon-Badat	Pa-001	Anadara shell	Ba et al., 1997	1555 ± 80	AD 420-600
Ndiamon-Badat	Beta-85886	Anadara shell	Ba et al., 1997	2370 ± 50	750-390
Mali					
Karkarichinkat south	N-1395	charcoal	Smith, 1974	4010 ± 160	2900-2300
Karkarichinkat north	N-1396	charcoal	Smith, 1974	4000 ± 90	2870-2390
Kobadi	Pa-221	burned hippo bone	Raimbault, 1986	3335 ± 100	1750-1510
Kobadi	Pa-222	burned animal bones	Raimbault, 1986	2880 ± 120	1260-920
Kolima Sud 1, level 10	GX-19814	cattle tooth	MacDonald and Van Neer, 1994	3084 ± 73	1440-1260
Windé Koroji-1 level 17	GX-19990	charcoal	MacDonald, 1996b	3635 ± 90	2140-1890
Windé Koroji-1 level 7	GX-19234	charcoal	MacDonald, 1996b	3115 ± 195	1650-1100
Zampia tumulus 7	GX-19232	charcoal	MacDonald, 1996b	2607 ± 86	900-550
Niger					
Chin Tafidet	Pa-292	cattle bone	Paris, 1997	3325 ± 260	2050-1300
Chin Tafidet	Pa-1054	human bone	Paris, 1997	3910 ± 150	2610-2140
In Tuduf 1	Pa-623	charcoal	Paris, 1997	3415 ± 200	2050-1500
In Tuduf 3	Pa-1048	human bone	Paris, 1997	3740 ± 200	2500-1900

cont...

Site	Lab number	Material	Reference	Date	Calibrated range
Arlit-Somair 4	Gif-1798	charcoal	Paris, 1997	4030 ± 110	2870-2460
Arlit	Gif-3057	charcoal	Paris, 1997	5380 ± 130	4350-4040
Adrar Bous	UCLA-1658	cattle bone	Carter and Clark, 1976	5760 ± 500	5200-4200
Adrar Bous	N-870	charcoal with bone	Smith, 1976	4910 ± 140	3940-3520
Adrar Bous, site 1	Pa-330	cattle bone	Roset, 1987	6350 ± 300	4750-4100
Adrar Bous, site 1	Pa-753	cattle bone	Paris, 1997	6200 ± 250	4550-4000
Burkina-Faso					
Ti-n-Akof (BF 94/133)	KN-4777	charcoal	Breunig and Neumann, this volume	3479 ± 45	1880-1750
Ti-n-Akof (BF 94/133)	UtC-6466	charcoal	Breunig and Neumann, this volume	3413 ± 37	1870-1840
Ti-n-Akof (BF 94/133)	KN-4776	charcoal	Breunig and Neumann, this volume	3380 ± 100	1880-1520
Ti-n-Akof (BF 94/133)	UtC-4906	seeds	Breunig and Neumann, this volume	2840 ± 49	1070-920
Pénténga (BF89/1), spit 4	KI-3350	charcoal	Breunig and Wotzka, 1991	890 ± 38	1040-1100
Pénténga (BF89/1), spit 17	KI-3351	charcoal	Breunig and Wotzka, 1991	7590 ± 90	5740-5540
Maadaga, layer 1, spit 6	KN-4299	charcoal	Wotzka, personal communication	2570 ± 120	840-520
Maadaga, layer 1, spit 3	UtC-2318	charcoal	Wotzka, personal communication	1700 ± 40	AD 250-400
Maadaga, layer 2, spit 20	UtC-1994	charcoal	Wotzka, personal communication	4250 ± 300	3350-2450
Maadaga, layer 2, spit 20	UtC-1995	charcoal	Wotzka, personal communication	1640 ± 50	AD 260-460
Ghana					
Ntereso	SR-52	charred wood	Davies, 1980	3580 ± 130	2140-1750
Ntereso	SR-6:	organic fill	Davies, 1980	3190 ± 120	1640-1310
Kintampo K6, layer 3	UCR-1692	palm nuts	Stahl, 1985	3550 ± 127	2130-1740
Nigeria					
Gajiganna (site 93/42)[a]	UtC-5515	charcoal	Breunig, personal communication	3690 ± 120	2290-1920

cont....

Gajigganna (site 93/42)[b]	UtC-5297	charcoal	3489 ± 34	1890-1760	Breunig, personal communication
Tuba Ajuz III[c]	UtC-6783	charcoal	3059 ± 50	1410-1260	Breunig, personal communication
Gajigganna B[d]	UtC-2330	charcoal	3150 ± 70	1520-1320	Breunig, personal communication
Gajigganna B II[e]	UtC-2796	charcoal	2730 ± 50	920-825	Breunig, personal communication
Gajigganna (site 93/10)[f]	UtC-3513	charcoal	2470 ± 70	770-510	Breunig, personal communication
Daima	Ul-6711	charcoal	2520 ± 110	810-520	Connah, 1968
Daima	Ul-6619	charcoal	2400 ± 95	760-390	Connah, 1968
Kursakata	UtC-3517	charcoal	2860 ± 60	1130-920	Gronenborn et al., 1995
Mege	UtC-4204	charcoal	2659 ± 36	890-800	Breunig, personal communication
Cameroon					
Déguesse mound	Ly-4177	livestock dung	3350 ± 270	2050-1300	Holl, 1998
Shum Laka upper grey layers	Beta-51836	charcoal	2120 ± 110	360-40	de Maret et al., 1993
Shum Laka upper grey layers	Hv-10587	charcoal	885 ± 55	AD 1040-1220	de Maret et al., 1987
Mbi, layer IVA	BM-2425	charcoal	4180 ± 160	3050-2500	Asombang, 1988
Mbi, layer VI	BM-2426	charcoal	2770 ± 120	1090-800	Asombang, 1988
Nkang, Pit 6	Lv-1940	charcoal	2580 ± 70	830-540	Mbida et al., 2000
Nkang, Pit 7bis	Lv-1941	charcoal	2340 ± 70	760-260	Mbida et al., 2000
Central African Republic					
Mondongué	Bdy-253	charcoal	140 ± 240	AD 1529-...	Koté, 1992
Lingbangbo	Bdy-255	charcoal	430 ± 180	AD 1300-1660	Koté, 1992
Lingbangbo	Bdy-582	charcoal	559 ± 77	AD 1300-1430	Koté, 1992
Gabon					
Oveng	Beta-14832	Anadara shell	1970 ± 70	90BC -AD110	Van Neer and Clist, 1991

cont...

Oveng	Gif-6424	shell	Van Neer and Clist, 1991	1650 ± 70	AD 250-530
Oveng	Beta-1483	charcoal and palm nuts	Van Neer and Clist, 1991	1740 ± 70	AD 220-400
Congo-Brazzaville					
Les Saras, Mayumbe	ARC-373	palm nuts	Schwartz et al., 1990	2110 ± 60	350-40
Congo-Kinshasa					
Ngovo	Hv-5258	charcoal	de Maret, 1986	2145 ± 45	360-110
Ngovo	Hv-5258	charcoal	de Maret, 1986	2035 ± 65	160 BC-AD 30
Dimba	Hv-5257	charcoal	de Maret, 1986	2035 ± 130	340 BC-AD 120
Mashita Mbanza	Hv-13451	charcoal	Pierot, 1987	265 ± 55	AD 1510-1800
Mashita Mbanza	Hv-13454	charcoal	Pierot, 1987	140 ± 55	AD 1670-...
Tongo I,1 200-300 cm	Gif-9006	charcoal	Van Neer, 2000	1620 ± 90	AD 260-550
Tongo I,1 200-300 cm	Gif-9010	charcoal	Van Neer, 2000	1690 ± 80	AD 240-430
Rwanda					
Akameru	GrN-7671	charcoal	Van Noten, 1983	1075 ± 95	AD 810-1040
Akameru	GrN-7672	charcoal	Van Noten, 1983	845 ± 75	AD 1050-1270
Gisagara II	GrN-9661	charcoal	Van Grunderbeek et al., 1983	925 ± 30	AD1030-1160
Remera I	GrN-9663	charcoal	Van Grunderbeek et al., 1983	1730 ± 30	AD 250-340
Angola					
Kamabanga I	Gif-6182	charcoal	de Maret, 1985	1120 ± 60	AD 850-990
Kitala II	Gif-5011	honey	de Maret, 1985	720 ± 60	AD 1220-1380

[a] oldest date phase I
[b] re-measurement of UtC-3515 context
[c] youngest date phase I
[d] oldest date phase II
[e] youngest date phase II
[f] abandonment

mentioned previously in Breunig and Neumann (1996), was wrong. A complete lower fourth premolar can be identified with certainty as African buffalo (*Syncerus caffer*), whereas the other large bovid teeth are too incomplete to allow identification. The only other identifiable fragment in the assemblage is a carapace fragment of soft-shelled turtle (*Trionyx triunguis*), indicating that standing water was present near the site. At the Iron Age mound BF94/45 of Oursi, about 30 km south of the previous site, large and small livestock occur from the beginning of the occupation, around 1900 bp. The only other site in the Oudalan province that yielded faunal remains is BF94/96 near Dori, at about 100 km south of Ti-n-Akof. It has a microlithic industry and is dated ca. 3000 bp (Breunig and Neumann, 1996). Only tooth remains, mainly enamel fragments, have been found. They all belong to a large bovid of the size class of domestic cattle and African buffalo, but the fragmentary nature of the finds excludes a species identification.

The archaeological sites in the Chaîne de Gobnangou, Burkina-Faso, are located in a savanna environment and are all cave sites or rock shelters. In contrast to the Oudalan region, habitation of the area goes back in time as far as ca. 7600 bp (Breunig and Wotzka, 1991). The faunal remains studied thus far (Frank *et al.*, in press) are in a good state of preservation, as is usually the case in cave sites. The fauna from Pénténga (previously BF 89/1) derives from levels dated between ca. 7600 and 900 bp. This site yielded only wild animals and the botanical analysis indicates a late introduction of domestic plants and cattle-keeping, after AD 1000 (Neumann and Ballouche, 1992). Maadaga rock shelter (previously BF 89/8 and BF 89/9) did not yield any domestic animals either, with the exception of a few dog finds in the upper layer. The dates for this site show that there have been a lot of perturbations in the Holocene section, which is about 40 cm deep, and it is not excluded that the dog finds are sub-recent. The upper layer (0-20 cm), dated by radiocarbon dates between ca. 2600 and 1700 bp, yielded pottery and microliths, whereas the second layer (20-40 cm), almost without pottery, has dates between ca. 4200 and 1600 bp (Ballouche *et al.*, 1993; Wotzka, personal communication). It is strange that dog is the only domestic species, since its introduction is usually associated with that of domestic herbivores. Maybe the rock shelter was inhabited by hunter-gatherers who only used dogs as an aid in hunting or as a source of meat and skin. The latter use is indicated by a phalanx showing cutmarks. It is unclear thus far whether pastoralism was really introduced so late in the region, or if we are dealing with an artifact related to the choice of the sites. Rock shelters and caves are not usually inhabited by pastoralists, but the absence of visible open air sites in the study area made it impossible to verify their possible presence on such locations.

The early colonization of the southern Chad area by pastoralists was until recently documented only by the finds from Daima in northeastern Nigeria (Connah, 1976, 1981; Gautier, 1987), and was dated between ca. 2500 and 2400 bp. New fieldwork in the Nigerian part of the Chad Basin (Breunig 1995; Breunig and Neumann, this volume; Breunig *et al.*, 1993, 1996; Gronenborn *et al.*, 1995) pushes the dates for the incursion of pastoralists back to ca. 3500 bp. As a result of increasing aridity, the Mega Chad basin shrank, and it was only around 3500 bp that the new land was accessible for people and their herds. The artifacts at the sites clearly point towards a Saharan origin. More than 120 archaeological sites of the Gajiganna Complex were recorded, of which 20 were excavated. A large series of radiocarbon dates and typological analysis of the pottery allowed the recognition of two phases in the settlement history. During Phase I, dated between ca. 3500 and 3100 bp, the majority of the settlements consisted of 50 to 100 m wide mounds with a maximum elevation of 1 m. They are all located close to clay-bottom depressions which, up to the present day, are filled with water during the rainy season. The small size of the sites and thickness of the deposits, as well as the rather poor yield of pottery, lithics and fauna, indicate seasonal occupation by mobile herder groups. Certain sites with thicker deposits of over 2 m seem to witness a recurrent seasonal use, whereas others with very few materials may reflect single, short-term occupation. There is no evidence for domestic plants during Phase I, but cattle and ovicaprines are present throughout the whole period. During Phase II, starting from about 3100 bp, the inhabitants from the region had apparently changed their lifestyle and formed sedentary communities. The settlement mounds become much larger and higher and the amount of excavated pottery and fauna per unit volume increases dramatically. The large amount of clay deposited on the mounds is due to the disintegration of houses. During Phase II, agriculture was practiced (Breunig et al., 1996).

About 3000 identifiable faunal remains from two settlement mounds (Gajiganna A and B) have been studied in detail. These sites belong to Phase II and show that throughout the occupation, from ca. 3150 to 2700 bp, ovicaprines, and especially cattle, were the most important food providers. The end of the Gajiganna culture is poorly documented and dates around 2500 bp. The reasons for this are unclear but increasing aridity may account for it. The botanical analysis from the Gajiganna sites A and B indicates that the initial dense vegetation was replaced by plants of the open savanna. No diachronic changes have been observed in the fish fauna that could point to increasing aridity, but among the terrestrial animals an increase in wild carnivores was noted towards the youngest levels of the upper cultural layer of Gajiganna mound A. Less phytomass, because of increasing aridity, overgrazing or both, results in a concentration of herbivores around water bodies where vegetation is more abundant. The carnivores, following the herbivores on which they preyed, may therefore have become more vulnerable to predation by man. The faunal remains from the new excavations in the Gajiganna area being so abundant, initial analysis involved in many cases only the verification of the domesticates' presence in the lower levels. Detailed analysis of all the material is on its way and will first concentrate on sites such as Bukarkurari, which were occupied during both Phases I and II. Special attention will be paid to aspects of seasonality which may help to elucidate the settlement-subsistence system through time.

The so-called firki region, east of the Gajiganna area, had been archaeologically explored by Connah (1981) but has recently been the subject of new fieldwork (Gronenborn 1996a,b; Gronenborn et al., 1995). Almost 70 sites have been recorded, of which three were excavated (Gronenborn, 1996a). The earliest settlements in the area date to about 3000 bp when waters from the Chad Lake had retreated farther. Comparison of the material culture from the firki and Gajiganna region shows that the pastoralists colonizing the firki area had no affinities with the herders from the west and that they therefore must have come from the east or the south. Faunal analysis of Mege (Lambrecht, 1997) and Kursakata (Gronenborn et al., 1995; Van Neer, unpublished) show that domestic cattle and ovicaprines were present from the very beginning (ca. 2900 bp for Kursakata; ca. 2700 bp for Mege). As in the Gajiganna area, it is noted that fishing and hunting were also regularly practiced.

Recently, a new find has been reported from the Cameroonian part of the Chad Basin, where livestock dung at the base of the Déguesse mound was dated to 3350 ± 270 (Ly-4177), representing the earliest proof for that region (Holl, 1998). The possible relation of these and associated finds to the early firki sites is to be evaluated after the complete publication of the results.

PROPAGATION OF DOMESTIC ANIMALS IN THE WOODED AREAS OF WEST AND CENTRAL AFRICA AND THE SOUTHERN SAVANNAS

Domestic stock has been reported from Ntereso and Kintampo rock shelter 6 in Ghana (Carter and Flight, 1972). The former site is located in dry savanna woodland, and was dated between about 3600 and 3200 bp. A bovine second phalanx was found but the authors hesitate in attributing it to domestic cattle. Gautier (1987) believes cattle is present at Ntereso on the basis of the synecological context and the occurrence of a slender horncore provisionally identified by Davies (1964) as 'Bos aegyptiacus' or long-horn cattle. Kintampo rock shelter 6 is situated at the fringe of the high forest and compares closely in date to Ntereso (Stahl, 1985). The bovine remains found in a level dated at 3550 bp were attributed to domestic cattle by Carter and Flight (1972) on the basis of a comparative metrical analysis of modern cattle and African buffalo. These identifications have been questioned (Stahl, 1985), but apparently the material has never been re-analyzed using the morphological criteria established by Peters (1988). The faunal remains from the new excavations at Kintampo (Stahl, 1985) were tentatively identified without adequate reference material. An analysis by Gautier (1987) of the metrical data published by Carter and Flight (1972) showed that the bovids from Kintampo had a shoulder-height of 1 m or less, which excludes an identification as African buffalo. The small size of the cattle breed at Kintampo seems to indicate that a race already existed that was comparable in stature to the modern ndama or muturu cattle. These are authochthonous breeds typically living in forested areas and resistant to trypanosomiasis. The only other archaeological indications for forest-adapted cattle comes from the Sine-Saloum in Senegal. Ndiamon-Badat shell mound dated between ca. 2400 and 1600 bp (Ba et al., 1997) is located in coastal savanna-forest mosaic and yielded remains of small breeds of cattle, sheep and goat (Van Neer, 1997). Small-sized cattle were also discovered at the nearby Faboura shell mound dated between 1900 and 1300 bp (Van Neer, unpublished).

At both Ntereso and Kintampo K6 goat bones have been found which were attributed to a dwarf breed. Carter and Flight (1972) showed that the length of the mandibular row of a Kintampo specimen compares closely with measurements of dwarf goats from Malakal, Sudan, but Gautier (1987) found the values observed for this measurement in normal goats to be similar to those found in the Kintampo specimen. Given the location of Kintampo, close to the rainforest, it is very likely that a dwarf breed of goat was indeed present, but a restudy of the material using other measurements and considering more modern material would be preferable to confirm this. The faunal report of Carter and Flight (1972) lists a number of wild species that were hunted, but quantitative data are missing.

Linguistic studies suggest that Bantu-speaking populations would have emigrated from their initial homeland in the Grassfields of Cameroon (Bastin *et al.*, 1979, 1982; Meussen, 1980). The migration of these people, who were not adapted to life in the forest, would have been from west to east through the savanna to finally reach the interlacustrine region. A further southward and then westward migration would have brought these people to the savanna south of the equatorial forest. However, certain groups would have been able to cross the rainforest belt through a more westerly route along the coastline or by traveling south along the waterways farther east (Bastin, 1978; David, 1980). There is scanty botanical and archaeological evidence that savanna corridors existed in the forest as a result of a late Holocene dry spell around 2100 bp (Schwartz *et al.*, 1990; Roche, 1991, but see Sowunmi, this volume), but more archaeological data will be needed to further document these hypothetical itineraries towards the southern savannas (de Maret, 1986). South of the rainforest both the eastern and western migratory waves would have met. No evidence has been found thus far for domestic animals in the sites of the Grassfields of Cameroon. They are lacking in the Iron Age layers of Shum Laka cave, dated between ca. 2150 and 900 bp, and the same was true for the undated pottery levels of Abeke rock shelter (de Maret *et al.*, 1987). Domestic stock is also absent from the cave at Mbi, where levels with pottery were dated at ca. 4200 and 2800 bp (Asombang, 1988). The absence of domestic animals may be related to the fact that the immediate surroundings of the aforementioned caves from the Grassfields are wooded. In addition, they may not be living sites, but rather represent places where hunters temporarily halted and buried their dead. The site of Nkang near Yaoundé, Cameroon, is located in a savanna-forest mosaic environment and is one of the rare sites documenting the early spread of domestic animals along the northern fringe of the equatorial forest. A few sheep and goat bones from a small breed were found in refuse pits and were securely dated between ca. 2600 and 2300 bp (Mbida *et al.*, 2000; Van Neer, 2000). The site belongs to the Obobogo tradition and yielded pottery, metal slag and lithic material. Nkang also yielded the earliest evidence for banana cultivation. It is important to note that no indication was found of a possible arid spell around 2500 bp (cf. Hassan, this volume). Both the archaeobotanical and archaeozoological data indicate that, at the time of occupation, the site was also located in an environment of the savanna-mosaic type. Subsistence was mainly based on fishing, hunting and mollusc collecting, whereas small livestock contributed little to the food provisioning. The small size of the ovicaprines shows again that a process of dwarfing has accompanied the introduction of ovicaprines in the forested areas of Africa. No other West Central African sites are available to illustrate the further early spread of domestic animals to the south. The few sites in the northern part of the equatorial forest with domesticates are located in the Central African Republic and all date to the second half of this millennium (Van Neer, 2000). At Lingbangbo, with dates between ca. 600 and 400 bp, dwarf goat was present, whereas unidentifiable ovicaprines occur at Mondongué, dated to 140 bp. Despite their young age, these sites are useful as a reference for food procurement strategies in forest areas since quantified faunal data are available. The only site, located along the coast and in the rainforest belt, that yielded a large faunal sample is Oveng, a shell midden near Libreville in Gabon (Van Neer and Clist, 1991). Pottery and iron slag were present and radiocarbon dates are between ca. 2000 and 1650 bp. The inhabitants relied mainly on fishing and shell collecting; mammal remains are heavily under-represented and comprise no domestic species. In the central part of the rainforest, only one site yielded domestic stock. At Nkile, the occupation of which is estimated to have covered the nineteenth century AD, possibly going back to the eighteenth or seventeenth centuries, ovicaprines, pig and domestic dog were found (Van Neer, 2000). Archaeological research by de Maret (1986) at the southern edge of the rainforest in Congo led to the definition of the Ngovo Group which would represent the first, or in any case very ancient, Bantu-speaking populations that would have crossed the rainforest and settled in the savannas to the south. Ngovo and Dimba caves are the only two sites with fauna preserved. Their deposits were dated between ca. 2100 and 2000 bp. Both assemblages lack domestic animals, but it should be emphasized that the excavated faunal samples are small and that the deposits may have been partly related to ritual practices.

The earliest occurrence of domestic stock in Central Africa south of the rainforest is at Kamabanga I, a shell midden in northern Angola. Domestic cattle were found in this site dated to ca. 1100 bp (Van Neer, 2000). The few other sites in the southern savannas with domestic stock are even more recent. Cattle found at Kitala, northern Angola, were dated to 720 bp, and at Mashita Mbanza, Congo, between 265 and 140 bp. Kamabanga, Kitala and Mashita Mbanza are all situated in a grassland environment where cattle keeping is still practiced today. In the wooded savanna belt further south, dwarf goat was found at Sanga in a human burial dated to the beginning of the second millennium.

THE INTERLACUSTRINE REGION

The interlacustrine region may have represented a more suitable corridor for the propagation of domestic animals than the western coastline or the equatorial forest. It has, however, not yet been established whether the first inhabitants of the area were Bantu-speaking populations coming from the west or if they were Nilotic pastoralists coming from the east, or a mixture of both populations. The area comprises extensive grassfields free of tsetse fly and trypanosomiasis as a result of the high elevation. The oldest finds of domestic cattle, sheep and goat from this region are from Tongo, 50 km north of Lake Kivu, and were found in early Iron Age levels dated between ca. 1700 and 1600 bp. The undated late Iron Age levels yielded cattle and goat remains (Van Neer, 2000). Palaeobotanical and geomorphological work in the region, summarized by Roche (1991), shows a humid phase between 2000 and 1700 bp, followed by an arid spell between 1600 and 1500 bp. After 1500 bp the climate turned again more humid and became comparable to the present-day conditions. The palynological data from Burundi and Rwanda show that, during the first centuries of our era, populations of the Urewe tradition (early Iron Age) had a profound effect on the environment. Around 1500 bp the occupation of the region would have been less dense, which would have allowed the environment to recover for about two centuries from the effects of the previous arid spell and the over-exploitation by the human inhabitants. This period of regeneration was followed by late Iron Age occupation. All the other interlacustrine sites with fauna yielded small samples only. In Rwanda, two cattle teeth were associated with a furnace dated to ca. 1700 bp at Remera I (Van Grunderbeek et al., 1983), and at Gisagara, near Butare, a faunal sample dated to ca. 900 bp comprised both small and large livestock (Gautier, n.d.). Test excavations in the caves of Akameru and Cyinkomane, a few kilometers west of Ruhengeri, yielded late Iron Age levels dated between ca. 1100 and 800 bp. The faunal assemblages were studied in their entirety and comprised cattle, sheep and goat as well as various wild species (Gautier, 1983).

REGIONAL DIFFERENCES IN THE RELATIVE IMPORTANCE OF DOMESTIC ANIMALS

In most articles dealing with the spread of early domestic animals emphasis has been laid on cattle. However, soon after an initial period of cattle exploitation in the Eastern Sahara, sheep and goat were also part of the herds that roamed the desert (Gautier, this volume). It appears, in addition, that the role of hunting and fishing in the food provisioning is often neglected. Maybe this narrow view results from parallels made to the present-day heavy reliance on livestock by certain pastoralist groups, such as the Turkana (Galvin, 1985; Soper, 1985). The Peul, living near the Niger today, do not practice fishing but consume fish obtained from Bozo fishermen (Sundström, 1972), whereas other groups such as the Nuer fish actively in the Nile on a seasonal basis (Evans-Pritchard, 1940). Archaeozoological data show that prehistoric herders in North and West Africa did not rely as heavily on their livestock as certain modern pastoralists. It is only in East Africa that sites have been reported in which the faunal remains comprise almost exclusively domestic stock, but this is a phenomenon occurring only from about 3000 bp onwards (Gifford-Gonzalez, 1998). Marshall and Stewart (1994) consider the high proportions of remains from wild animals and fish at the 2000 year old Elmenteitan site of Gogo Falls in Kenya—as opposed to the livestock-dominated samples from other Elmenteitan and savanna pastoral Neolithic sites—as a possible indication of environmental stress. In North and West Africa, hunting and fishing always contributed significantly to the procurement of animal protein, and this may be seen as a continuation of strategies that were important before livestock keeping was practiced. Diversification may have made prehistoric populations less vulnerable to the effects of climatic stress or disease of the herds. However, it still remains to be verified for the considered region if the degree of reliance on fishing and hunting among past pastoralist groups, as inferred from faunal analyses, can be linked to adverse periods.

On the basis of quantitative data given in faunal reports, an estimation can be given of the former relative importance of hunting, fishing and stock keeping. Unfortunately such data are lacking for the majority of the sites in the region reviewed here. For many sites, the presence of cattle or small livestock is simply mentioned in passing. In certain cases the relative importance of other taxa is indicated in very general terms. However, detailed species lists with a quantification of the remains are very rare, thus hampering the investigation of the importance of herding versus hunting and fishing. Even in cases where quantitative data are available, caution is needed when intersite comparisons are carried out. The contribution of each strategy will be evaluated here on the basis of the number of identified specimens (NISP) found on each site where such data are provided. Alternative ways of establishing proportions are the calculation of minimum number of individuals (MNI) or the use of bone weight. The calculation of MNIs depends heavily on sample size (Poplin, 1976), and the number of stratigraphic and horizontal units distinguished on a site, and is, in addition, difficult to reproduce by different, individual researchers. The MNI method has more limitations than fragment counts (Gautier, 1984) and its application is now mostly limited to special find contexts where there is evidence for the disposal of complete carcasses. The bone weight method (Kubasiewicz, 1973) has been recommended in the past since it would be a good measure of the amount of meat corresponding to the deposited bone refuse. The major drawback of this method is that there is no straightforward relationship between these two parameters (Reitz et al., 1987). Although certain researchers still systematically provide bone weights and/or MNIs, such data for the region under consideration are very scanty. Working with those units would significantly reduce the number of fauna that can be taken into account. The use of fragment counts for a comparison of the relative importance of stock keeping, hunting and fishing is, therefore, the most suitable strategy for the region considered here. A major limitation of any faunal analysis, independent from the quantification method, is that the recovered skeletal remains have been subjected to the effects of refuse disposal phenomena, differential preservation, sampling strategies, and sample size. Excavations being most often limited to a small part of a site, a distorted picture of the average consumption pattern is possible when there was not random disposal of the refuse from mammals, fish and molluscs over the site. The degree of surface weathering and diagenetic processes vary considerably from site to site but can also show variation within one site. Differential destruction of small remains can be responsible for the low incidence of remains from fish and other small species. An over-representation of cattle is, therefore, often observed in sites with poor preservation of bone. Also, sampling strategies can influence to a large extent the relative abundance of small and larger faunal remains. Surface collecting and lack of sieving result, amongst others, in the under-representation of fish and other small creatures, whereas larger animals such as bovids tend to be over-represented. Even in cases where fish remains are collected without sieving, bias exists, as reflected by the abundance, or exclusive presence, of large species such as Nile perch and clariid catfish.

In Figure 15.2, the relative abundance of domestic animals versus wild species and fish is expressed in terms of the percentage of identified remains available at each considered site. It is obvious that domestic animals at certain sites were only a secondary resource, not the staple food. This is the case for sites in heavily wooded environments (Ndiamon-Badat, Nkile, Lingbangbo and Nkang), where bone samples comprise only a minor portion of ovicaprines, and, sometimes, small-sized cattle. In open environments the contribution of domesticates is also low, such as at Dhar Tichitt. The same can be seen on the sites in the Méma and Gourma region where there is a heavy reliance on aquatic resources. Quantitative data are available for Kolima Sud I, whereas for the other sites from those areas the contribution of fish to the diet cannot be properly estimated since their analysis was limited to an on-site identification of taxa without any quantification. The proportion of fish was, however, said to be large (MacDonald, 1994). This points to a continuation of the food procurement strategies similar to those practiced during the wet Holocene phase in the regions of Hassi-el-Abiod and Erg Ine Sakane, where fish remains abound. Anthropological data have been used to link the earlier occupants of the Méma region to those areas (Georgeon *et al.*, 1992). For the sites in the desert of northern Mali, however, a quantified approach is difficult due to the sampling strategies which were carried out by different specialists for mammals and fish remains. Sampling involved the surface collection of the most diagnostic pieces of each animal group (Gayet, personal communication). An important question in the interpretation of sites with abundant fish bones and few livestock remains is whether the bones represent the food refuse of a single group. The margins of large water bodies inhabited by fishermen and hunters may have been visited by pastoralist groups occasionally or on a seasonal basis.

The highest contribution of livestock remains to the total faunal assemblages is noted in the Gajiganna and firki area of northeastern Nigeria and it is clear that the emphasis here was on pastoralism. The advantage of these newly excavated sites is that sieving was practiced consistently and the fauna was sampled in a similar way at all the mounds. This means that intersite comparisons in this region do not suffer from differences in sampling strategies that hamper the inclusion of several other North and West African faunas in the discussion. The proportions of livestock seen at Gajiganna A and B, Kariari C and Mege are high, but it is also obvious that amongst this group of sites near Lake Chad significant variation occurs. This is especially remarkable for Gajiganna A and B, two settlement mounds situated close to each other and which are, in addition, partly contemporaneous. In this case, the high proportion of fish at Gajiganna B is explained by the presence of a refuse pit with remains in a very good state of preservation. This example illustrates that caution is needed when employing faunal data for palaeoeconomic or palaeoenvironmental reconstructions. Once all the fauna from the Gajiganna and firki areas are identified, it will be ascertained whether a relationship can be found between the proportions of the different food procurement strategies and the local topography and the proximity of open water, the age of the deposits, the size of the mounds and other variables such as state of preservation. It also remains to be seen to what extent the mounds were inhabited; on a year round or on a seasonal basis.

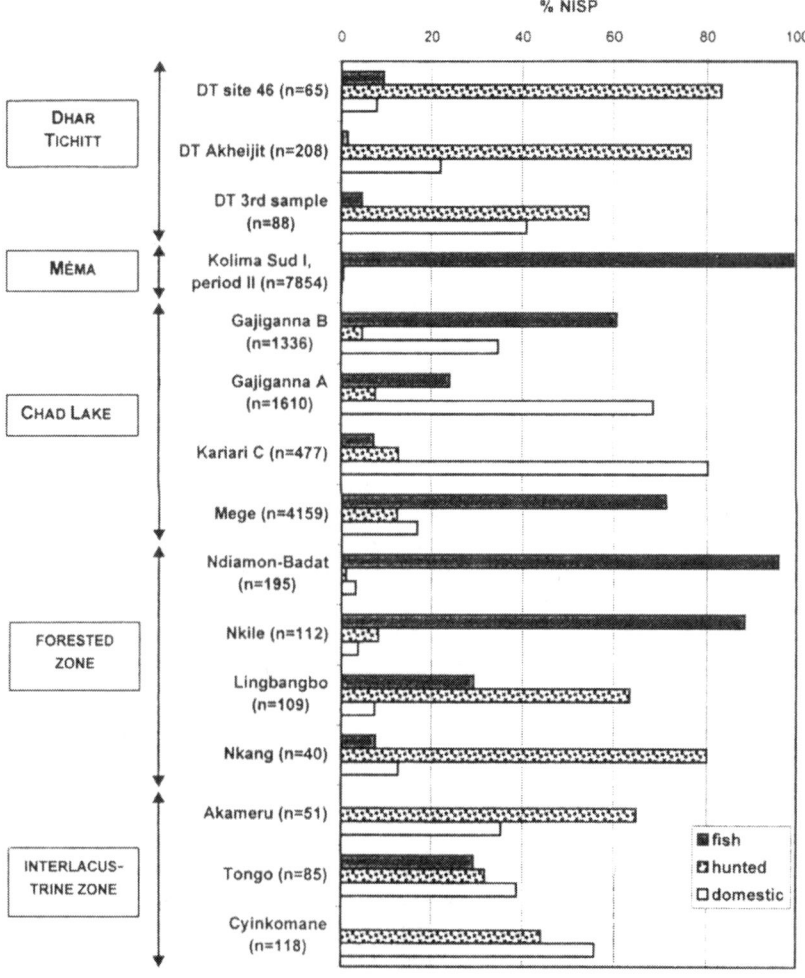

Figure 15.2. Relative importance of fishing, hunting and herding on sites in the considered region (calculated on the basis of number of identified specimens).

The foregoing comparisons are based on the proportions of the number of fragments of domestic animals, fish and hunted species. It is obvious that the figures are only an approximation of the relative frequency with which an activity was exercised and that a conversion to meat yield is needed to make a better estimation of the contribution to the diet. Here, again, different methodological approaches are possible (for a review see Reitz *et al.*, 1987), but the simplest method consists of multiplying the number of fragments by an arbitrary, average total weight. When a mean total weight is taken of 300 kg for cattle, 25 kg for ovicaprines (cf. Vigne, 1991), 50 kg for each hunted species and 1 kg for fish, then it is obvious that cattle is the major source of animal protein. Theoretically, the total weight of each hunted species should be considered separately and, the spectrum being different at each site, an average value should not be given. However, with the exception of the Dhar Tichitt sites, where the contribution of large-sized hunted species is high, probably as a result of the sampling techniques, the hunted faunas considered here are predominated by medium-sized to small mammals. Conversion to total weights (Figure 15.3) confirms that the dietary contribution of small livestock in the forest sites is small (Nkang, Lingbangbo, and Nkile), and it appears that also in the Méma region domestic animals were of less importance than the wild terrestrial and aquatic fauna (Kolima Sud I, period II). The Dhar Tichitt samples show that both hunting and cattle keeping were the major strategies for obtaining animal protein. Due to the small sample sizes and the different recovery techniques for each sample, it would be speculative to try and explain the differences among the three assemblages. At the shell midden of Ndiamon-Badat, cattle seems to have been the major contribution to the diet, but it should be underlined that the molluscs have not been taken into account and that the fish remains must be heavily under-represented since no sieving was practiced. The large average size of the fish remains and the low number of unidentifiable pieces show that the recovery of the ichthyofauna was incomplete. The sites Akameru, Tongo and Cyinkomane show that large livestock was the major food source in the interlacustrine area, but it is obvious that hunting also contributed significantly to the diet. The heaviest reliance on livestock, especially cattle, is seen in the Gajiganna region. In terms of total weight, the contribution of fishing and hunting is negligible. At Mege, the only site from the firki region, hunted animals provided slightly more proteins, but livestock remained the major source.

Cattle have a high meat yield compared to other animal resources that are usually exploited. However, it should be underlined that, after slaughtering, cattle have to be consumed rather rapidly, even when methods for the preparation of dried or smoked meat were known. The meat yield of individual fish or small and medium-sized game is limited, but since consumption usually follows immediately after capturing, losses due to bacterial decay or insect pests are restricted. Hunting, snaring and fishing may therefore constitute a more regular food procurement strategy than the slaughtering of cattle. It is unclear when the use of dairy products and blood started among the early pastoralists. The use of secondary products can be indirectly inferred from age profiles obtained from the analysis of tooth wear and epiphyseal fusion of long bones. In herds primarily kept for their meat, slaughtering will usually be practiced when the individuals are sub-adult, i.e. when they have more or less reached mature size. Only a limited number of adult specimens are kept for reproduction. When attention is focused on secondary products, the individuals from a herd are slaughtered at a much older age, possibly with the exception of young males that may have been removed from the herds at a younger age. In order to obtain significant age profiles that can be informative about the uses of the livestock herds, sufficient data are needed. These can only be obtained from large samples since the proportion of age-diagnostic pieces is limited. Even from the relatively large samples from the Chad area, data are insufficient at the moment to make a

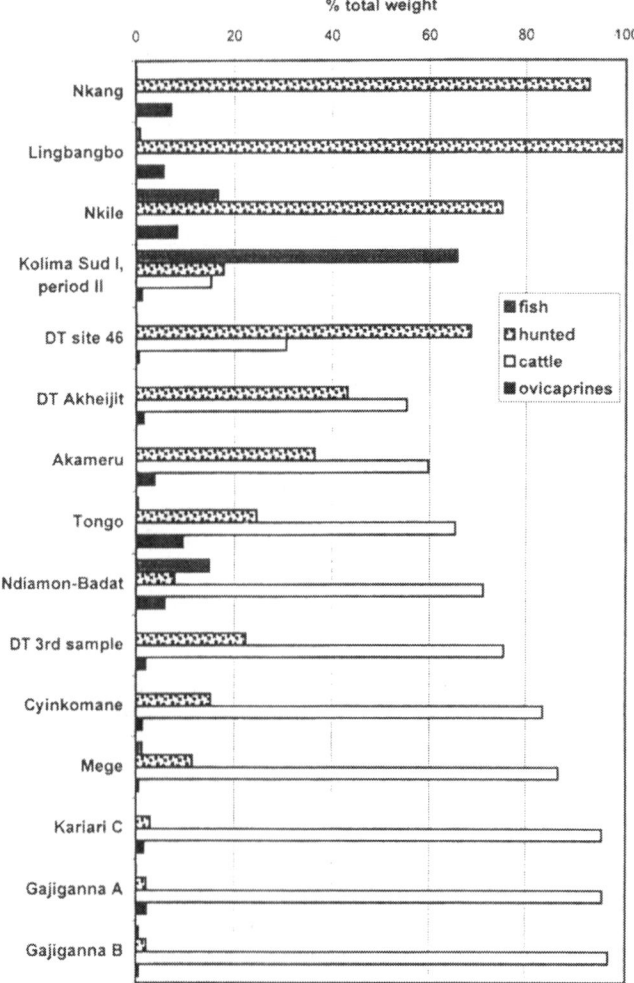

Figure 15.3. Dietary contribution of fishing, hunting and herding on sites in the considered region (calculated on the basis of number of identified specimens multiplied by an average total weight). The sites are arranged in order of increasing importance of cattle.

reasonable statement about the age profiles. Moreover, it should be kept in mind that age profiles can be distorted as a result of poor preservation. Due to differences in bone density, young individuals can be absent or heavily under-represented when conditions for preservation were poor. The foregoing consider-ations start from the assumption that the herders were maintaining the size of their herds. In societies where the size of the herds is a measure of wealth, slaughtering may have been restricted to very old individuals, no longer able to follow the seasonal movements, and to young and sub-adult males that were difficult to handle or no longer desired for reproduction.

CONCLUSIONS

The faunal data show that a fast dispersal of livestock occurred, during the fourth millennium bp, in the Sahel and the northern savannas as far as the fringe of the rainforest. It is clear from the overall faunal composition that the southward migrating populations were not exclusively searching for suitable pasture for their herds, but that they preferred regions that also allowed them to continue to hunt and fish. The degree to which these additional resources were exploited also varied. This is well illustrated in the Méma and Gourma regions where stock keeping was minimal. At the southern edge of the Chad Lake, cattle and ovicaprines were the main food resource, but exploitation of aquatic species remained an important activity. The almost exclusive reliance on livestock as it is seen in several modern pastoralists groups is a phenomenon that can be observed archaeologically only from about 3000 bp onwards, and this only in East Africa. The diverse food procurement strategies of prehistoric people in West and Central Africa may have allowed them to overcome periods of environmental stress. However, the data available thus far do not permit us to infer periods of stress on the sole basis of the faunal composition.

The small size of the cattle at Kintampo and of the ovicaprines at Nkang, and possibly also at Ntereso and Kintampo, demonstrates that during the fourth millennium bp breeds were already adapted to forest regions. The degree of reliance on domestic stock in forested areas was poor. Maybe this was related to less demographic pressure compared to that in open environments? (There is no hard evidence for less demographic pressure; the smaller number of sites may be related to poorer archaeological visibility and to the low number of teams working in those areas. The inhabitants of these sites practiced broad spectrum exploitation; long-term exploitation of this kind by growing numbers of sedentary people may have led ultimately to the more frequent use of domestic animals). Despite the rapid distribution of the concept of animal keeping, it seems that the ideas were not adopted everywhere at such an early date. Late occurrence of livestock is known from southern Burkina Faso and from the savannas just south of the rainforest, but it is unclear thus far whether this is a factor related to the choice of the excavated sites, poor preservation of bone finds and small sample size, or if this is an illustration of the reluctance of certain groups to adopt pastoralism.

Information on the role of stock keeping, hunting and fishing in western and central Africa is still fragmentary as a result of the low number of excavations carried out in this vast region. Moreover, faunal preservation depends heavily on soil conditions, which differ regionally. In general, the acid soils are responsible for the low incidence of faunal remains in forested regions where bones are almost exclusively preserved in very young sites and in caves or rock shelters. Further advances in the knowledge of the mechanisms allowing prehistoric people to deal with the threats of aridity on food security will depend on the discovery of well preserved faunal assemblages which, in addition, need to be sampled carefully, analyzed in a quantitative way, and published fully.

ACKNOWLEDGMENTS

This text presents research results of the Belgian Program on Interuniversity Poles of Attraction initiated by the Belgian State, Prime Minister's Office, Science Policy Programming. I also thank Achilles Gautier (Gent) and Peter Breunig (Frankfurt) for their comments on an earlier version of this paper.

REFERENCES

Asombang, R. (1988). *Bamenda in Prehistory: the Evidence from Fiye Nkwi, Mbi Crater and Shum Laka Rockshelters*, PhD dissertation, University of London. [unpublished]

Ba, C., Descamps, C., and Thilmans, G. (1997). Fouille d'un tumulus à Ndiamon-Badat. In Thilmans, G. (ed.), *Fouille et Dégradations dans les Îles du Saloum*, IFAN Dakar, pp. 1-14.

Ballouche, A., Küppers, K., Neumann, K., and Wotzka, H.-P. (1993). Aspects de l'occupation humaine et de l'histoire de la végétation au cours de l'holocène dans la région de la Chaîne de Gobnangou, S.E. Burkina Faso. *Berichte des Sonderforschungsbereichs 268 'Kulturentwicklung und Sprachgeschichte im Naturraum Westafrikanische Savanne'* 1: 13-31.

Bastin, Y. (1978). Statistique grammaticale et classification des langues bantoues. *Linguistics in Belgium* 2: 17-37.

Bastin, Y., Coupez, A., and de Halleux, B. (1979). Statistique lexicale et grammaticale pour la classification des langues bantoues. *Bulletin de l'Académie royale des Sciences d'Outre-Mer* 3: 375-387.

Bastin, Y., Coupez, A., and de Halleux, B. (1982). Classification lexicostatistique des langues bantoues (214 relevés). *Bulletin de l'Académie Royale des Sciences d'Outre-Mer* 27: 173-199.

Breunig, P. (1995). Gajiganna und Konduga. Zur frühen Besiedlung des Tschadbeckens in Nigeria. *Beiträge zur Allgemeinen und Vergleichenden Archäologie* 15: 3-48.

Breunig, P., and Neumann, K. (1996). Archaeological and archaeobotanical research of the Frankfurt University in a West African context. *Berichte des Sonderforschungsbereichs 268 'Kulturentwicklung und Sprachgeschichte im Naturraum Westafrikanische Savanne'* 8: 181-191.

Breunig, P., and Neumann, K. (1999). Archäologische und archäobotanische Forschung in Westafrika. *Archäologisches Nachrichtenblatt* 4: 336-357.

Breunig, P., and Wotzka, H.-P. (1991). Archäologische Forschungen im Südosten Burkina Fasos 1989/90: Vorbericht über die erste Grabungskampagne des Frankfurter Sonderforschungsbereichs 268 'Westafrikanische Savanne'. *Beiträge zur Allgemeinen und Vergleichenden Archäologie* 11: 145-187.

Breunig, P., Ballouche, A., Neumann, K., Rösing, F.W., Thiemeyer, H., Wendt, P., and Van Neer, W. (1993). Gajiganna. New data on early settlement and environment in the Chad basin. *Berichte des Sonderforschungsbereichs 268 'Kulturentwicklung und Sprachgeschichte im Naturraum Westafrikanische Savanne'* 2: 51-74.

Breunig, P., Neumann, K., and Van Neer, W. (1996). New research on the Holocene settlement and environment of the Chad Basin in Nigeria. *African Archaeological Review* 13: 111-145.

Carter, P.L., and Clark, J.D. (1976). Adrar Bous and African cattle. In Abébé B., Chavaillon, J., and Sutton, J.E.G. (eds.), *Proceedings of the Seventh Panafrican Congress of Prehistory and Quaternary Studies - Addis Ababa 1971*, Addis Ababa, pp. 487-493.

Carter, P.L., and Flight, C. (1972). A report on the fauna from the sites of Ntereso and Kintampo Rock Shelter 6 in Ghana: with evidence for the practice of animal husbandry during the second millennium B.C. *Man* **7**: 277-282.

Chenal-Vélardé, I. (1997). Les premières traces de boeuf domestique en Afrique du Nord: état de la recherche centré sur les données archéozoologiques. *Archaeozoologia* **9**: 11-40.

Connah, G. (1968). Radiocarbon dates for Benin city and further dates for Daima, N.E. Nigeria. *Journal of the Historical Society of Nigeria* **4**: 313-320.

Connah, G. (1976). The Daima sequence and the prehistoric chronology of the Lake Chad region of Nigeria. *Journal of African History* **17**: 321-352.

Connah, G. (1981). Man and a lake. In *Le Sol, la Parole et l'Écrit. Mélanges en Hommage à Mauny*, Société Française d'Histoire d'Outre-Mer, Paris, pp. 161-178.

Corridi, C. (1997). Some new archaeozoological data from Tadrart Acacus, Libya (9th to 5th millennium B.P.). *Archaeozoologia* **9**: 41-48.

David, N. (1980). Early Bantu expansion in the context of Central African prehistory: 4.000-1 B.C. Colloque du C.N.R.S. *L'Expansion Bantoue*, n.s. **2**: 265-278.

Davies, O. (1964). *The Quaternary in the Coastlands of Guinea*, Jackson, Glasgow.

Davies, O. (1980). The Ntereso cultures in Ghana. In Swartz, B.K., and Dumett, R. (eds.), *West African Culture Dynamics: Archaeological and Historical Perspectives*, Mouton, The Hague, pp. 205-225.

de Maret, P. (1985). Recent archaeological research and dates from Central Africa. *Journal of African History* **26**: 129-148.

de Maret, P. (1986). The Ngovo group: an industry with polished stone tools and pottery in Lower Zaïre. *African Archaeological Review* **4**: 103-133.

de Maret, P., Clist, B., and Van Neer, W. (1987). Résultats des premières fouilles dans les abris sous roche de Shum Laka et Abeke au nord-ouest du Cameroun. *L'Anthropologie* **91**: 559-584.

de Maret, P., Asombang, R., Cornelissen, E., Lavachery, P., Moeyersons, J., and Van Neer, W. (1993). Preliminary results of the 1991-1992 field season at Shum Laka, Northwestern Province, Cameroon. *Nyame Akuma* **39**: 13-15.

Descamps, C., Thilmans, G., Thommeret, J., Thommeret, Y., and Hauptmann, E.F. (1977). Données sur l'âge et la vitesse d'édification de l'amas coquillier de Faboura (Sénégal). *Bulletin ASEQUA (Dakar)* **51**: 27-32.

Evans-Pritchard, E.E. (1940). *The Nuer: a Description of the Modes of Livelihood and Political Institutions of a Nilotic People*, Oxford University Press, Oxford.

Frank, T., Breunig, P., Müller-Haude, P., Van Neer, W., Neumann, K., Vogelsang, R., and Wotzka, H.-P. (in press). The Chaîne de Gobnangou, SE Burkina Faso - archaeological, archaeobotanical, archaeozoological and geomorphological studies. *Beiträge zur Allgemeinen und Vergleichenden Archäologie* **21**: 127-190.

Galvin, K. (1985). *Food Procurement, Diet, Activities and Nutrition of Ngisonyoka, Turkana Pastoralists in an Ecological and Social Context*, PhD dissertation, State University of New York. [unpublished]

Garcea, E.A.A. (1993). *Cultural Dynamics in the Sahara-Sudanese Prehistory*, Gruppo Editoriale Internazionale, Roma.

Gautier, A. (1983). Les restes osseux des sites d'Akameru et de Cyinkomane (Ruhengeri, Rwanda). In Van Noten, F. (ed.), *L'Histoire Archéologique du Rwanda*, Annales du Musée Royal de l'Afrique Centrale, Sciences Humaines 112, Tervuren, pp. 104-120.

Gautier, A. (1984). How do I count you, let me count the ways? Problems of archaeozoological quantification. In Grigson, C., and Clutton-Brock, J. (eds.), *Animals and Archaeology: 4. Husbandry in Europe*, British Archaeological Reports International Series 227, Oxford, pp. 237-251.

Gautier, A. (1987). Prehistoric men and cattle in north Africa: a dearth of data and a surfeit of models. In Close, A.E. (ed.), *Prehistory of Arid North Africa, Essays in Honor of Fred Wendorf*, Southern Methodist University Press, Dallas, pp. 163-187.

Gautier, A. (n.d.) *Les Restes de Mammifères de Gisagara près de Butare*, Unpublished report, Gent.

Georgeon, E., Dutour, O., and Raimbault, M. (1992). Paléoanthropologie du gisement lacustre néolithique de Kobadi (Mali). *Préhistoire et Anthropologie Méditerranéennes* **1**: 85-97.

Gifford-Gonzalez, D. (1998). Early pastoralists in East Africa: ecological and social dimensions. *Journal of Anthropological Archaeology* **17**: 166-200.

Gronenborn, D. (1996a). Kundiye: archaeology and ethnoarchaeology in the Kala-Balge area of Borno State, Nigeria. In Pwiti, G., and Soper, R. (eds.), *Aspects of African Archaeology. Papers from the 10th Congress of the PanAfrican Association for Prehistory and Related Studies*, University of Zimbabwe Publications, Harare, pp. 449-459.

Gronenborn, D. (1996b). Beyond Daima: recent excavations in the Kala-Balge region of Borno State. *Nigerian Heritage* 5: 34-46.

Gronenborn, D., Van Neer, W., and Skorupinksi, T. (1995). Kleiner Vorbericht zur archäologischen Feldarbeit südlich des Tschad-Sees. *Berichte des Sonderforschungsbereichs 268 'Kulturentwicklung und Sprachgeschichte im Naturraum Westafrikanische Savanne'* 5: 27-39.

Hassan, F.A. (1996). Abrupt Holocene climatic events. In Pwiti, G., and Soper, R. (eds.), *Aspects of African Archaeology. Papers from the 10th Congress of the PanAfrican Association for Prehistory and Related Studies*, University of Zimbabwe Publications, Harare, pp. 83-89.

Hassan, F. A. (2000). Climate and cattle in north Africa. In Blench, R., and MacDonald, K. C. (eds.), *The Origins and Development of African Livestock: Archaeology, Genetics, Linguistics, and Ethnography*, University College London Press, London, pp. 61-86.

Holl, A. (1998). The dawn of African pastoralism: an introductory note. *Journal of Anthropological Archaeology* 17: 81-96.

Joleaud, L. (1936). Gisements de vertébrés quaternaires du Sahara. *Bulletin de la Société d'Histoire Naturelle d'Afrique du Nord* 26(bis): 23-39.

Koté, L. (1992). *Naissance et Développement des Économies de Production en Afrique Centrale*, Doctoral thesis, Paris.

Kubasiewicz, M. (1973). Spezifische Elemente der pölnische archäologischen Forschungen des letzten Vierteljahrhunderts. In Matolcsi, J. (ed.), *Domestikationsforschung und Geschichte der Haustiere*, Kaido, Budapest, pp. 371-376.

Krzyzaniak, L. (1992). The later prehistory of the upper (main) Nile: comments on the current state of research. In Klees, F., and Kuper, R. (eds.), *New light on the Northeast African Past*, Heinrich-Barth Institut, Köln, pp. 239-248.

Lambrecht, S. (1997). *Archeozoölogische Studie van het Site Mege (1e millennium BC - 1983 AD). Aanduidingen voor de eerste Veeteelt in het ZW-Tchaadbekken*, Licentiate Thesis, Katholieke Universiteit Leuven, Leuven.

MacDonald, K.C. (1994). *Socio-Economic Diversity and the Origins of Cultural Complexity Along the Middle Niger (2000 BC to AD 300)*, Ph.D. dissertation, Cambridge. [unpublished]

MacDonald, K. C. (1996a). Tichitt-Walata and the Middle Niger: evidence for cultural contact in the second millennium BC. In Pwiti, G., and Soper, R. (eds.), *Aspects of African Archaeology. Papers from the 10th Congress of the PanAfrican Association for Prehistory and Related Studies*, University of Zimbabwe Publications, Harare, pp. 429-440.

MacDonald, K. C. (1996b). The Windé Koriji complex: evidence for the peopling of the eastern Inland Niger Delta (2100-500 BC). *Préhistoire et Anthropologie Méditerranéenne* 5: 147-165.

MacDonald, K. C., and MacDonald, R.H. (2000). The origins and development of domesticated animals in arid West Africa. In Blench, R., and MacDonald, K. C. (eds.), *The Origins and Development of African Livestock: Archaeology, Genetics, Linguistics, and Ethnography*, University College London Press, London, pp. 127-162.

MacDonald, K. C., and Van Neer, W. (1994). Specialised fishing peoples in the later Holocene of the Méma region (Mali). In Van Neer, W. (ed.), *Fish Exploitation in the Past. Proceedings of the Seventh Meeting of the ICAZ Fish Remains Working Group*, Annales du Musée Royal de l'Afrique Centrale, Sciences Zoologiques 274, Tervuren, pp. 243-251.

Marshall, F., and Stewart, K. (1994). Hunting, fishing and herding pastoralists of western Kenya: the fauna from Gogo Falls. *Archaeozoologia* 7: 7-27.

Mbida, C., Van Neer, W., Doutrelepont, H., and Vrydaghs, L. (2000). Evidence for banana cultivation and animal husbandry during the first millennium BC in the forest of Southern Cameroon. *Journal of Archaeological Science* 27: 151-162.

Meeussen, A. (1980). Apports nouveaux en matière de classification et du degré d'archaïsme des langues bantoues. Colloque du C.N.R.S. *L'Expansion Bantoue*, n.s. 2: 457-472.

Neumann, K., and Ballouche, A. (1992). Die Chaîne de Gobnangou in SE Burkina Faso - ein Beitrag zur Vegetationsgeschichte der Sudanzone W-Afrikas. *Geobotanische Kolloquien* 8: 53-68.

Neumann, K., and Vogelsang, R. (1996). Paléoenvironnement et préhistoire au Sahel du Burkina Faso. *Berichte des Sonderforschungsbereichs 268 'Kulturentwicklung und Sprachgeschichte im Naturraum Westafrikanische Savanne'* 7: 177-186.

Paris, F. (1996). *Les Sépultures du Sahara Nigérien du Néolithique à l'Islamisation: Coutumes Funéraires, Chronologie, Civilisation*, Orstom (Mémoires et thèses), Paris .

Paris, F. (1997). Les inhumations de Bos au Sahara méridional au Néolithique. *Anthropozoologia* 9: 113-122.

Peters, J. (1988). Osteomorphological features of the appendicular skeleton of African buffalo, *Syncerus caffer* (Sparrman,1779) and of domestic cattle, *Bos primigenius* f. taurus Bojanus 1827. *Zeitschrift für Säugetierkunde* 53: 108-123.

Pierot, F. (1987). *Etude Ethnoarchéologique du Site de Mashita Mbanza (Zaïre)*, Licentiate thesis, Université Libre de Bruxelles.

Poplin, F. (1976). Remarques théoriques et pratiques sur les unités utilisées dans les études d'ostéologie quantitative, particulièrement en archéologie préhistorique. *Union Internationale des Sciences Préhistoriques et Protohistoriques, IXe Congrès, Nice. Thèmes spécialisés*: 124-141.

Raimbault, M. (1986). Le gisement néolithique de Kobadi (Sahel malien) et ses implications paléoclimatiques. *INQUA Symposium 86 Dakar 'Changements Globaux en Afrique'*, Dakar, pp. 393-397.

Raimbault, M. Guérin, C., and Faure, M. (1987). Les vertébrés du gisement néolithique de Kobadi (Mali). *Archaeozoologia* 1: 219-238.

Reitz, E.J., Quitmyer, I.R., Hale, H.S., Scudder, S.J., and Wing, E.S. (1987). Application of allometry to zooarchaeology. *American Antiquity* 52: 304-317.

Roche, E. (1991). Evolution des paléoenvironnements en Afrique centrale et orientale au Pléistocène supérieur et à l'Holocène. Influences climatiques et anthropiques. *Bulletin de la Société Géographique de Liège* 27: 187-208.

Roset, J.P. (1987). Néolithisation, Néolithique et post-Néolithique au Niger nord-oriental. *Bulletin de l'Association Française pour l'Etude du Quaternaire* 4: 203-214.

Schwartz, D., de Foresta, H., Dechamps, R., and Lanfranchi, R. (1990). Découverte d'un premier site de l'Age du Fer ancien (2.110 B.P.) dans le Mayumbe congolais. Implications paléobotaniques et pédologiques. *Comptes Rendus Hebdomadaires des Séances de l' Académie des Sciences Paris* 310(2): 1293-1298.

Shaw, T. (1981). The Late Stone Age in West Africa and the beginnings of African food production. In Roubet, C., Hugot, H.J., and Souville, G. (eds.), *Préhistoire Africaine*, Recherche sur les Grandes Civilisations 6, Editions ADPF, Paris, pp. 213-235.

Smith, A.B. (1974). Preliminary report of excavations at Karkarichinkat Nord and Karkarichinkat Sud, Tilemsi Valley, Republic of Mali, Spring 1972. *West African Journal of Archaeology* 4: 33-55.

Smith, A.B. (1976). A microlithic industry from Adrar Bous, Ténéré desert, Niger. In Abebe, B., Chavaillon, J., and Sutton, J.E.G. (eds.), *Proceedings of the Seventh Panafrican Congress of Prehistory and Quaternary Studies - Addis Ababa 1971*. Addis Ababa, pp. 181-196.

Smith, A.B. (1980a). The Neolithic tradition in the Sahara. In Williams, M.A.J., and Faure, H. (eds.), *The Sahara and the Nile*, Balkema, Rotterdam, pp. 451-465.

Smith, A.B. (1980b). Domesticated cattle in the Sahara and their introduction into West Africa. In Williams, M.A.J., and Faure, H. (eds.), *The Sahara and the Nile*, Balkema, Rotterdam, pp. 489-501.

Soper, R.C. (1985). *Socio-Cultural Profile of Turkana District*, Institute of African Studies, University of Nairobi and The Ministry of Finance and Planning, Nairobi.

Stahl, A.B. (1985). Reinvestigation of Kintampo 6 rock shelter, Ghana: implications for the nature of culture change. *African Archaeological Review* 3: 117-150.

Sundström, L. (1972). *Ecology and Symbiosis: Niger Water Folk*, Studia Ethnographica Upsaliensia 35, Uppsala.

Van Grunderbeek, M.C., Roche, E., and Doutrelepont, H. (1983). *Le Premier Age du Fer au Rwanda et au Burundi. Archéologie et Environnement*, Institut National de Recherche Scientifique, Publication 23, Butare.

Van Neer, W. (1997). Etude des ossements animaux de l'amas coquillier de Ndiamon-Badat (Delta du Saloum). In Thilmans, G. (ed.), *Fouille et Dégradations dans les Îles du Saloum*, Dakar, pp. 15-21.

Van Neer, W. (2000). Domestic animals from archaeological sites in Central and West-Central Africa. In Blench, R., and MacDonald, K. C. (eds.), *The Origins and Development of African Livestock: Archaeology, Genetics, Linguistics, and Ethnography*, University College London Press, London, pp. 163-190.

Van Neer, W., and Clist, B. (1991). Le site de l'age du fer ancien d'Oveng (Province de l'Estuaire, Gabon), analyse de sa faune et de son importance pour la problématique de l'expansion des locuteurs bantu en Afrique Centrale. *Comptes Rendus Hebdomadaires des Séances de l' Académie des Sciences Paris* **312**(II): 105-110.

Van Noten, F. (1983). *Histoire Archéologique du Rwanda*, Annales du Musée Royal de l'Afrique Centrale, Sciences Humaines 112, Tervuren.

Vigne, J.-D. (1991). The meat and offal weight (MOW) method and the relative proportion of ovicaprines in some ancient meat diets of the north-western Mediterranean. *Revista di Studi Liguri* A **LVII**(1-7): 21-47.

16
BOVINES IN EGYPTIAN PREDYNASTIC AND EARLY DYNASTIC ICONOGRAPHY

S. Hendrickx

INTRODUCTION

The earliest possible evidence for the symbolic importance of bovines dates to the terminal Palaeolithic. In a few Qadan tombs at Tushka, the deceased were accompanied by horncores, which obviously must have had symbolic significance (Wendorf, 1968, p. 875). In two cases, an almost complete horncore was placed directly over the skeletons, near the head. A third example was near the head of a burial. Furthermore, bovine bones were scattered over the surface. The excavators suggest the possibility that the horncores served as grave markers.

The importance of bovines for Neolithic subsistence hardly needs any explanation. With regard to the Egyptian case, domesticated cattle is most probably already attested in the Western Desert from about 9300 cal BC onwards (Gautier, 1984; Wendorf and Schild, 1995). The recent find at Nabta Playa of buried cattle bones dating to the middle and late Neolithic (McKim Malville *et al.*, 1998) strongly confirms the importance of cattle, not only from the economic point of view, but especially as an element of great religious importance. This is particularly well illustrated by the cattle tumulus E-94-1 at Nabta Playa, dated to 6470 ± 270 bp, where a young cow was buried in a roofed, clay-lined chamber (McKim Malville *et al.*, 1998, p. 488). It seems obvious that for the Neolithic/ Ceramic cultures in the Western Desert a very important interdependence existed between man and cattle (see, for example, Close, 1996), which was also religiously expressed.

This strong relationship may have become somewhat less marked with the introduction of agriculture, which can be attested for the first time with the Fayum Neolithic culture, about 5000 cal BC. Through its lithic technology, the Fayum Neolithic shows strong resemblances to the late Neolithic of the Western Desert, but agriculture, introduced most probably from the Levant, was obviously the basis of subsistence. Cattle are attested at the Neolithic settlements from Lower Egypt at Merimde Beni Salama (von den Driesch and Boessneck, 1985), Maadi (Boessneck *et al.*, 1989) and Buto (Boessneck and von den Driesch, 1997). The presence of a number of pottery figurines of bovines in all layers at Merimde seems to support their importance other than just economically (Eiwanger, 1984, pp. 53-54, table 63, 1988, p. 40, table 47, 1992, p. 60, tafel 89-90). All of these figurines show little detail and the horns always stand out as the most important characteristic. Also, it is remarkable that no other animals but bovines occur among the pottery figurines from Merimde (Eiwanger, 1992, p. 60). It is less obvious whether similar figurines also occurred at Maadi (Rizkana and Seeher, 1989, pp. 11-12, plate 1) and Buto (von der Way, 1997, p. 112, tafel 56).

The Predynastic sequence in Upper Egypt started with the appearance of the Badarian culture, attested with certainty for the period between 4400 and 4000 BC, but which probably already existed before that time. A few bovines have been found buried among the human burials in Badarian cemeteries, illustrating their socio-religious importance (Brunton and Caton-Thompson, 1928, p. 12, tombs 5422, 5434). Strangely, figurative representations of bovines are not attested for the Badarian. Most probably this is only a consequence of our limited knowledge of this period. More information is available for the subsequent Nagada units, which are subdivided into several sub-units (Hendrickx, 1989, 1996).

REPRESENTATIONS OF BOVINES

Representations of bovines in Predynastic art[1] are frequently attested from Nagada I times onwards. We will first turn our attention to the more realistic images, although a clear distinction between 'realistic' and 'stylized' is not always possible. As will be discussed further, stylized elements may be combined with overall realistic representations. The earliest examples, probably all of them bulls, occur on White Cross-lined pottery, typical of Nagada IA-IIA times. Because of the rarity of White Cross-lined pottery with figurative decoration, it is not surprising that the corpus of examples is limited (Appendix A, nos. 1-5). Modeled figurines of bulls were also attached to the rims of White Cross-lined pots (Appendix A, nos. 6-8). Although one could still claim that the fundamental reason for the depiction of bovines is to be found in the economic importance of the animals, these are evidently not merely representations illustrating economic wealth. Indeed, animals such as the hippopotamus and the crocodile figure more frequently on White Cross-lined pottery than bovines, and exceptionally even in combination with them. Although crocodile and hippopotamus could have been hunted for their meat, they are not of economic importance to farmers, but on the contrary extremely harmful for their crops. Therefore, a more symbolic, probably religious and/or sociological interpretation for the bovines must be taken into consideration. Contemporaneous with the White Cross-lined pottery, a number of clay figurines of bovines is also known. The examples found at el-Amra (MacIver and Mace, 1902, plates V, IX) are from a funerary context and date mainly to the Nagada I and early Nagada II period. Many of these figurines represent cows and calves and are probably not of great relevance for the present study, which will place the emphasis on bulls. A clay statuette of a bull has recently been found in the elite tomb U-235 at Abydos (Hartung, 1998, p. 83, tafel 4, c). A very nice example from Gebel Tarif (Quibell, 1905, no. 14709) is unfortunately undated. At first view, these statuettes could be considered substitutes for real animals which would be at the disposal of the deceased in his eternal life. This is however contradicted by the presence of similar figurines of, for example, hippopotamus in other tombs of the same period. Once again a symbolic meaning for at least a number of these statuettes can be supposed.

On the Decorated pottery, characteristic of Nagada IIC-D, bovines are extremely rare; only two examples are known to me (Appendix B), although other types of animals are frequently depicted on this type of pottery. There are important iconographic and compository differences between White Cross-lined and Decorated pottery. The absence of bovines on Decorated pottery must refer to the particular intention at which the decorative schemes of these vessels are aimed. Since this is still a much debated topic (for recent interpretations, see Adams, 1988, pp. 48-53; el-Yahky, 1985; Midant-Reynes, 1992, pp. 180-182; Smith, 1993), for which by no means a consensus exists, it will not be brought into consideration here. It has however been suggested that the cult of a goddess related to the cow could be recognized on the Decorated jars in the women with raised arms and some types of standards (Baumgartel, 1955, p. 81, 1960, pp. 146-147; Adams, 1988, p. 48; Hassan, 1992). A number of these standards certainly represent bovine horns (Petrie, 1920, plate XXIII, 5, 3-8; Newberry, 1913). On one occasion, the complete figure of a bull is depicted on a standard, inscribed on a pottery fragment from the region of Asfun el-Mata'na (Weigall, 1907, p. 49).

Among the figurative graffiti occurring occasionally on pottery, bovines seem to be present (e.g., Petrie, 1896, plate LI, 14-15; Petrie and Mace, 1901, plate XX, 28; Quibell, 1905, no. 11733; Brunton, 1948, plate XXII, 2), but their number remains limited and the little detail given in this kind of representation makes them unsuitable for the purpose of the present study.

Another important type of object frequently found in Predynastic tombs are the greywacke or mudstone palettes, many of them animal shaped. Bovines however seem to be very rare among them, the most frequently occurring animals being fish, different kinds of birds and tortoises. Furthermore, the identification of certain palettes as representing bovines is debatable, because of the simplified manner in which the animals are rendered. An important characteristic for the representation of bulls is the hump (cf. Grigson, 1991: figure 1c), which probably allows us to recognize a number of rather uncharacteristic animal shaped palettes also as bovines (Appendix C).

On the wall painting from Hierakonpolis, tomb 100, dating to Nagada IIC, bovines can most probably be recognized in one of the labels referring to power and control, which are located below the row of boats. The label in question shows a man overthrowing a bull, of which the feet have been tied together. In this case, the power of the bull is used to reinforce the importance of the overthrower. Here, the animal is probably to be taken literally and not as a symbol. A similar phenomenon can also be observed with regard to the lion. On the Battlefield palette, the lion represents the victorious king, while on the Hunter's palette, the lion is hunted by man.

In general, it can be said that representations of bovines can only be found very exceptionally on objects found in Predynastic or Early Dynastic tombs. Therefore, it can be conjectured that the symbolic meaning of bovines, and especially of the bull, had no direct relationship with the funeral world.

Cattle are frequently represented in rock art. However, these will not be discussed extensively here given the difficulties in dating and other specific problems related to rock art, such as the difficulty in defining the absence or presence of relationships between individual drawings. In general, it seems that the representations of bovines are mainly to be placed chronologically during the Nagada III period and the Old Kingdom. At Elkab, none of the representations of bovines could be dated prior to the Nagada III period (Huyge, 1995, 2001). This is all the more remarkable because the florescence of the rock art in Upper Egypt, and probably also in Nubia, is to be placed during Nagada II and III. The Nagada III rock art at Elkab is characterized by numerous images referring to royal ideology, among them the 'victorious' bull (Huyge, 1995, 2001). However, Huyge regards the majority of bovines dating to the Old Kingdom as representing sacrificial beasts, related to the temple cult.

Bulls occur frequently, among many other animals, on the well known group of late Predynastic, presumably mainly Nagada IIIA-B, ivory and bone carvings, for which unfortunately the provenance is generally unknown (Appendix D), and on the 'ivories' from the main deposit at Hierakonpolis (Quibell and Petrie, 1900, plate XII, 1-2, XIII, XVII; Adams, 1974a, pp. 60-61, no. 326). These representations seem strongly related to those on the decorated palettes, dating probably also mainly from the same period. Bulls appear on the Hunter's palette, the Oxford palette, the Louvre palette, the Narmer palette and the Libyan palette (cf. Cialowicz, 1991). On all of these objects, the bull can appear either individually or as one of the animals occurring in the characteristic rows of animals. With regard to the latter, the bull does not stand out among the other animals. The meaning of these animal rows is not totally clear, although they most probably represent the idea of order being imposed by socio-religious powers which, however, are not clearly identified (cf. Kemp, 1989, pp. 46-53; Baines, 1995, p. 111). Obviously the representation of individual bovines are of greater interest for the present study.

On the Narmer palette and the Louvre palette, as well as on the Koptos colossi (Dreyer, 1995), the bulls are clearly meant to be 'victorious bulls', representing the king as triumphant (cf. Störk, 1984, col. 258; Davis, 1992, p. 169).

A number of bovines can also be found on a group of objects which are, in general, a little more recent, at Nagada IIIB-C2, than the previous group. On these objects, bovines seem to appear in a religious context, the exact meaning of which remains unfortunately most problematic. On the Narmer macehead (Cialowicz, 1987, pp. 38-41; Friedman, 1996) the figure of a bull is used for identifying 400,000 pieces of cattle among the captured booty, but the image of two bovines, possibly a cow and a calf, within some kind of enclosure is obviously more interesting, but far more difficult to explain. The animals seem related to the carrying chair and its occupant, above which they are located. However, it remains unclear whether they identify the person concerned in some manner or merely indicate his provenance or some other circumstance. The related depiction of a bull in what might also be some kind of enclosure, on a tablet from the time of Hor-Aha, found in tomb B 18-19 [Nagada IIIB-C1] at Abydos (Petrie, 1901, plate X, 2), suggests a locality. This is confirmed by another tablet from cemetery B dating also to the reign of Hor-Aha, on which the figure of a bull is used for the identification of a building (Petrie, 1901, plate XI, 1). On a tablet from the tomb of Hemaka [Nagada IIIC2] finally, a standard with a bull is carried in a procession (Emery, 1938, plate 18).

The religious context of the bull is also confirmed by a few statuettes found in the earliest temple levels at Abydos. These include pottery statuettes from the Osiris temenos (Petrie, 1902, p. 26, plate LIII, 40-42), and an ivory (Petrie 1903, p. 27, plate IX, 204) and a quartz statuette (Petrie, 1903, p. 25, plate VI, 63), found respectively in the rooms M 64 and M 69 of the temple area. The provenance of Early Dynastic statuettes in de Kofler-Truniger Sammlung is generally also considered to have been the earliest temple levels at Abydos, but this has been contradicted by Dreyer (1986, pp. 54-55). Among these statuettes, there is one of a bovine (Luzern, Sammlung Kofler-Truniger K 9643 J, cf. Müller, 1964, p. 22, A 18, color photograph in Wildung, 1981, Abb. 39).

On a fragmentary ivory tablet from the tomb of Den at Abydos [Nagada IIIC2], a bull, who is most probably hunted by the king, is depicted on the symbol for mountains (Dreyer, 1998b, p. 163, tafel 12, e). This is probably not to be considered as a reference to the location of the hunt but rather as an indication that a wild bull is meant. Discounting the detail from the Hierakonpolis painting discussed previously, this is the oldest representation of the wild bull hunt, which was to become a classic element of the royal iconography.

Finally, the very beautiful drawing of a bull on an ostrakon from the tomb of Hemaka [Nagada IIIC2] must be mentioned (Emery, 1938, p. 40, no. 431, plate 19D). It is generally considered as an artist's sketch, without further implications.

The stylistic manner in which bovines are rendered differs in no way from the rest of the Predynastic and Early Dynastic art. This means that two dimensional images are always shown in profile, while three dimensional representations are symmetrical and in the majority of cases do not attempt to suggest movement or any kind of action. The horns of the animals are always prominently represented, and certainly functioned as an obvious characteristic for identifying the animals. Nearly all of the bovines have incurved horns, turned towards the spectator, as became typical for Dynastic Egyptian art. In a few exceptional cases, the horns are rendered more or less in perspective. The most obvious examples are the White Cross-lined cup from Naqa ed-Deir (cf. Appendix A, no. 2) and the Libyan palette. In only a few cases, lyre-shaped horns occur, but it is also possible that the hartebeest was being represented and not the bull. The most remarkable examples are to be found on the Carnarvon knife handle (see Appendix D, no. 4), a decorated ostrich egg (Oxford 1895.990; Payne, 1993, p. 253, no. 2104, figure 85; Friedman, 1995, p. 60, 'gazelles or hartebeest') from tomb 1480 at Nagada (Nagada IC) and a figurative flint of unrecorded provenance (Berlin 15774; Scharff, 1929, no. 97, plate 21, '*Kuhantilope*', Boessneck, 1988, figure 10 '*Kuhantilope*'; Osborn and Osbornova, 1998, p. 4, 'hartebeest'). However, the actual horns of bulls which are known for early Egypt, and especially the large number of them which were found around mastaba S 3504 at Saqqara (Nagada IIIC2, reign of Djed) show us that there was a great variability in shape (Emery, 1954, pp. 8-9, plate I, VI-VII). Therefore, the rendering of the horns in Predynastic and Early Dynastic art has clearly become idealized and uniformed. Although this is a normal feature for

Egyptian art, even at this early stage, this can also imply that the incurved shape of the horns had symbolic value, especially if one considers the emphasis which is placed upon them in the representations.

A most particular rendering of the bull is the so-called 'double bull' found on the Hunter's palette (Cialowicz, 1991, pp. 55-56; Baines, 1995, p. 151). The 'double bull' clearly relates to the building next to it, the meaning of which unfortunately remains open for discussion. Baines (1995, p. 112) suggests the group of the building and the 'double bull' represent royalty with a device that later disappeared, but he also leaves the possibility open that the 'double bull' could identify the building. Whatever the interpretation, the symbolic significance of the 'double bull' cannot be denied.

The importance of bovines is also illustrated by the burial of a small group of them at Hierakonpolis, Locality 6, tomb 7, dating probably to the Early Dynastic period (Hoffman et al., 1982, pp. 55-56, Adams, 2000, pp. 33-34). The burial of these animals has been considered by the excavators, in a very speculative manner, to represent "...either a royal bull and its family or Hathor and her family" (idem; see also Adams, 1996, pp. 6-7).

REALISTIC REPRESENTATIONS OF BOVINE FEATURES

We will now turn our attention to the realistic representations of parts of bovines, among which horns, heads and legs are the most important. The detailed image of a bull's head can be found on a small block of limestone from the Early Dynastic temple at Hierakonpolis (London UC.14859. Quibell and Petrie, 1900, p. 7, plate II; Adams, 1974a, p. 4, no. 3/3, plate 3, 3). Several pendants in the shape of bull's heads are known. They are most probably to be considered as amulets and can be compared with the more complex bull's head amulets which will be discussed later. In the same manner a macehead decorated with two foreparts of bulls, from the main deposit at Hierakonpolis (Quibell and Petrie, 1900, plate XIX, 3), is to be compared with maceheads decorated with bull's head amulets still to be discussed. Of particular interest are three ivory double bull's head amulets, said to come from Abydos (Brussels E.3381a-c; ex Hilton Price Collection no. 4551; Price, 1897, no. 890; Price, 1900; Capart, 1904, figure 139, 1905, figure 154; Figure 16.1), which can be related to the double bull mentioned previously, although their shape is also related to the so-called 'pelta' palettes (cf. infra).

A most interesting image occurs on a decorated hippopotamus tusk from the main deposit at Hierakonpolis (Quibell and Petrie, 1900: plate XIV; Adams, 1974a, p. 75, no. 384, University College 14875), on which can be seen the typical facade of a niched building of the type which is generally interpreted as a palace, but which could also represent the enclosure wall of a temple. Above each of the doors hangs a bull's head, or more probably a bucranium, with the horns curving downwards. Obviously these are meant as an identification of the building, or as a general indication of its importance. A row of strongly stylized bull's heads or bucrania on an unpublished jar preserved at Brussels (E. 6400) may have had a similar meaning. The provenance of the jar is not known, but it belongs to Petrie's Predynastic type L 36 (Petrie, 1920, plates XLVII-XLVIII), and therefore probably dates to the beginning of the Nagada III period.

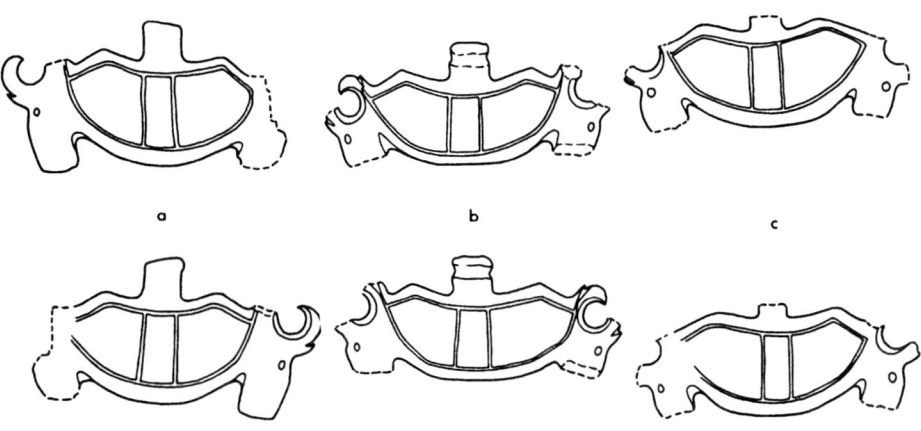

Figure 16.1. 'Double bull's head amulets'. Abydos (?), formerly Hilton-Price collection (Brussels E.3381a-c). Scale 1:2.

Actual horns of bulls have been found in a limited number of Predynastic tombs. The best documented examples are from Naqa ed-Deir (tomb 7097 (Nagada IID?), tomb 7174 (date unknown), tomb 7525 [Nagada IIC-D1?] Lythgoe and Dunham, 1965, pp. 53-54, figure 21a, p. 100, figure 41h, pp. 339-341, figures 151-152). Because the horns had been removed from the skull, they were not intended as food offerings, but must have had symbolic meaning. A most interesting object in this respect is a pottery horn from tomb 20 at Gerza (SD 58 = Nagada IIC-D?), which is terminated by a realistic head of a bovine, most probably a bull (Petrie *et al.*, 1912, p. 23, plate VII, 13). Although this object is clearly symbolic, its exact meaning is, once again, unclear.

Standards with different types of emblems can be seen on the large majority of the boats figuring on Decorated pottery. Among them, standards with bull's horns occur frequently (Petrie, 1920, plate XXIII, 5, 4-8; Newberry, 1913). Two main types can be distinguished. The first type shows two pairs of horns, one within the other; the second type show a number of horns attached to some kind of pole. It is very tempting to recognize in some standards of the first type at the same time a schematized human figure with raised arms, although it is impossible to prove this. Nevertheless, these standards clearly illustrate the religious, and perhaps already the political, relevance of the bull's horns.

An indisputable expression of the symbolic importance of bovines are the legs of chairs, beds and other furniture shaped as bovine legs, most probably bulls. The earliest example can be dated to Nagada IIIA2 and comes from Hierakonpolis Hk. 6 tomb 11 (Adams, 1996, p. 13, 2000, pp. 109-111, plates xxxiiib-xxxivb). Most examples known are from the royal tombs at Abydos

(Amélineau, 1899, plate XXXII, 1904, plate XIV; Petrie, 1900, plates XII, XXVII, 1901, plates XXXII, XXXIV, XXXVI-XLI) and the mastabas of the highest officials at Saqqara (Emery, 1938, p. 40, plate 19, 1939, pp. 63-64, 1949, pp. 57-59, 1954, pp. 38-55, plates XXVI-XXVII, XXIX, 1958, p. 84, plate 102, 1961, plate 48, figure 130). Similar objects are known from Nagada (de Morgan, 1897, p. 189; Quibell, 1905, CG 14045-14051; Kahl and Engel, 2001, Abb. 17), Tarkhan (Petrie, 1913, plates VIII-IX, XIV), Helwan (Saad, 1969, plate 45) and Abu Roash (Klasens, 1959, p. 60, figure 10). Nearly all of them have been found in elite tombs and it is most obvious that furniture in the funerary equipment illustrates the high social status of the tomb owner. There are however also a few examples from the main deposit at Hierakonpolis (Quibell and Green, 1902, plate XVI). During more recent periods of Egypt's history, besides bull's legs, those of lions were also frequently used for furniture. There can hardly be any doubt that these two impressive and powerful animals have been chosen as protectors, or originally perhaps even as personifications of the individuals using this furniture.

A strange greywacke bull's leg with three perforations (Brussels E.6154;[2] Figure 16.2) cannot have been used as a palette and was probably some sort of amulet.

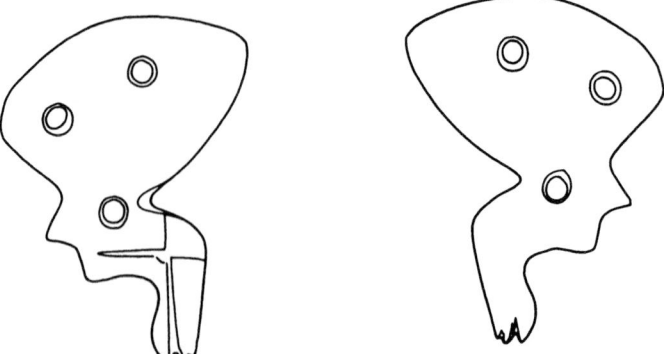

Figure 16.2. Bull's leg amulet. Provenance not recorded, formerly MacGregor collection (Brussels E.6154). Scale 1:2.

As was the case for the image of the whole animal, the bull's head may also have been used as an identification for the king during the early Nagada III period. In tomb U-j at Abydos, dated to Nagada IIIA1, inscriptions in ink on Wavy-handled jars have been found which represent a bull's head, or once again more probably a bucranium, on a pole (Dreyer, 1998a, pp. 65-67). A large number of similar inscriptions has been found in tomb U-j, each time showing an animal in combination with a plant. These inscriptions have been interpreted by Dreyer (1995, 1998a, pp. 84-86) as royal domains, where the animals, or once a shell, are the names of late Predynastic kings. This explanation is however less obvious for the inscription with the bucranium, which is of particular interest because in that case the plant is absent. Also, the head or bucranium on a pole is not a living animal, in contrast to the other representations. Recently another example of the same emblem has been found as a part of a large rock art tableau at Gebel Tjauti (Darnell and Darnell, 1995-1996, p. 66, figure 8), on the desert road which cuts the bend of the Nile between the Luxor and Abydos areas (Darnell and Darnell, 1997). This tableau has been interpreted as related to the victory of a king from Abydos, probably even the Scorpion of tomb U-j, over another Upper Egyptian king, who is probably to be looked for at Nagada because of the location of Gebel Tjauti (Friedman and Hendrickx, in press). On this tableau, the bucranium on a pole occurs between the victorious king, who has a macehead in front of him, and a bound prisoner. Unfortunately, it is not obvious whether the bucranium identifies the king, the prisoner, or represents an idea in itself.

STYLISTIC REPRESENTATIONS

Besides the easily recognizable, more or less realistic representations (Appendix E), there are also far more complex Predynastic images which include bovines or some of their most important characteristics. A key document in this regard is a very delicately worked flint (Brussels E.6185a; Figure 16.3) from a small cache of flints in the so-called 'royal mastaba' at Nagada. This piece was found in 1904-1905 by Garstang (1905) and until now only a photograph of it has been published (Charron, 1990, p. 87, 105, no. 432). The date of the object is less obvious than its provenance might suggest. Among the other flints from the cache are two very fine rhomboidal knives (Hendrickx 1994, pp. 52-55), which are generally considered to date to the Nagada I and early Nagada II period (Baumgartel, 1960, p. 32). Furthermore, despite the fact that the objects have been found *in situ*, all of them are damaged, and it is therefore quite possible that they predate the reign of Hor-Aha, during whose reign the tomb at Nagada was constructed (Kahl and Engel, 2001). In that case, they would have been included in the funerary equipment as heirlooms referring to tradition, a custom which is not unknown for Early Dynastic Egypt (Sowada, 1999). At first view the piece under discussion strongly resembles the well known human figurines from the Nagada I-II period (Ucko, 1968). Most of the known figurines are female, although male examples also exist. On the Decorated pottery, the figures with raised arms are always female, while on the extremely rare White Cross-lined pots with human representations there are only male figures with raised arms (Hendrickx, 1994; 1995, 1999; Köhler, 1998). The flint object from Nagada can also easily be considered to represent a bull's head. Indeed, it obviously is not a human figure because there is no head and, from the finishing of the piece, it can be seen that there has never been one. The delicate craftsmanship of the piece shows that the artisan who produced it was definitely capable of adding a head, if he had wanted to. Most unfortunately the ends of the horns/arms are broken off and therefore it is impossible to identify whether they are horns or arms. However, this problem is probably irrelevant, because it is quite obvious that characteristics from both human and bovine representations are combined. The head of the bull can also be regarded as the trunk of a human (female) body, the

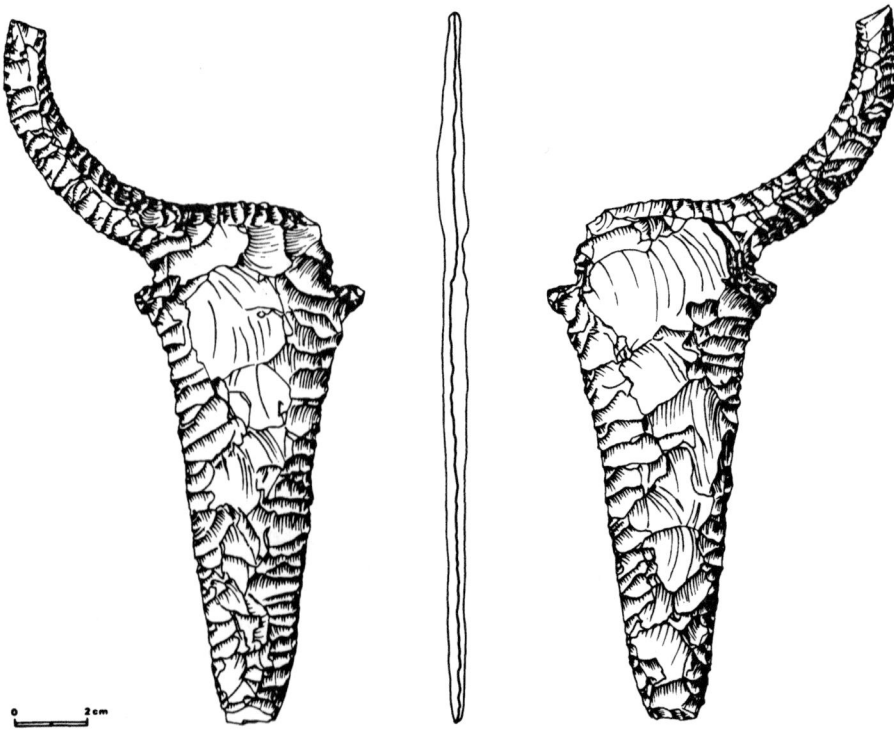

0 ____ 2 cm

Figure 16.3. Figurative flint. Nagada, 'royal tomb', formerly MacGregor collection (Brussels E.6185a). Scale indicated.

horns as arms and finally the triangular extension on both sides as the eyes of the bull or the breasts of a woman. When compared to the modeled bull's heads around mastaba S 3504 at Saqqara, it is indeed likely that the extensions represent the eyes of the bull and not his ears. The interchangeability between the eyes of bovines and female breasts will also be recognized in a number of objects still to be discussed. It seems obvious that this confusion was a deliberate aim. The bovine element in the flint object from Nagada is confirmed by another flint, unfortunately without provenance, showing a bull's head with naturalistic horns but also with the triangular extensions underneath the horns (London BM EA.32124; Capart, 1905, figure 115).

Admittedly, the described flint figure remains unique and is also made using a very particular technique which certainly had its restrictions.[3] There are, however, other documents which show the combination of stylized elements from different contexts. With regard to the combination of elements representing both a bull and a woman, the case of the so-called bull's head amulets is the most remarkable. This type of pendant is the most common one during Predynastic times and continues to be used until the Early Dynastic period. On a few occasions, bull's head amulets have been found as parts of necklaces (Stanton and

Hoffman, 1988, p. 82, no. 70; Payne, 1993, figure 72, no. 1713). Petrie already recognized both the chronological importance as well as the formal evolution of this type of objects, and according to him they were "the oldest form of amulet. It begins at S.D. 46 or earlier, and continues in use till S.D. 67, when it is very degraded" (Petrie and Mace, 1901, p. 26). This corresponds with Nagada IIC (or earlier) to the beginning of the Nagada III period, which should certainly be expanded until Nagada IIIC2 (cf. Appendix F). Most of the examples known come from tombs, but they have also been found in living areas, the earliest levels of the temples at Abydos and Hierakonpolis and even at sites in Palestine (Appendix F). There are also a few objects which show a close relation with the bull's head amulets. The shape of the amulet has been used on two stone vases from the main deposit at Hierakonpolis (London UC.15010, Quibell and Green, 1902, plate XLVIIIa; Adams, 1974a, p. 22, no. 109, plate 15; Cambridge, Fitzwilliam Museum E.13.1898, Quibell and Petrie, 1900, plate XVII; Vassilika, 1995a, pp. 12-13) and a few maceheads, one from tomb 1051 at Abusir el-Meleq (Nagada IIC?; Cairo JdE 38743, Scharff, 1926, p. 49) and another probably from Badari (Berlin 15142, Scharff, 1931, pp. 80-81, no. 152, Abb. 21, tafel 8, 152). For two other examples the provenance is unknown (Scharff, 1926, p. 49, 1931, pp. 80-81). Furthermore, there is a unique example of a stone vase of unrecorded provenance resembling such an amulet (Oxford 1948.18; Baumgartel, 1960, plate VI, 3; Payne, 1993, p. 144, no. 1201, figure 57).

The general appearance of the bull's head amulets is very characteristic and consists of two parts (Figure 16.4). The first is a 'disc', which is the front part of the objects and is delineated by the downwards curved 'horns', while the second is a 'cylinder' which comes out of the back part of the disk and serves as a 'foot' to the object. The manner in which both parts are integrated can differ greatly, from an almost complete separation to an integration into a new shape, more or less oval with a straight base. Perforations have always been made through the upper part of the 'cylinder', showing clearly that this is the back part of the objects. However, at least two types can be distinguished, which also have chronological implications. They show an evolution towards a more oval shape and stronger integration between the two parts distinguished before (Figure 16.5). In some cases, the 'horns' are detached from the body of the objects, but this is probably merely a matter of the quality of the craftsmanship.

These amulets were originally identified by Petrie (1914, p. 44, plate XXXVIII, 212a-m) as representing ram's heads. Baumgartel (1960, pp. 73-74) was the first to note the ambiguous character of the objects, with both human and bovine characteristics, but considered the amulets themselves to represent the mother goddess, while Hoffman (1989, p. 321) hesitated between a bucranium or an elephant amulet. Most recently, an identification as an elephant amulet has also been presented by Van Lepp (1999). Needler (1984, pp. 317-318) considered the amulet to refer to the prototype of a bull god or cow goddess. She rejected the identification by Baumgartel of a mother goddess because, "such an abstraction would be unprecedented and inconsistent with the prehistoric artist's unfailing preference for easily recognized forms" (Needler, 1984, p. 318), an interpretation which of course is not shared at all in the present article. Both Adams (1995) and Vassilika (1995b) recently described this type of amulet purely as symbolizing bulls or bovines in a more general sense. The identification as a ram's head never

Figure 16.4. 'Bull's head' amulet. Provenance not recorded (Brussels E.2335). Scale 3:4.

Figure 16.5. 'Bull's head' amulet, late type. Provenance not recorded, formerly Scheurleer collection (Brussels E.7126). Scale 3:4.

gained general acceptance and is most unlikely because there are no related depictions of rams known for this period, but it is nevertheless still accepted by Otto (1986, p. 140). It is also most improbable that the amulets represent elephants, the 'horns' being the elephant's tusks and the cylindrical part his trunk, because the cylindrical part is on the back of the object and especially because this part is very short and inconspicuous for the earlier type of amulets. Also, for a number of amulets, the cylindrical part ends in a circular thickening which can also be found at the snout of some bovine shaped palettes, but never at the trunk of elephant shaped palettes.

The identification of the bull's head amulet as bovine is strongly supported by an unpublished rock drawing at Gebel Faradi, about 8 km south of Elkab, which shows a bucranium with the horns curved downwards. Considering the surrounding drawings, the bucranium is to be dated to the Early Dynastic period (Huyge, personal communication).

A final argument for the identification of the bovine element in the bull's head amulet is to be found in the bucrania surrounding mastaba S 3504 at Saqqara, dating to the time of Djed. This mastaba was surrounded by a low bench on which had been placed about 300 bull's heads modeled in clay, with real horns sticking out of them (Emery, 1954, pp. 8-9, plates I, VI-VII). The remarkable point is that the heads have been modeled, but unfortunately it is not clear from the excavation report if the actual skulls are present under the clay modeling. The shape of the heads shows a strong resemblance to the more or less oval type of bull's head amulets with a straight base, but with, of course, the notable difference that the horns are pointing outwards. Nevertheless, the general outline of the heads, as well as the position and shape of the eyes, which are in reality small bulbs of clay, leave no doubt concerning their relation with the bull's head amulet. The bucrania around mastaba S 3504 can therefore be considered definite proof of the fact that the bull is indeed the animal behind the bull's head amulet.

The human elements distinguished by Baumgartel for the bull's head amulets are most obvious if one compares them to a particular type of statuette representing women with their arms curved underneath their breasts (Appendix G). The manner in which the arms, and especially the hands, are curved is physically impossible. Also the hands have never been worked in detail; on the contrary, the arms are generally finished by blunt points. The manner in which the 'hands' are curved inwards can be found identically on the already mentioned human statuettes with raised arms (Ucko, 1968). It is to be noted that no male statuettes with the arms in front of the chest are known. The combination of arms and breasts seems therefore essential for these statuettes. Also, the heads are never shown in much detail, and it is clear that these statuettes do not aim at depicting the individuality of some particular person. For all of these reasons and also for the fact that the upper legs are represented in a manner which places emphasis on the pubic area, Baumgartel's interpretation as fertility figures is not illogical. Nevertheless, the fact that the arms have been shaped in a manner allowing them to be considered as horns also remains obvious.

The limited amount of information available does not allow us to make chronological distinctions in the statuettes with raised or lowered arms. Both already occur during the Nagada I period. The very well known ivory Badarian statuette from tomb 5107 at Badari (London BM EA.58648; Brunton and Caton-Thompson, 1928, plates XXIV, 2, XXV, 3-4; Ucko, 1968, pp. 2, 70) already shows the arms below the breast. However, as the relative position of the arms and breasts is not identical to the Nagada culture examples, this statuette will not be taken into consideration here. It is therefore impossible to say whether the statuettes with the arms underneath the breasts are an adaptation of the type with raised arms, or directly inspired by the shape of the horns of a bull. Anyhow, this question is not of great importance because the resemblance to the horns is clear in both cases.

All in all, there seems to be little doubt that the bull's head amulet is a combination of female elements, with the emphasis on fertility, and the bull. Another example of this combination can probably be found in a few jars with applied decoration, which are at present in the Ashmolean Museum at Oxford (1895.1220, Nagada tomb 1449, Payne, 1993, no. 105; E.3195, Abadiya tomb B 101, Payne, 1993, no. 106; E.2952, Hu, tomb U 179, Payne, 1993, no. 107). This was already recognized by Baumgartel (1960, pp. 31-32, see however Midant-Reynes, 1992, pp. 169-171).

The bull's head amulet disappears during the Early Dynastic period, and its shape was at that moment so much degraded that the original meaning may already have been lost. There is no trace of a similar type of amulet during the Old Kingdom, but there are two possibilities for the bull's head amulet having been integrated into formal Egyptian art as defined by Kemp (1989, pp. 19-107).

First there is the intriguing formal resemblance between the bull's head amulet and the representation of the false beard as a hieroglyphic sign, such as that found on a very nicely worked tablet from the tomb of Djed at Abydos (Petrie, 1900, plates X, 9, XIII, 2; Schott, 1951, p. 27, Textabb. 3; see also Kahl 1994, p. 720, s 26). The sign continues to be used during the Old Kingdom and is supposed to represent the deified royal beard. Very little is known of the origin and meaning of the false beard, although Wildung (1984, p. 973), when trying to explain the white crown of Upper Egypt through a visual similarity with the Imiut-emblem, suggested a relation with the tail of a calf hanging from the Imiut-emblem. However, the formal relation between the Imiut-emblem and the white crown is far from evident, especially for the Early Dynastic representations of these two objects, which are far more different from each other than they are during more recent times. This can easily be seen by comparing the white crown on the Scorpion macehead (Oxford E.3632, see most recently Gautier and Midant-Reynes, 1995; Cialowicz, 1997) with the Imiut-emblem on a tablet from Abydos (Philadelphia 9396, Logan, 1990). Also, Wildung's explanation does not take into account that the king also wears the false beard with the red crown. Therefore, it seems logical to consider the false beard as an independent symbol and not merely as something which goes with the white crown. Even though the words *hbsw.t* (beard) and *hbs.t* (animal tail) are closely related (Wildung, 1984, pp. 973-974), it cannot be denied that there is little formal resemblance between the tail of a calf and the false beard of the king. Finally, another reference to bovines might have been at the origin of the false beard. In a very tentative manner, it could be suggested that the purpose of the false beard was to create an image of the king as a personification of the bull's head amulet. In that case, the most important element would not have been the beard itself, but the strings by which it was attached, imitating the horns/arms of the amulet.

The second element is the emblem of the goddess Bat, several examples of which are known from the Nagada III period (Appendix H). The horns of Bat show a very strong curve, the tips reaching back almost to the top of her head. The curve is completely similar to that of the horns/arms of the bull's head amulets, but upwards instead of downwards. Also, the round face of Bat is very similar to the shape of the amulets. Obviously Bat, as a cow goddess, fits very well with the above proposed interpretation of the bull's head amulets being a combination of human and bovine elements.

In the end, one can only try to point out a few elements which may be related to the meaning of the bull's head amulets. Their use as pendants, in combination with their numeric importance, and to a certain extent also the attitude of the female statuettes with the arms 'supporting' the breasts, point to a general prophylactic characteristic of the bull's head amulet. This is also confirmed by the great symbolic importance of the bull, and especially his horns, referring to power and strength in general, as already discussed. Finally, there is the female element in the bull's head amulet, which does not necessarily refer to fertility only, but could also indicate the regeneration of life. Tentatively, the prophylactic value of the bull's head amulet can be situated within the context of the power and strength needed for the regeneration of life after death.

BOVINES AND BIRDS

Another combination occurring in Predynastic art is that between the bull and a bird, which often has a long neck and originally may have been an ostrich. For the large majority of these birds, their exact identity seems of little importance because very little detail is given in the representations. The most obvious examples showing a combination of bovine and bird elements are two palettes (Figure 16.6), unfortunately both of them without provenance (Brussels E.4992, unpublished; private collection Kilchberg, Switzerland, Page-Gasser and Wiese, 1997, pp. 30-31, no. 12), where the necks and heads of birds, in a shape which also occurs separately as an amulet (Appendix I; Figure 16.7), are used for the legs of bovines. Similarly, on another palette of unrecorded provenance, the horns of a bovine have been replaced by the bird amulet (Leipzig 2886, Onasch and Steinmann, 1997, p. 21), for which a theriomorph limestone vase of equally unrecorded provenance (Cairo JdE 66628, Saleh and Sourouzian, 1986, no. 6) can be mentioned as a parallel. An element of great importance is the shape of the horns and ears of the animals on the first two palettes, which can also be found on several of the other bovine shaped palettes (cf. Appendix C). The tips of the horns are curved outwards, while the ears are of a particular triangular or trapezoidal shape. This type of horns and ears also occurs for other animal shaped palettes and other objects, (e.g., Payne, 1993, nos. 1808, 1903, 1904) and should therefore rather be considered as a standardized element than a naturalistic rendering. The horns with curved tips show a great resemblance with the heads and necks of birds of the type just mentioned. This relationship can be elaborated further through the non-figurative amulets which will be discussed later. For a number of them the bird-like shape of the horns seems beyond discussion.

The combination of the bovine and the bird can most probably also be recognized in the so-called 'pelta-palette'.[4] Two general classes of pelta palettes are to be distinguished. The first one (Petrie, 1920, plate XLIV, type 30-32), which is not of interest for the present study, most probably refers to boats, some of them with bird shaped prows. These palettes have also been identified as bats (Tupinier, 1984; Osborn and Osbornova, 1998, p. 30), which is, however, most unlikely because of the asymmetric shape of many examples, among other reasons. The second group (Petrie, 1920, plate XLIV, type 100-101; Appendix J; Figure 16.8) shows two birds' heads and necks in such a manner that they also resemble a pair of outward curved horns. The relationship to bovine horns can be made not only because of the general shape, but also by the already mentioned amulets of pelta shape with bull's heads (Figure 16.1). Furthermore, there is an example of a pelta palette on which the animal head, unfortunately broken off, but apparently of the normal bird type, is surmounted with the typical horns ending in bird heads (Appendix J, no. 19). It has been argued that the boat type pelta palette was the earliest, and the origin of the second type (Vandier, 1952, pp. 384-388), but the chronological evidence only allows us to say that both types occur mainly during Nagada IIA-IIB. There is, on the other hand, limited evidence that the pelta palette with only one bird's head (Petrie, 1920: plate XLIV, type 100; Figure 16.9), of which all dated examples can be attributed to Nagada IIA, could be the origin of the type with two birds' heads (see also Regner, 1996, plate 21).

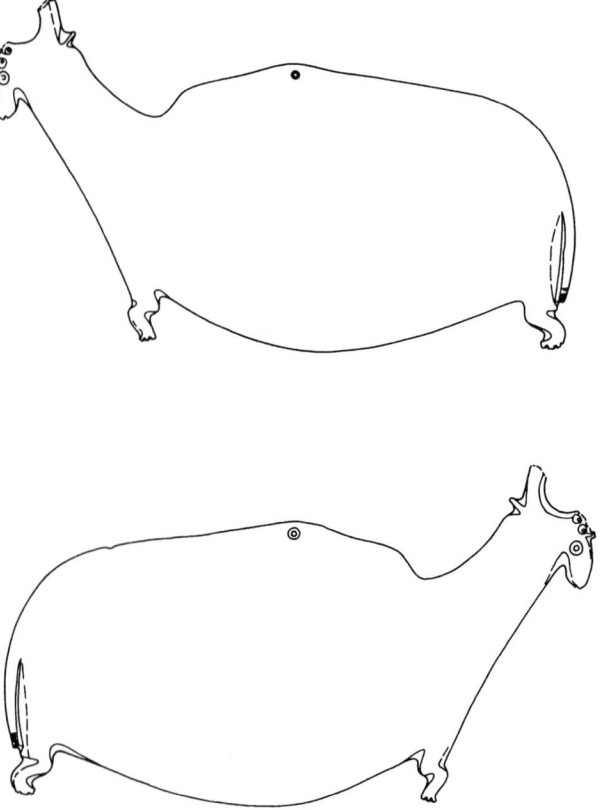

Figure 16.6. Palette in the shape of bovid with bird amulet as leg. Provenance not recorded (Brussels E.4992). Scale 1:3.

Figure 16.7. Bird amulet. Provenance not recorded (Brussels E.2179). Scale 1:2.

Figure 16.8. 'Pelta' palette with two bird heads. Provenance not recorded (Brussels E.421). Scale 1:2.

Figure 16.9. 'Pelta' palette with one bird head. Provenance not recorded (Brussels E.422). Scale 1:2.

This can be regarded as evidence, be it inconclusive, for the suggestion already made that the iconographic elements concerning bovines, birds and humans originally developed independently.

This kind of direct relationship between the bull and the bird remains nevertheless rare and, admittedly, some examples are still open for discussion. Both animals, however, can be linked with human figures, as was demonstrated above for the bull. The relationship between the bird and human figurines is obvious from the female figurines already mentioned several times, which often have heads shaped like those of birds. This however will not be discussed in detail here, because it falls beyond the aims of the present study. It should nevertheless be mentioned that the bird occurring so frequently on Decorated pottery (Hendrickx, 2000) and palettes during the Nagada II period, and especially during Nagada IIC-D, would gradually be replaced by standardized representations of falcons, typical for Nagada III. Once again, the original iconography of the birds was not assimilated into formal Egyptian art. The image of the falcon, on the contrary, probably already had its formal shape by the very end of the Nagada II period and certainly from the beginning of the Nagada III period.

DEVELOPED SYMBOLS

The type of art discussed so far combines into one image two or more highly stylized elements which originally must have represented individual ideas. In this manner, human representations with raised arms, the horns of a bull and the necks and heads of birds are combined. Next, we will see that the shapes which were obtained by combining different figurative elements can also be used as symbols on their own, without further figurative context, implying that these shapes had a recognizable cognitive value of their own. A few of the most remarkable examples will be presented here.

There exist a number of rhomboidal palettes (Petrie, 1920, plate XLIV, 1921, plate LVIII, type 91 T-U) which, on one of their pointed ends, all have a similar decoration (Appendix K). It consists of a ring which is not closed at its top, clearly representing horns, underneath which are two small, horizontally protruding triangles. Two types can be distinguished. For the first type, the horns consist of one continuous unbroken curve with tipped ends indicated, while the ends of the second type are incurved. The relationship with the flint bull's head from Brussels (Figure 16.3), especially for the second type with the particular type of horns and ears mentioned above, is extremely pertinent. The relationship with the bull is furthermore confirmed by a rhomboidal palette of unknown provenance, on top of which two antithetic bull's heads could originally be found (Brussels E.2182; Figure 16.10). Another interesting example without provenance clearly shows the interchangeability of the horns and the bird's heads (Cairo CG 14172, Quibell, 1905, p. 226). The rhomboidal shape of the palettes, however, does not show figurative resemblance and occurs very frequently without this decoration.

Similar decorations, both with horns in one continuous curve and with horns with incurved ends, also occur on other types of palettes and different kinds of objects in bone and ivory. It is impossible to differentiate the examples with horns in one curve (Appendix L; Figure 16.11) chronologically from those with incurved horns, the large majority of them dating to Nagada I and early Nagada II, and in at least one case they occur together on the same object (Appendix L, no. 6). Another interesting example is a palette on top of which a similar decoration can be found which, however, through the addition of a line separating the horns from the palette itself, shows strong resemblance to the Bat emblem (Appendix L, no. 25; Figure 16.12). This resemblance was already noted for a number of other objects with horns in one unbroken curve by Scharff (1926, p. 53, no. 339, 1929, pp. 140-141, no. 270, 1931, p. 263, no. 854; Appendix L, nos. 2, 8, 14) and their shape identified as 'Hathorkopf'.

The examples with incurved horns however are far more numerous (Figure 16.13), although not all of them have the triangular extensions (Appendix M; Figure 16.14). As previously stated, nearly all of them date to Nagada I and early Nagada II, but the most recent example dates to Nagada IIIA2. The large number of objects indicates the importance of this type of symbol, while on the other hand the wide typological variety of objects indicates that the symbol was not dependent on a particular type of object as regards content, but had a meaning by itself. It should also be mentioned that a few of the amulets (Petrie, 1920, plate

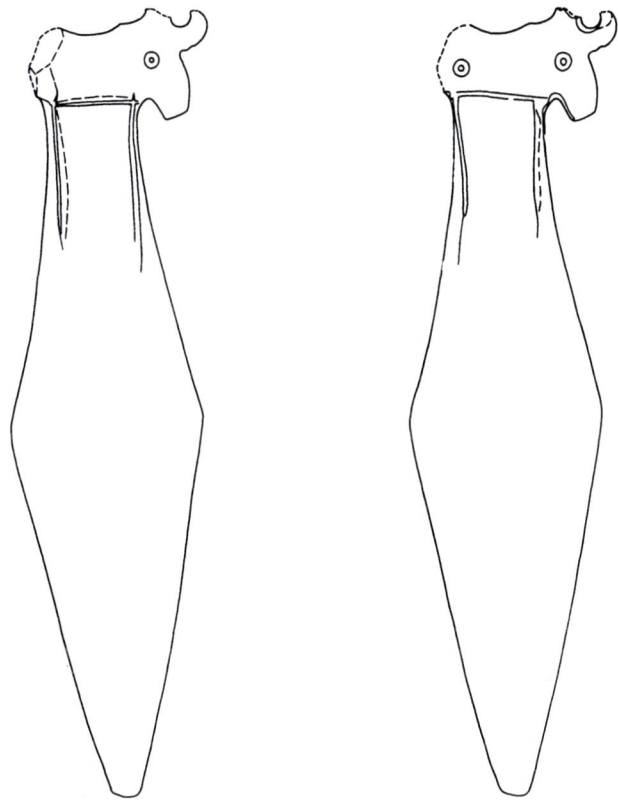

Figure 16.10. Rhomboidal palette decorated with double (?) bull's head. Provenance not recorded (Brussels E.2182). Scale 1:3.

XLIV; type 103 C, H; Figure 16.13) are decorated with inlayed eyes/breasts, confirming the relationship with the bull's head and probably at the same time also with the female figurines.

On the objects mentioned, slight individual differences can be recognized in the combination of the 'horns', triangular extensions, 'eyes/breasts'. Apparently, their significance was so familiar that they could be adapted with decorative purposes. The manner in which they can be combined is, for example,

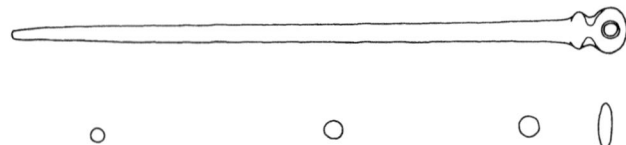

Figure 16.11. Greywacke needle. Provenance not recorded (Brussels E.2187). Scale 1:3.

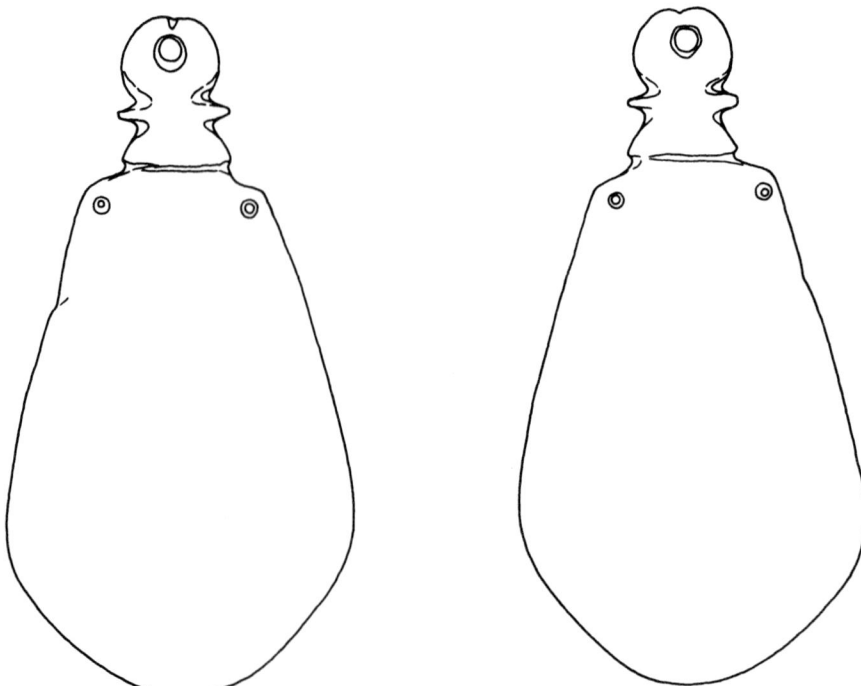

Figure 16.12. Palette with simplified Bat emblem. Provenance not recorded, formerly Scheurleer collection (Brussels E.7129). Scale 1:2.

also clear in a double bird head palette for which the provenance has not been recorded, showing the triangular extensions which in themselves were apparently considered sufficient as reference to the bovine element (Brussels E.2886; Figure 16.15).

There seems to be little doubt that these symbols are prophylactic because they occur on objects such as pendants, hair combs and needles, which

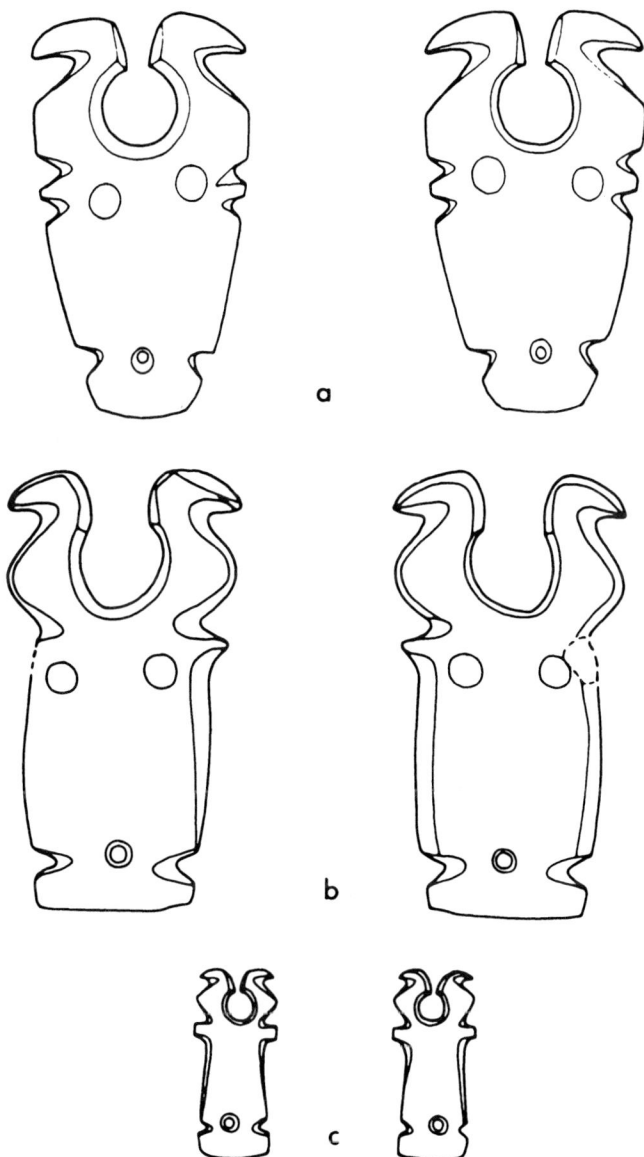

Figure 16.13. Amulets: a. provenance not recorded (Brussels E.2882); b. provenance not recorded (Brussels E.2880); c. Ballas-Zawaida (bought), formerly MacGregor collection (Brussels E.6188b). Scale 3:4.

were worn or meant to be clearly visible, and which are in a general manner most suited for such a purpose. But since we find here the same basic elements as for the bull's head amulets, they may refer, in more or less the same manner, to 'power' and eventually to regeneration in the afterlife.

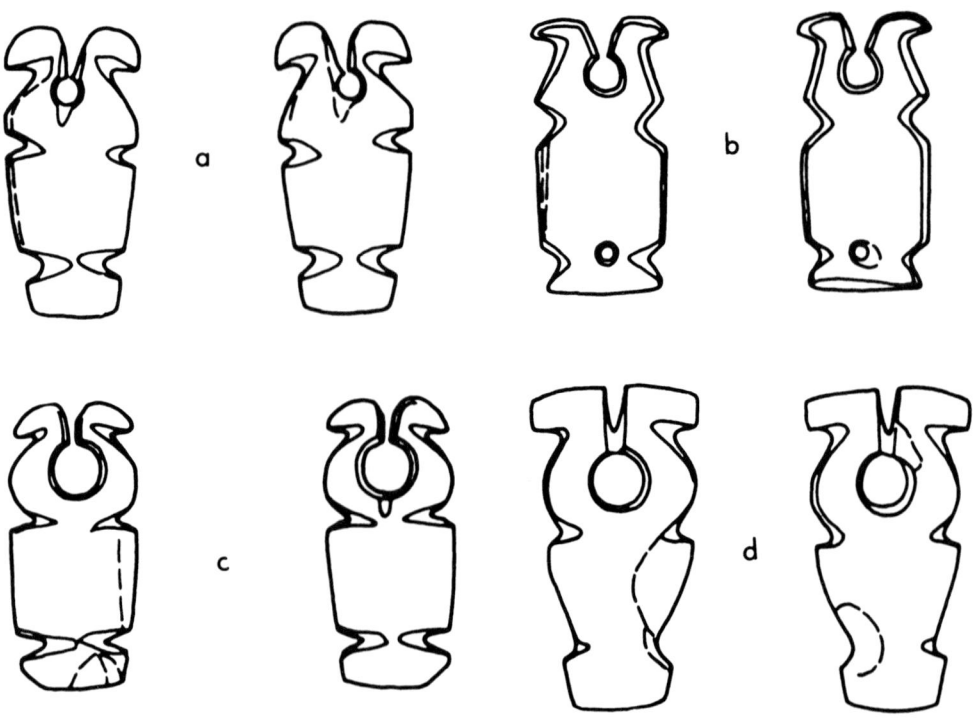

Figure 16.14. Amulets: a. provenance not recorded (Brussels E.2881); b. provenance not recorded, formerly MacGregor collection (Brussels E.6188e); c. Ballas-Zawaida (bought), formerly MacGregor collection (Brussels E.6188c); d. unidentified Petrie excavation (Brussels E.1231). Scale 3:4.

CONCLUSIONS

In this article it has been argued that an important part of Predynastic and Early Dynastic art used stylized representations with symbolic values. This stylization was by no means the result of inadequacy among the early Egyptian artisans. On the contrary, if one looks at the astonishing craftsmanship with which some objects were decorated,[5] it is quite obvious that these artisans were capable of producing almost any kind of representation desired. Therefore, if a representation is stylized, it should be regarded as intentional. One of the most important reasons for the stylization would certainly have been that the artisans did not want to render the exact image of one individual animal but, on the contrary, the general idea and characteristics of the animal. Indeed, we are not dealing with a kind of art which tries to illustrate particular moments or events in a realistic manner. Another reason for this mode of representation was to allow multiple interpretations, or more exactly to combine originally independent ideas into new symbols. They would become a kind of label, which could be used in

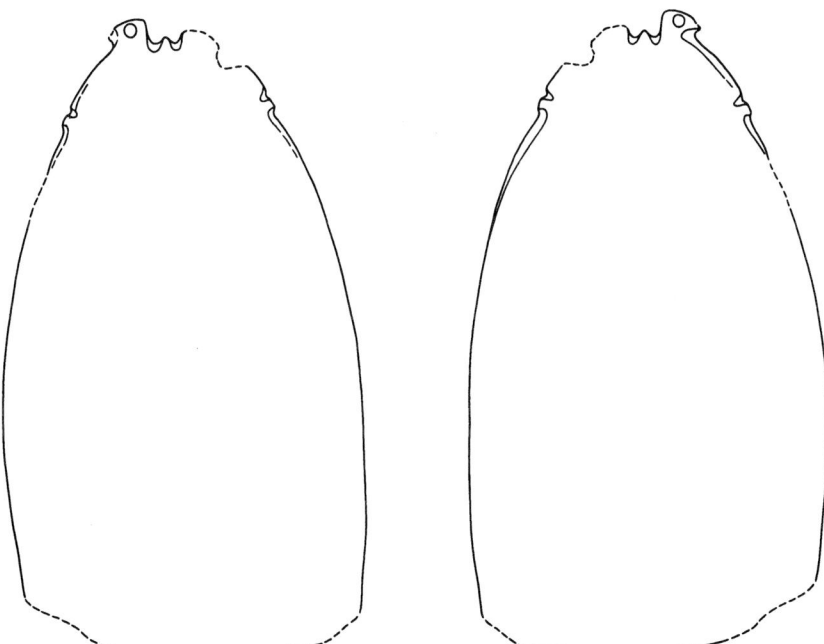

Figure 16.15. 'Double bird head' palette. Provenance not recorded (Brussels E.2886). Scale 1:3.

different contexts, both political and religious, which are always intimately related in Egyptian culture.

Because of this stylization, zoological identifications of the animals depicted are in many cases a perilous undertaking. Not only are the animals often rendered with little detail, but one also has to take into consideration that characteristic elements of different animals may have been amalgamated into one representation. The shape of the horns, which is so often used for the identification of animals, was of great symbolic importance and seems to have become highly standardized.

This mode of representation occurs from the very beginning of the Nagada period and continues throughout the whole of Predynastic and Early Dynastic culture. Changes will certainly have occurred during this period of nearly a thousand years, but a number of basic iconographic elements continued to be used. Some of them were integrated into formal Egyptian art, others were not and disappeared from the artistic record.

As for the subject discussed in the present article, a close relationship can be observed between bovines (principally the bull), birds (originally probably the ostrich but from the Nagada III period onwards mainly the falcon), and human beings provided with 'strength'. During the historical period, this is reflected in the identification of the Egyptian kings with both the falcon and the bull.

The emergence of kingship from the Nagada IIC period onwards brought about the development of a standardized art, with the religious and political confirmation of divine kingship as the main objective. A number of previously existing iconographic elements became integrated into formal Egyptian art, but many others disappeared from the repertoire. As already mentioned, the bull's head was probably used for the formal representation of the goddess Bat from the very beginning of the Nagada III period onwards. Apparently, Bat was a very important goddess during the early period of Egypt's history (Fischer, 1962, 1975). She was the principal goddess of the seventh Upper Egyptian nome, but her position in the Egyptian pantheon was, from the beginning of the Middle Kingdom, to be taken over by Hathor (Haynes, 1996).

Concerning royal titles and iconography, it has long been attested that the bull is one of the favorite animals with which the king identifies himself. Among the most important titles of the kings can be found 'Strong Bull, Great of Strength' and 'Bull of Horus'. The most characteristic iconographic element is the bull's tail attached to the king's girdle (Jéquier, 1918; Staehelin, 1982, p. 615, no. 34), as can already be seen on the Narmer palette.

NOTES

The present article is part of a larger study on Predynastic art which is still in progress. The objects used for illustrations are from the Egyptian collection of the Royal Museums for Art and History in Brussels. I wish to thank Luc Limme, curator of the collection, for permission to publish these artifacts; Ilona Regulski for inking the drawings of these objects and Françoise Roloux for making the drawing of E.6185a (Figure 16.3). Most useful comments and information on various subjects were furnished by Barbara Adams, Renée Friedman, Achilles Gautier, Dirk Huyge, Dagmar Kleinsgütl and Wim Van Neer.

(1) The word 'art' is not used here in its modern sense, where the artist himself decides about the subject, style etc. of his work, but more in the sense of artisanat (see Davis, 1983; Junge, 1990).

(2) This piece has been attributed by Baumgartel (1970, plate LVI) to tomb 1758 at Nagada because of an ink inscription on the object. However, the same number can be found on three more palettes from Brussels (E.6155, E.6186A-B) and refers to the number they originally had in the MacGregor collection. The provenance of these palettes remains unknown.

(3) With regard to the technical possibilities of the Nagada flint knappers, see however the extremely fine figure of an hippopotamus found at Hierakonpolis (Friedman and Adams, 1992, p. 68) and more generally the well known ripple flake knives (Midant-Reynes, 1987).

(4) 'Pelta', Latin, 'small crescent shaped shield'. The name was given by Petrie when looking for foreign relations between Egypt and the Mediterranean world.

(5) See, for example, the ivory fragments discovered a few years ago at Abydos, tomb U-127, which date s back to the Nagada IID period (Dreyer, 1993, table 6). Similar objects are not known for Nagada I and early Nagada II, but for those periods great craftsmanship is evident in the production of basalt vases or some of the animal shaped palettes.

REFERENCES

Adams, B. (1974a). *Ancient Hierakonpolis*, Aris and Phillips Ltd, Warminster.
Adams, B. (1974b). *Ancient Hierakonpolis, Supplement*, Aris and Phillips Ltd, Warminster.
Adams, B. (1988). *Predynastic Egypt*, Shire Egyptology 7, Aylesbury.
Adams, B. (1995). Bull's Head. In Phillips, T. (ed.), *Africa. The Art of a Continent*, Royal Academy of Arts, London, p. 63.
Adams, B. (1996). Elite tombs at Hierakonpolis. In Spencer, A.J. (ed.), *Aspects of Early Egypt*, British Museum Press, London, pp. 1-15.
Adams, B. (2000). *Excavations in the Locality 6 Cemetery at Hierakonpolis 1979-1985*, Egyptian Studies Association Publication no. 4, British Archaeological Reports Internatinal Series 903, Archaeopress, Oxford.
Amélineau, E. (1899). *Les Nouvelles Fouilles d'Abydos I*, (1895-1896), Ernest Leroux, Paris.
Amélineau, E. (1904). *Les Nouvelles Fouilles d'Abydos III*, (1897-1898), Ernest Leroux, Paris.
Andelkovic, B. (1995). *The Relations Between Early Bronze Age I Canaanites and Upper Egyptians*, Centre for Archaeological Research, Belgrade.
Ayrton, E.R., and Loat, W.L.S. (1911). *Pre-dynastic Cemetery at El-Mahasna*, Egypt Exploration Fund 31, London.
Baines, J. (1995). Origins of Egyptian kingship. In O'Connor, D., and Silverman, D.P. (eds.), *Ancient Egyptian Kingship. Probleme der Ägyptologie*, Bd. 9, E. J. Brill, Leiden, pp. 95-156.
Baumgartel, E.J. (1955). *The Cultures of Prehistoric Egypt* I, Oxford University Press, London.
Baumgartel, E.J. (1960). *The Cultures of Prehistoric Egypt* II, Oxford University Press, London.
Baumgartel, E.J. (1970). *Petrie's Naqada Excavation: A Supplement*, Bernard Quaritch, London.
Boessneck, J. (1988). *Die Tierwelt des Alten Ägypten*, Verlag C. H. Beck, München.
Boessneck, J., and von den Driesch, A. (1997). Tierknochenfunde. In von der Way, T., *Tell el-Fara'în - Buto I. Ergebnisse zum frühen Kontext. Kampagnen der Jahre 1983-1989*, AV 83, Philipp von Zabern, Mainz, pp. 206-216.
Boessneck, J., von den Driesch, A., and Ziegler, R. (1989). Die tierreste von Maadi und dem Friedhof am Wadi Digla. In Rizkana, I., and Seeher, J., *Maadi III. The Non-Lithic Small Finds and the Structural Remains of the Predynastic Settlement*, Philipp von Zabern, Mainz am Rhein, pp. 87-128.
Bonnet, C. (1997). Le groupe A et le pré-Kerma. In Wildung, D. (ed.), *Soudan, Royaumes sur le Nil*, Flammarion, Paris, pp. 36-47.
Brunton, G. (1927). *Qau and Badari* I, British School of Archaeology in Egypt and Egyptian Research Account 44, Bernard Quaritch, London.
Brunton, G. (1937). *Mostagedda and the Tasian Culture*, Bernard Quaritch, London.
Brunton, G. (1948). *Matmar*, Bernard Quaritch, London.
Brunton, G., and Caton-Thompson, G. (1928). *The Badarian Civilisation and Prehistoric Remains near Badari*, British School of Archaeology in Egypt and Egyptian Research Account 46. London.
Burgess, E.M., and Arkell, A.J. (1958). The reconstruction of the Hathor Bowl. *Journal of Egyptian Archaeology* 44: 6-11.
Capart, J. (1904). *Débuts de l'Art en Egypte*, Vromant and Co., Brussels.
Capart, J. (1905). *Primitive Art in Egypt*, Grevel, London.
Chantre, E. (1907). La nécropole memphite de Khozan (Haute Egypte) et l'origine des Egyptiens. *Bulletin de la Société d'Anthropologie de Lyon* 26: 229-246.
Charron, A. (1990). L'époque thinite. In *L'Egypte des Millénaires Obscurs*, Hatier et Musées de Marseille, Paris, pp. 77-98.
Churcher, C.S. (1984). Zoological study of the ivory knife handle from Abu Zaidan. In Needler, W., *Predynastic and Archaic Egypt in the Brooklyn Museum*, Brooklyn Museum, Brooklyn, pp. 152-169.
Cialowicz, K.M. (1987). Les têtes de massues des périodes prédynastique et archaïque dans la vallée du Nil. *Prace Archeologiczne* 41: 7-68.
Cialowicz, K.M. (1991). *Les Palettes Égyptiennes aux Motifs Zoomorphes et sans Décoration. Etudes de l'Art Prédynastique*, Studies in Ancient Art and Civilization 3, Uniwersytet Jagiellonski, Krakow.
Cialowicz, K.M. (1992). La composition, le sens et la symbolique des scènes zoomorphes prédynastiques en relief. Les manches de couteaux. In Friedman, R., and Adams, B. (eds.),

The Followers of Horus. Studies Dedicated to Michael Allen Hoffman 1944-1990, Egyptian Studies Association Publication 2, Oxbow Monograph 20, Oxford, pp. 247-258.

Cialowicz, K.M. (1997). Remarques sur la tête de massue du roi Scorpion. In Sliwa, J. (ed.), Studies in Ancient Art and Civilization 8, *Prace Archeologiczne* **59**: 11-27.

Cleyet-Merle, J.-J., and Vallet, F. (1982). Egypte. In Beck, F., Cleyet-Merle, J.J., Duval, A., Mohen, J.-P., and Vallet, F., *Archéologie Comparée. Afrique, Europe Occidentale et Orientale. Catalogue Sommaire Illustré des Collections du Musée des Antiquités Nationales de Saint-Germain-en-Laye*, 1, Editions de la Réunion des musées nationaux, Paris, pp. 68-165.

Close, A.E. (1996). Carry that weight: The use and transportation of stone tools. *Current Anthropology* **37**: 545-553.

Darnell, J.C., and Darnell, D. (1995-1996). The Theban Desert Road Survey (The Luxor-Farshût Desert Road Survey), *The Oriental Institute 1995-1996 Annual Report*, The Oriental Institute, Chicago, pp. 62-70.

Darnell, J.C., and Darnell, D. (1997). Exploring the 'narrow doors' of the Theban Desert. *Egyptian Archaeology* **10**: 24-26.

Davis, W.M. (1983). Artists and patrons in Predynastic and early Dynastic Egypt. *Studien zur Altägyptischen Kultur* **10**: 119-139.

Davis, W.M. (1992). *Masking the Blow. The Scene of Representation in Late Prehistoric Egyptian Art*, University of California Press, Berkeley / Los Angeles.

de Morgan, J. (1896). *Recherches sur les Origines de l'Egypte. I. L'age de la Pierre et des Métaux*, Ernest Leroux, Paris.

de Morgan, J. (1897). *Recherches sur les Origines de l'Egypte II. Ethnographie Préhistorique et Tombeau Royal de Negadah*, Ernest Leroux, Paris.

Dreyer, G. (1986). *Elephantine VIII. Der Tempel der Satet. Die Funde der Frühzeit und des Alten Reiches*, AV 39, Philip von Zabern, Mainz am Rhein.

Dreyer, G. (1993). Umm el-Qaab. Nachuntersuchungen im frühzeitlichen Königsfriedhof. 5./6. Vorbericht. *Mitteilungen des Deutschen Archäologischen Instituts, Abteilung Kairo* **49**: 23-62.

Dreyer, G. (1995). Die Datierung der Min-Statuen aus Koptos. In *Deutsches Archäologisches Institut Abteilung Kairo, Kunst des Alten Reiches. Symposium im Deutschen Archäologischen Instituts, Abteilung Kairo am 29. und 30. Oktober 1991*, Deutschen Archäologischen Instituts, Abteilung Kairo Sonderschrift 28, Mainz am Rhein, pp. 49-56.

Dreyer, G. (1998a). *Umm el-Qaab I. Das Prädynastische Königsgrab U-j und seine frühen Schriftzeugnisse*, AV 86, Philipp von Zabern, Mainz am Rhein.

Dreyer, G. (1998b). Grab des Dewen. In Dreyer, G., Hartung, U., Hikade, T., Köhler, E.C., Umm el-Qaab. Nachuntersuchungen im frühzeitlichen Königsfriedhof 9./10. Vorbericht. *Mitteilungen des Deutschen Archäologischen Instituts, Abteilung Kairo* **54**: 141-164.

Drouot-Richelieu. (1996). *Collection Jean-Marie Talleux*, Antiques, Mercredi 6-Jeudi 7 décembre 1996, Paris.

Eisenberg, J.M. (1995). *Art of the Ancient World*, Royal-Athena Galleries n° 71, January 1995, New York.

Eiwanger, J. (1984). *Merimde-Benisalâme I. Die Funde der Urschicht*, AV 47, Philip von Zabern, Mainz am Rhein.

Eiwanger, J. (1988). *Merimde-Benisalâme II. Die Funde der Mittleren Merimdekultur*, AV 51, Philip von Zabern, Mainz am Rhein.

Eiwanger, J. (1992). *Merimde-Benisalâme III. Die Funde der Jüngeren Merimdekultur*, AV 59, Philip von Zabern, Mainz am Rhein.

el Yahky, F. (1985). Clarifications on the Gerzean Boat Scenes. *Bulletin de l'Institute Française d'Archéologie Orientale* **85**: 187-195.

Emery, W.B. (1938). *The Tomb of Hemaka*. Excavations at Saqqara, Government Press, Cairo.

Emery, W.B. (1939). *Hor-Aha. Excavations at Saqqara 1937-1938*, Government Press, Cairo.

Emery, W.B. (1949). *Great Tombs of the First Dynasty I. Excavations at Saqqara*, Government Press, Cairo.

Emery, W.B. (1954). *Great Tombs of the First Dynasty II. Excavations at Sakkara*, Egypt Exploration Society 46, London.

Emery, W.B. (1958). *Great Tombs of the First Dynasty III. Excavations at Sakkara*, Egypt Exploration Society 47, London.

Emery, W.B. (1961). *Archaic Egypt*, Harmondsworth, England.

Firth, C.M. (1912). *The Archaeological Survey of Nubia: Report for 1908-1909*, Government Press, Cairo.

Firth, C.M. (1915). *The Archaeological Survey of Nubia: Report for 1909-1910*, Government Press, Cairo.

Fischer, H.G. (1962). The cult and the nome of the Goddess Bat. *Journal of the American Research Center in Egypt* 1: 7-24.

Fischer, H.G. (1975). Bat. *Lexicon der Ägyptologie* 1: col. 630-632.

Forrer, R. (1901). *Ueber Steinzeit Hockergräber zu Achmim, Naqada etc. in Ober Ägypten und Über Europaïsche Parallelfunde*, Achmim Studien 1, Verlag Karl J. Trübner, Strasbourg.

Friedman, R.F. (1995). Decorated ostrich egg. In Phillips, T. (ed.), *Africa. The Art of a Continent*, Royal Academy of Arts, London, pp. 52, 54-58, 60-62, 64-65.

Friedman, R.F. (1996). The ceremonial centre at Hierakonpolis locality HK 29A. In Spencer, A.J. (ed.), *Aspects of Early Egypt*, British Museum Press, London, pp. 16-35.

Friedman, R.F., and Adams, B. (eds.) (1992). *The Followers of Horus. Studies Dedicated to Michael Allen Hoffman 1944-1990*, Egyptian Studies Association Publication 2, Oxbow Monograph 20, Oxford.

Friedman, R.F., and Hendrickx, S. (in press). A Protodynastic tableau at Gebel Tjauti. In, Darnell, J.C., and Darnell, D. (eds.), *Theban Desert Road Survey, Volume I, The Rock Inscriptions of Gebel Tjauti in the Theban Western Desert*, The Oriental Institute, Chicago.

Garstang, J. (1903). *Mahâsna and Bêt Khallaf*, Egyptian Research Account 7, London.

Garstang, J. (1905). The So-called Tomb of Mena at Negadeh in Upper Egypt. *British Association for the Advancement of Science, Report of the 74th Meeting, Cambridge, 1904*, London, pp. 711-712.

Gautier, A. (1984). Archaeozoology of the Bir Kiseiba region, Eastern Sahara. In Wendorf, F., and Schild, R. (ass.), Close, A.E. (ed.), *Cattle-Keepers of the Eastern Sahara: The Neolithic of Bir Kiseiba*. Southern Methodist University Press, Dallas, pp. 49-72.

Gautier, P., and Midant-Reynes, B. (1995). La tête de massue du roi Scorpion. *Archéo-Nil* 5: 87-127.

Gophna, R. (1980). Excavations at En Besor, 1976. *Atiqot* 14: 9-16.

Grigson, C. (1991). An African origin for African cattle? Some archaeological evidence. *African Archaeological Review* 9: 119-144.

Hartung, U., and Friedhof U. (1998). In Dreyer, G., Hartung, U., Hikade, T., Köhler, E.C., Umm el-Qaab. Nachuntersuchungen im frühzeitlichen Königsfriedhof 9./10. Vorbericht. *Mitteilungen des Deutschen Archäologischen Instituts, Abteilung Kairo* 54: 79-100.

Hassan, F.A. (1992). Primeval goddess to divine king. The mythogenesis of power in the early Egyptian state. In Friedman, R., and Adams, B. (eds.), *The Followers of Horus. Studies Dedicated to Michael Allen Hoffman*, Egyptian Studies Association Publication 2, Oxbow Monograph 20, Oxford, pp. 307-322.

Haynes, J.L. (1996). Redating the Bat capital in the Museum of Fine Arts, Boston. In der Manuelian, P. (ed.), *Studies in Honor of William Kelly Simpson*, vol. I., Museum of Fine Arts, Boston, pp. 399-408.

Hendrickx, S. (1986). Predynastische objecten uit Naqada en Diospolis Parva (Boven Egypte). *Bulletin des Musées Royaux d'Art et d'Histoire* 57(2): 31-44.

Hendrickx, S. (1989). *De Grafvelden der Naqada-cultuur in Zuid-Egypte, met Bijzondere Aandacht voor het Naqada III Grafveld te Elkab. Interne Chronologie en Sociale Differentiatie*, Ph.D. dissertation, Leuven. [unpublished]

Hendrickx, S. (1994). *Antiquités Préhistoriques et Protodynastiques d'Egypte*, Guides du Département Égyptien 8, Musées Royaux d'Art et d'Histoire, Bruxelles.

Hendrickx, S. (1995). Vase decorated with Victory Scene. In Phillips, T. (ed.), *Africa. The Art of a Continent*, Royal Academy of Arts, London, pp. 59-60.

Hendrickx, S. (1996). The relative chronology of the Naqada culture: problems and possibilities. In Spencer, A.J. (ed.), *Aspects of Early Egypt*, British Museum Press, London, pp. 36-69.

Hendrickx, S. (1999). Peaux d'animaux comme symboles prédynastiques. A propos de quelques représentations sur les vases White Cross-lined. *Chronique d'Egypte* 74: 203-230.

Hendrickx, S. (2000). Autruches et flamants – les oiseaux représentés sur la céramique prédynastique de la catégorie Decorated. *Cahiers Caribéens d'Egyptologie* 1: 21-52.

Hoffman, M.A. (1989). A stratified Predynastic sequence from Hierakonpolis (Upper Egypt). In Krzyzaniak, L., and Kobusiewicz, M. (eds.), *Late Prehistory of the Nile Basin and the Sahara*, Poznan Archaeological Museum, Poznan, pp. 317-323.

Hoffman, M.A., Lupton, C., and Adams, B. (1982). Excavations at Locality 6. In Hoffman, M.A., *The Predynastic of Hierakonpolis. An Interim Report*, Macomb, Illinois, pp. 38-61.

Huyge, D. (1995). *De Rotstekeningen van Elkab (Boven-Egypte): Registratie, Seriatie en Interpretatie*, Ph.D. dissertation, Leuven. [unpublished]

Huyge, D. (2001). Cosmology, ideology, and personal religious practice in ancient Egyptian rock art. In Friedman, R. F. (ed.), *Egypt an dNubia: Gifts of the Desert*, British Museum Press, London.

Jéquier, G. (1918). La queue de taureau insigne des rois d'Egypte. *Bulletin de l'Institute Française d'Archéologie Orientale* 15: 165-168.

Junge, F. (1990). Versuch zu einer ästhetik der ägyptischen kunst. In Eaton-Krauss, M., and Graefe, E. (eds.), *Studien zur ägyptischen Kunstgeschichte*, Hildesheimer Ägyptologische Beiträge 29, Hildesheim, pp. 1-38.

Junker, H. (1912). *Bericht Über die Grabungen der Kaiserlichen Akadamie der Wissenschaften in Wien, auf dem Friedhof in Turah. Winter 1909-1910*, Denkschriften der Akademie der Wissenschaften in Wien 56, Wien.

Kahl, J. (1994). *Das System der Ägyptischen Hieroglyphenschrift in der 0.-3. Dynastie*, Göttinger Orientforschungen, IV, Reihe Ägypten, Band 29, Wiesbaden.

Kahl, J., and Engel, E. –M. (2001). Vergraben, verbrannt, verkannt und vergessen. In *Funde aus dem "Menesgrab"*, Münster.

Kemp, B.J. (1989). *Ancient Egypt. Anatomy of a Civilization*, Routledge, London.

Klasens, A. (1958). The excavations of the Leiden Museum of Antiquities at Abu-Roash: Report of the first season 1957. Part II. *Oudheidkundige Mededelingen uit het Rijksmuseum van Oudheden te Leiden* 39: 20-31.

Klasens, A. (1959). The excavations of the Leiden Museum of Antiquities at Abu-Roash: Report of the second season 1958. Part II. Cemetery 400. *Oudheidkundige Mededelingen uit het Rijksmuseum van Oudheden te Leiden* 40: 41-61.

Köhler, E.C. (1998). Frühe keramik. In Dreyer, G., Hartung, U., Hikade, T., Köhler, E.C., Umm el-Qaab. Nachuntersuchungen im frühzeitlichen Königsfriedhof 9./10. Vorbericht. *Mitteilungen des Deutschen Archäologischen Instituts, Abteilung Kairo* 54: 100-115.

Kroeper, K., and Wildung, D. (1985). *Minshat Abu Omar. Münchner Ostdelta - Expedition. Vorbericht 1978-1984*, Schriften aus der Ägyptischen Sammlung 3, Munich.

Kroeper, K., and Wildung, D. (2000). *Minshat Abu Omar II, Ein vor- und Frühgeschichtlicher Friedhof im Nildelta*, Gräber, Philipp von Zabern, Mainz, pp. 115-204.

Logan, T.J. (1990). The Origins of the Jmy-wt Fetish. *Journal of the American Research Center in Egypt* 27: 61-69.

Lythgoe, A.M., and Dunham, D. (1965). *The Predynastic Cemetery N7000. Naqa-ed-Dêr*, Part IV, University of California Publications, Egyptian Archaeology 7, Berkeley / Los Angeles.

McIver, R.D., and Mace, A.C. (1902). *El Amrah and Abydos. 1899-1901*, Egypt Exploration Fund 23, London.

McKim Malville, J., Wendorf, F; Mazar, A.A., and Schild, R. (1998). Megaliths and Neolithic astronomy in southern Egypt. *Nature* 392: 488-491.

Midant-Reynes, B. (1987). Contribution à l'étude de la société prédynastique: le cas du couteau 'ripple-flake'. *Studien zur Altägyptischen Kultur* 14: 185-224.

Midant-Reynes, B. (1992). *Préhistoire de l'Egypte. Des Premiers Hommes aux Premiers Pharaons*, Armand Colin, Paris.

Mond, R.L., and Myers, O.H. (1937). *Cemeteries of Armant* I, Egypt Exploration Society 42, London.

Müller, H.W. (1964). *Ägyptische Kunstwerke, Kleinfunde und Glas in der Sammlung E. und M. Kofler-Truniger, Luzern*, Münchner Ägyptologische Studien 5, München.

Needler, W. (1984). *Predynastic and Archaic Egypt in The Brooklyn Museum*, Wilbour Monographs 9, Brooklyn.

Newberry, P.E. (1913). List of vases with cult-signs. *Liverpool Annals of Archaeology and Anthropology* 5: 137-142.

Nordström, H.-A. (1972). *Neolithic and A-Group Sites. Scandinavian Joint Expedition to Sudanese Nubia* 3, Scandinavian University Books, Stockholm.

Onasch, A., and Steinmann, F. (1997). Die frühdynastische zeit. um 2960-2640 v. Chr. In Krauspe, R. (ed.), *Das Ägyptische Museum der Universität Leipzig*, Philipp von Zabern, Mainz, pp. 22-25.

Osborn, D.J., and Osbornova, J. (1998). *The Mammals of Ancient Egypt*, Aris and Phillips, Warminster.

Otto, B. (1986). Der narrative und der hierographische stil im frühen Ägypten und im Zweistromenland. In *Actes du Congrès 'Symmetrie'*, Darmstadt, pp. 137-149.

Page-Gasser, M., and Wiese, A.B. (eds.) (1997). *Ägypten. Augenblicke der Ewigkeit. Unbekannte Schätze aus Schweizer Privatbesitz*, Philipp von Zabern, Mainz.

Payne, J.C. (1993). *Catalogue of the Predynastic Egyptian Collection in the Ashmolean Museum*, Oxford University Press, Oxford.

Petrie, W.M.F. (1896). *Naqada and Ballas*, Bernard Quaritch, London.

Petrie, W.M.F. (1900). *The Royal Tombs of the First Dynasty*, Part I, Egypt Exploration Fund 18, Kegan Paul, Trench, Trübner and Co., London.

Petrie, W.M.F. (1901). *The Royal Tombs of the Earliest Dynasties*, Part II, Egypt Exploration Fund 21, Kegan Paul, Trench, Trübner and Co., London.

Petrie, W.M.F. (1902). *Abydos*, Part I, 1902, Egypt Exploration Fund 22, Kegan Paul, Trench, Trübner and Co., London.

Petrie, W.M.F. (1903). *Abydos*, Part II, 1903, Egypt Exploration Fund 24, Kegan Paul, Trench, Trübner and Co., London.

Petrie, W.M.F. (1913). *Tarkhan I and Memphis V*, British School of Archaeology in Egypt and Egyptian Research Account 23, London.

Petrie, W.M.F. (1914). *Amulets*, Constable, London.

Petrie, W.M.F. (1920). *Prehistoric Egypt*, British School of Archaeology in Egypt and Egyptian Research Account 31, London.

Petrie, W.M.F. (1921). *Corpus of Prehistoric Pottery and Palettes*, British School of Archaeology in Egypt and Egyptian Research Account 32, London.

Petrie, W.M.F. (1925). *Tombs of the Courtiers and Oxyrhynkhos*, British School of Archaeology in Egypt and Egyptian Research Account 37, London.

Petrie, W.M.F., and Mace, A.C. (1901). *Diospolis Parva. The Cemeteries of Abadiyeh and Hu. 1898-1899*, Egypt Exploration Fund 20, Kegan Paul, Trench, Trübner and Co., London.

Petrie, W.M.F., Wainwright, G.A., and Mackay, E. (1912). *The Labyrinth, Gerzeh and Mazguneh*, British School of Archaeology in Egypt and Egyptian Research Account 21, Bernard Quaritch, London.

Price, F.G.H. (1897). *A Catalogue of the Egyptian Antiquities in the Possession of F.G. Hilton Price*, Bernard Quaritch, London.

Price, F.G.H. (1900). Some ivories from Abydos. *Proceedings of the Society of Biblical Archaeology* 22: 160-161.

Priese, K.-H. (ed.) (1991). *Ägyptische Museum. Staatliche Museen zu Berlin. Ägyptisches Museum und Papyrussamlung*, Philipp von Zabern, Mainz am Rhein.

Quibell, J.E. (1905). *Catalogue Général des Antiquités Egyptiennes. nos. 11.001 - 12.000 et 14.001 - 14.754. Archaic Objects*, Government Press, Cairo.

Quibell, J.E., and Green, F.W. (1902). *Hierakonpolis II*, Egyptian Research Account 5, London.

Quibell, J.E., and Petrie, W.M.F. (1900). *Hierakonpolis I*, Egyptian Research Account 4, London.

Regner, C. (1996). *Schminkpaletten*, Bonner Sammlung von Aegyptiaca 2, Wiesbaden.

Reisner, G.A. (1908). *The Early Dynastic Cemeteries at Naqa-ed-Dêr*, Part I, University of California Publications, Egyptian Archaeology 2, Leipzig.

Reisner, G.A. (1910). *The Archaeological Survey of Nubia. Report for 1907-1908, Vol. I, Archaeological Report*, National Printing Department, Cairo.

Rizkana, I., and Seeher, J. (1989). *Maadi III. The Non-Lithic Small Finds and the Structural Remains of the Predynastic Settlement*, AV 80, Philipp von Zabern, Mainz am Rhein.

Saad, Z.Y. (1951). *Royal Excavations at Helwan (1945-1947)*, Annales du Service des Antiquités de l'Egypte, supplément cahier 14, Cairo.

Saad, Z.Y. (1969). *The Excavations at Helwan: Art and Civilization in the First and Second Egyptian Dynasties*, University of Oklahoma Press, Oklahoma.

Saleh, M., and Sourouzian, H. (1986). *Die Hauptwerke im Ägyptischen Museum Kairo*, Verlag Philip von Zabern, Mainz am Rhein.

Schäfer, H. (1896). Neue altertümer der 'new race' aus Negadeh. *Zeitschrift für Ägyptische Sprache und Altertumskunde* 3: 158-161.

Scharff, A. (1926). *Das Vorgeschichtliche Gräberfeld von Abusir el-Meleq*, Wissenschaftliche Veröffentlichung der Deutschen Orient Gesellschaft 49, Leipzig.

Scharff, A. (1929). *Die Altertümer der Vor- und Frühzeit Ägyptens. II. Bestattung, Kunst, Amulette und Schmuck, Geräte zur Körperpflege, Spiel- und Schreibgeräte, Schnitzereien aus Holz und Elfenbei, Verschiedenes*, Staatliche Museen zu Berlin, Mitteilungen aus der Ägyptischen Sammlung 5, Berlin.

Scharff, A. (1931). *Die Altertümer der Vor- und Frühzeit Ägyptens. I. Werkzeuge, Waffen, Gefässe*, Staatliche Museen zu Berlin. Mitteilungen aus der ägyptischen Sammlung 4, Berlin.

Schlögl, H. (ed.), (1978). *Geschenk des Nils. Ägyptische Kunstwerke aus Schweizer Besitz*, Schweizerischer Bankverein, Basel.

Schott, S. (1951). *Hieroglyphen: Untersuchungen zum Ursprung der Schrift*, Abhandlungen der Akademie der Wissen-schaften in Mainz 24, Mainz.

Smith, A.L. (1993). Identification d'un potier prédynastique. *Archéo-Nil* 3: 23-33.

Smith, W.S. (1960). *Ancient Egypt as Represented in the Museum of Fine Arts, Boston*, Museum of Fine Arts, Boston.

Sowada, K.N. (1999). Black-topped ware in early Dynastic contexts. *Journal of Egyptian Archaeology* 85: 85-102.

Spencer, A.J. (1980). *Catalogue of Egyptian Antiquities in the British Museum. V. Early Dynastic Objects*, British Museum Publications, London.

Staehelin, E. (1982). Ornat. In Helck, W., and Westendorf, W., *Lexikon der Ägyptologie*, Band V. Otto Harrassowitz, Wiesbaden, cols. 613-618.

Stager, L.E. (1992). The periodization of Palestine from Neolithic through early Bronze times. In Ehrich, R. (ed.), *Chronologies in Old World Archaeology*, 3rd revised edition, University of Chicago Press, Chicago, pp. 22-41.

Stanton, E.B., and Hoffman, M.A. (1988). Description of the exhibition artifacts. In Hoffman, M.A., Willoughby, K.L., and Stanton, E.B. (eds.), *The First Egyptians*, University of South Carolina, Columbia, pp. 59-112.

Störk, L. (1984). Rind. In Helck, W., and Westendorf, W., *Lexikon der Ägyptologie*, Band V. Otto Harrassowitz, Wiesbaden, cols. 257-263.

Tupinier, D. (1984). Plaque à fard de l'Egypte prédynastique representant une chauve-souris. *Bulletin Mensuel de la Société Linnéenne de Lyon* 9: 296-299.

Ucko, P.J. (1968). *Anthropomorphic Figures of Predynastic Egypt and Neolithic Crete with Comparative Material from the Prehistoric Near East and Mainland Greece*, Royal Anthropological Institute, Occasional Papers 24, London.

Vandier, J. (1952). *Manuel d'Archéologie Égyptienne. I. Les Époques de Formation*, A. Jean Piccard, Paris.

Van Lepp, J. (1999). The misidentification of the Predynastic bull's head amulet. *Göttinger Miszellen* 168: 101-111.

Vassilika, E. (1995a). *Egyptian Art*, Fitzwilliam Museum Handbooks, Cambridge.

Vassilika, E. (1995b). Bull's head. In Phillips, T. (ed.), *Africa. The Art of a Continent*, Royal Academy of Arts, London, p. 63.

von den Driesch, A., and Boessneck, J. (1985). *Die Tierknochenfunde aus der neolithischen Siedlung von Merimde-Beni Salâme am Westlichen Nildelta*, Institut für Palaeoanatomie, Domestikationsforschung und Gestichte der Tiermedizin, Munich.

von der Way, T. (1997). *Tell el-Fara'în - Buto I. Ergebnisse zum Frühen Kontext. Kampagnen der Jahre 1983-1989*, AV 83, Philipp von Zabern, Mainz.

Weigall, A.E.P. (1907). A report on some objects recently found in Sebakh and other diggings. Prehistoric drawing from Maala. *Annales du Service des Antiquités de l'Egypte* 8: 49-50.

Welsby, D.A. (1995). Female figurine. In Phillips, T. (ed.), *Africa. The Art of a Continent*, Royal Academy of Arts, London, pp. 104-105.

Wendorf, F. (1968). Late Paleolithic sites in Egyptian Nubia. In Wendorf, F. (ed.), *The Prehistory of Nubia*, vol. II, Fort Burgwin Research Centre and Southern Methodist University Press, Dallas, pp. 791-953.

Wendorf, F., and Schild, R. (1995). Are the Early Holocene cattle in the Eastern Sahara domestic or wild? *Journal of Evolutionary Anthropology* 3: 118-128.

Wildung, D. (1981). *Ägypten vor den Pyramiden*, Münchner Ausgrabungen in Ägypten, Mainz am Rhein.

Wildung, D. (1984). Zur formgeschichte der landeskronen. In *Studien zur Sprache und Religion Ägyptens. Band 2: Religion*, Fs. Westendorf. Göttingen, pp. 967-980.

Wildung, D. (ed.) (1997). *Soudan, Royaumes sur le Nil*, Flammarion, Paris.

APPENDIX A. White Cross-lined pottery with representations of bovines

Painted Decoration

(1) Mahasna, tomb H 97 Nagada IC. Type C 53 (London BM EA.49025), Ayrton and Loat, 1911, plate xiv, 1

(2) Naqa ed-Deir, tomb 7014 Nagada IC. Type C 49 (Berkeley), Lythgoe and Dunham, 1965, figure 3, d-e

(3) Provenance not recorded [date unknown]. Type C 6 L (London UC.15331), Petrie, 1920, plate XVI, 66

(4) Provenance not recorded [date unknown]. Type C 96 L (London UC.15334), Petrie, 1920, plate XVI, 69

(5) Provenance not recorded [date unknown]. Type C 95 (London UC.15335), Petrie, 1920, plate XVI, 70

Applied Decoration

(6) Nagada (?) [date unknown]. Type unknown, loose figure (Berlin 13805), Scharff, 1929, p. 39, no. 60, figure 13; Priese, 1991, pp. 4-5

(7) Provenance not recorded [date unknown]. Type C 75 (Genève, Roland Cramer), Page-Gasser and Wiese, 1997, pp. 24-25

(8) Provenance not recorded [date unknown]. Type C 79 (Boston 041814), Smith, 1960, p. 19, plate 2

APPENDIX B. Decorated pottery with representations of bovines

(1) Dakka, cem. 102, t. 140 [Nagada IIC-D?]. Type D (München ÄS.2728), Firth, 1915, p. 65; Wildung, 1997, p. 43, no. 34

(2) Provenance not recorded [date unknown]. Type W 19 (Cairo CG 11733), Quibell, 1905, p. 149

APPENDIX C. Palettes representing probable bovines. After Petrie, 1920, 1921

(1) el-Amra, tomb A 63 [date unknown]. Palette type 3 D (not located), Petrie, 1921, plate LII

(2) Ballas North Town (?) [date unknown]. Fragment of a palette (Oxford 1895.870), Payne, 1993, no. 1807

(3) Gebel Tarif, tomb unknown [date unknown]. Palette cf. type 4 N (Cairo CG 14145), Quibell, 1905, p. 223

(4) Hu, tomb U 247 [SD 39 = Nagada IIA-B?]. Palette type 4 k (not located), Petrie and Mace, 1901, plate XI, 1; Petrie, 1921, plate LII

(5) Khor Bahan, cemetery 17, tomb 56 [Nagada IB-IIA]. Palette type 4 J (not located), Reisner, 1910, I, pp. 120-121, II, plate 63b, 10; Petrie,

1921, plate LII (Petrie, 1920, p. 37 identifies the animal as hartebeest, which however seems highly unlikely)

(6) Nagada, tomb 241 [Nagada IIA]. Palette type 4 S (London UC.4243), Petrie, 1921, plate LII; Baumgartel, 1970, p. X

(7) Nagada, tomb 271 [Nagada IIA]. Palette type 9 D, tortoise with two bovid heads (Oxford 1895.841), Petrie, 1921, plate LII; Payne, 1993, p. 222, no. 1809

(8) Nagada, tomb 1515 [Nagada IIA]. Romboidal palette type 92 d, decorated with incised bovid (Oxford 1895.825), Petrie, 1896, plate LI, 15a; Payne, 1993, p. 227, no. 1868

(9) Nagada, tomb T 4 [Nagada IIB]. Palette type 3 D (not located), Petrie, 1921, plate LII

(10) Nagada (?) [date unknown]. Palette type 4 N (Brussels E.6187), Hendrickx, 1994, pp. 44-45 (previously identified as sheep)

(11) Provenance not recorded [date unknown]. Palette type 3 J (London UC.15769), Petrie, 1920, plate XLIII, 3 J

(12) Provenance not recorded [date unknown]. Palette type 4 N (London UC.15770), Petrie, 1920, plate XLIII, 4 N

(13) Provenance not recorded [date unknown]. Realistic palette of a bovine, eventually a cow (previously in private collection in Switzerland), Eisenberg, 1995, no. 226

(14) Provenance not recorded [date unknown]. Palette (not located), Drouot-Richelieu, 1996, no. 274

APPENDIX D. Ivory and bone carvings with representations of bovines (bibliography cf. Cialowicz, 1992)

(1) Abu Zeidan, tomb 32 [Nagada IIIB?]. Knife handle (Brooklyn 09.889.118), Churcher, 1984

(2) Abydos, tomb U-127 [Nagada IID1]. Knife handle (Abydos storeroom), Dreyer, 1993, tafel 6, f

(3) Sheikh Hamada (?)[date unknown]. Pitt-Rivers knife handle (London BM EA.68512)

(4) Provenance not recorded [date unknown]. Carnarvon knife handle (New York 1926 (26.7.1281)

(5) Provenance not recorded [date unknown]. Davis comb (New York 1915 (30.8.224)

(6) Provenance not recorded [date unknown]. Fragment of ivory plaque (London BM EA.66953), Spencer, 1980, no. 451

(7) Provenance not recorded [date unknown]. Shell plaque (Berlin 13797), Scharff, 1929, p. 83, no. 113, tafel 22 (It is strongly questionable whether this object really dates to the Early Dynastic period, its style being considerably different from the other decorated ivories)

APPENDIX E. Naturalistic bovine heads, mainly amulets

(1) Abydos, Royal tombs [Nagada IIIC-D]. Fragment of a statuette? (not located), Amélineau, 1904, plate XXV

(2) Gerza, tomb 205 [SD 64 = Nagada IID2-IIIA2?]. White limestone (not located), Petrie *et al.*, 1912, plate V

(3) Hemamiya, tomb 1809 [Nagada IIIA2]. Black steatite (not located), Brunton, 1927, plate XVII (= Adams, 1988, p. 44, g)

(4) Hierakonpolis [date unknown]. Pottery (Brooklyn 09.889.327), Needler, 1984, pp. 364-365, no. 291

(5) Matmar, tomb 2001 [Nagada IIIA2]. Black limestone (not located), Brunton, 1937, plate XXII, 30

(6) Nagada, tomb 1289 [Nagada IIIA2]. Carnelian (not located), Petrie, 1896, plate LVIII, 1289, (not mentioned in Baumgartel, 1970)

(7) Nagada, tomb 1759 [Nagada IC]. Glass (sic; not located), Petrie, 1920, p. 43, plate IX, 47 (not mentioned in Baumgartel ,1970)

(8) Nagada (bought) [date unknown]. Pottery (Berlin 13810), Schäfer, 1896, p. 160; Scharff, 1929, pp. 39-40, no. 61, figure 13

(9) Turah south, no tomb number [Nagada IIIB?]. Unidentified stone (not located), Junker, 1912, Abb. 86, tafel L, b

(10) Provenance not recorded [date unknown]. Limestone (London BM EA.32134), Spencer, 1980, no. 560, plate 59

(11) Provenance not recorded (?) [date unknown]. Green serpentine (London UC?), Petrie, 1914, p. 19, plate V, no. 62d

APPENDIX F. Catalogue of bull's head amulets with a known provenance

Tombs

(1) Abadiya, tomb B 378 B [SD 52 = Nagada IIC-D?]. Serpentine (Oxford E.E.34), Payne, 1993, p. 210, no. 1713, figure 72

(2) Abusir el-Meleq, tomb 60 d 7 [date unknown]. Limestone (Berlin 18663), Scharff, 1926, p. 57, no. 369

(3) Abusir el-Meleq, tomb 60 k 6 [Nagada IIC-III]. Ivory (Berlin 18620), Scharff, 1926, p. 57, no. 370

(4) Abusir el-Meleq, tomb 4 d 2 [Nagada IIC-III]. Calcite (Berlin 19266), Scharff 1926, p. 57, n° 371

(5) Abusir el-Meleq, tomb C [Nagada IIIA2?]. Unidentified stone (Berlin 19617), Scharff, 1926, p. 57, no. 372

(6) Abydos, tombs of the courtiers, tomb 618 [Nagada IIIC1, Djer]. No description (not located), Petrie, 1925, plate XX

(7) Ballas, tomb Q 709 [date unknown]. Serpentine (not located), Petrie, 1896, p. 45, plate LVIII, Q709/5

(8) Debod, cemetery 23, tomb 47 [Nagada IIIA-B]. Unidentified stone (not located), Reisner, 1910, I, p. 158, II, plate 70a, 13

(9) Gerza, tomb 229 [SD 47-57 = Nagada IIC-IID2?]. Gypsum (not located), Petrie *et al.*, 1912, plate V

(10) Helwan, tomb 597H5 [Nagada IIIC-D]. Green diorite (Cairo), Saad, 1951, p. 36, plates XLI, XLIII

(11) Hemamiya, tomb 1620 [Nagada IIIC2]. Black limestone (not located), Brunton, 1927, plate XVII

(12) Hemamiya, tomb 1629 [Nagada II]. Limestone (not located), Brunton and Caton-Thompson, 1928, plate XLIX

(13) Hemamiya, tomb 1773 [Nagada II-IIIA]. Limestone (not located), Brunton and Caton-Thompson, 1928, plate XLIX

(14) Hu, tomb U 379 [SD 67 = Nagada IIIA1-IIIA2?]. Two examples (stone and ivory) part of a necklace (London UC.10834-8), Stanton and Hoffman, 1988, p. 82, no. 70

(15) Matmar, tomb 236 [Nagada IIIA2]. Amethyst (not located), Brunton, 1948, plate XXII, 31

(16) Matmar, tomb 5109 [Nagada IIC]. Malachite (not located), Brunton, 1948, plate XV, 2

(17) Mediq, cemetery 79, tomb 76 [date unknown]. Serpentine (?) (not located), Firth, 1912, I, p. 136, II, plate 37a, 17

(18) Mediq, cemetery 79, tomb 117 [Nagada IIIA1-IIIA2]. Green stone (not located), Firth, 1912, I, pp. 139-140, II, plate 37a, 16

(19) Minshat Abu Omar. Kroeper and Wildung, 1985, p. 90, Abb. 308. Several examples have been found at Minshat Abu Omar (no details have been published yet), all in tombs of children.

(20) Nagada, tomb 1788 [SD 34-46 = Nagada IC-IIB?]. Elephant ivory, eyes filled in with black paste (Oxford 1895.908), Petrie, 1896, plate LXI, 4; Baumgartel, 1960, plate VI, 2; Payne, 1993, p. 207, no. 1693, figure 72; Adams, 1995, p. 63, no. 1.16b

(21) Nagada, tomb 1788 [SD 34-46 = Nagada IC-IIB?]. Hippopotamus ivory (London UC.6005), Adams, 1988, p. 51, 1995, p. 63, no. 1.16a; Petrie, 1920, plate IX, 4

(22) Naqa ed-Deir, tomb N 3031 [2nd dyn. according to Reisner]. Limestone (Berkeley?), Reisner, 1908, p. 76, plate 73c

(23) Shellal, cemetery 7, tomb 317 [Nagada IIIA2?]. Green steatite (not located), Reisner, 1910, I, p. 23, II, plate 70a, 7

(24) Sialy, cemetery 40, tomb 14 [Nagada IIIA-B]. Green stone (not located), Reisner, 1910, I, p. 236, II, plate 70a, 2

(25) Tarkhan, tomb 1256 [Nagada IIIA2-IIIC2]. Green serpentine (London UC.15236), Petrie, 1914, p. 44, plate XXXVIII, 212d

Living Sites

(26) Hemamiya, North Spur [1 ft. to 1 ft. 6 in. = Nagada IIC-IID?]. White limestone (not located), Brunton and Caton-Thompson, 1928, p. 108, plate LXXI, no. 63

(27) Hierakonpolis, Nekhen town, 10N5W, level 3 [= Nagada II-IIIA2]. Diorite (Hierakonpolis store room), Hoffman, 1989, p. 321, figure 1, 2

(28) Hierakonpolis, Nekhen town, 10N5W, level 3 [= Nagada II-IIIA2]. Serpentine (Hierakonpolis store room), Hoffman, 1989, p. 321, figure 1, 4

Temple Sites

(29) Abydos, Osiris temple. Green serpentinite (not located), Petrie, 1902, p. 23, plate LI, 4

(30) Abydos, Osiris temple. Green serpentinite (not located), Petrie, 1902, p. 23, plate LI, 5 ('probably under Mena')

(31) Abydos, Osiris temple, level 177. Unidentified stone (not located), Petrie, 1903, p. 30, plate XIV, 281 ('1st dynasty')

(32) 'Abydos, Osiris temple' (?). Graywacke (Luzern, Sammlung Kofler-Truniger K 9646 A), Müller, 1964, p. 29, A 30; Schlögl, 1978, p. 27, no. 80

(33) Hierakonpolis, Main Deposit. Limestone (London UC.15002), Quibell and Green, 1902, plate XLVIIIb; Adams, 1974a, p. 22, no. 110, plates 15, 17; Adams 1974b, pp. 6, 132; Stanton and Hoffman, 1988, p. 83, no. 71

Palestinian Sites

(34) Azor, tomb. (unpublished, cf. Gophna, 1980, p. 15, no.13.)

(35) En Besor, stratum III. Carbonate rock, metamorphic marble (not located), Gophna, 1980, p. 15, figure 5, plates 1, 4; Stager, 1992, figure 7, 68; Andelkovic, 1995, figure 11, 18

(36) Tell el-Asawir, tomb. Unidentified stone (not located), Andelkovic, 1995, p. 27, figure 3, 3

APPENDIX G. Female statuettes with arms curved underneath the breasts

(1) Halfa Degheim, site 277, tomb 16B [Nagada IID2-IIIA1]. Pottery (Khartoum, Sudan National Museum SNM 13729), Nordström, 1972, pp. 27, 127, plates 56, 3, 197; Welsby, 1995, p. 105, no. 1.74; Bonnet, 1997, pp. 40-41.

(2) Nagada, tomb 1611 [SD 36-38 = Nagada IA-IB?]. Baked clay (Oxford 1895.125), Ucko, 1968, pp. 28-30, 89, no. 38; Payne, 1993, no. 39

(3) Nagada or Ballas [date unknown]. Baked clay (Oxford 1895.126), Petrie, 1896, plate VI, 4; Ucko, 1968, pp. 22-23, 85, no. 30; Payne, 1993, no. 38

(4) Provenance not recorded [date unknown]. Baked clay (London UC.15156), Petrie, 1896, plate XXXVI, 96, 1920, plate IV, 5; Ucko, 1968, p. 140, no. 172

(5) Provenance not recorded [date unknown]. Clay (London UC.15153), Petrie, 1920, plate IV, 9; Ucko, 1968, p. 139, no. 171

(6) Provenance not recorded [date unknown]. Baked clay (London UC.15160), Petrie, 1920, plate IV, 10, V, 6, plate VI; Ucko, 1968, p. 141, no. 175

(7) Provenance not recorded [date unknown]. Baked clay (London UC.15162), Petrie, 1920, plate V, 4-5, plate VI; Ucko, 1968, p. 146, no. 187

(8) Provenance not recorded [date unknown]. Baked clay (London UC.15813), Ucko, 1968, p. 147, nos. 190, 474

(9) Provenance not recorded [date unknown]. Baked clay (London UC.15814), Ucko, 1968, p. 148, nos. 191, 475

(10) Provenance not recorded [date unknown]. Clay (New York 07.228.71), Ucko, 1968, p. 155, no. 204

APPENDIX H. Objects decorated with the head of Bat (examples with known provenance only)

(1) Abydos, tomb O [Djer, Nagada IIIC1]. Fragment of an ivory vase; probably same vase as following (not located), Petrie, 1901, plate VI, 22

(2) Abydos, tomb O [Djer, Nagada IIIC1]. Fragment of an ivory vase; probably same vase as previous (Berlin 18140), Scharff, 1929, pp. 82-83, no. 112, Abb. 58, table 22

(3) Abydos, tomb X [Adjib, Nagada IIIC2]. Ivory tablet (not located), Petrie, 1900, plate XXVII, 71

(4) Abydos (?) [date unknown]. Statuette of Bat (?) in the sanctuary of the goddess Repit (Luzern, Kofler-Truniger K 9643 R), Müller, 1964, p. 29, A31; Schlögl, 1978, p. 27, no. 81

(5) Abu Roash, tomb 389 [Nagada IIIC1]. Fragment of ivory box (Cairo?), Klasens, 1958, p. 53, figure 20, y, plate XXV, 2

(6) Gerza, tomb 59 [SD 47-77 = Nagada IIC-IIIC1?]. 'Gerza palette' (Cairo), Petrie *et al.*, 1912, plate VI, 7

(7) Hierakonpolis, Temple area [date unknown]. Fragments of stone vase (Oxford E.132, E.3645 and London UC.16245), Quibell and Petrie, 1900, plate XVIII, 21; Quibell and Green, 1902, plate LIX, 4-7; Burgess and Arkell, 1958; Adams, 1974a, p. 50, no. 272/37

(8) Hierakonpolis, Main Deposit. Narmer palette (Cairo, CG 14716)

(9) Nagada, tomb 218 [date unknown]. 'Pot mark' (not located), Petrie, 1896, plate LIII, 116

(10) Nagada, tomb 584 [Nagada IIIA1]. 'Pot mark' (not located), Petrie, 1896, plate LII, 77a

(11) Naqa ed-Deir, tomb N 1532 [Nagada IIID]. Gold pendant of a bull with the Bat amulet at its neck (Cairo CG 204A1), Reisner, 1908, plate 6, 9

APPENDIX I. Bird amulets, Petrie, 1920, pl. XLIV, 102 N-P (examples with known provenance only)

(1) el-Mahasna, tomb H 25 [Nagada IIA]. Ivory, 3 examples (not located), Ayrton and Loat, 1911, plate XV, 2

(2) el-Mamariya, tomb unknown [date unknown]. Greywacke, 2 examples (Brooklyn 07.447.615-616), Needler, 1984, pp. 98-99

(3) Matmar, cem. 3000 [date unknown]. Greywacke (Oxford 1932.894), Brunton, 1948, plate XVI, 23; Payne, 1993, p. 240, no. 1985, figure 82

(4) Matmar, tomb 3123 [Nagada IIB]. Ivory, 2 examples (Cairo JdE 57433A-B), Brunton, 1948, plate XVI, 21-22, plate XVII, 70-71

(5) Nagada, tomb 146 [date unknown]. Greywacke (London UC.5667), Petrie, 1920, plates XLIV, XLV, 37, 1914, plate XLII, 251e; Baumgartel, 1970, p. VI

(6) Nagada, tomb 1590 [Nagada IA]. Greywacke (London UC.4121), Petrie, 1920, plates XLIV, XLV, 38

(7) Nagada, tomb 1781 [SD 47 = Nagada IIC?]. Greywacke, 3 examples (London UC.5664-5666), Petrie, 1896, plate LXIV, 90, 1914, plate XLII, 251d, 1920, plate XLV, 34-36; Baumgartel, 1970, LVII

(8) Nagada (?) [date unknown]. Greywacke (Cairo CG 14156), Quibell, 1905, p. 224

(9) Naqa ed-Deir 7453 [Nagada IIC?]. Greywacke (Berkeley), Lythgoe and Dunham, 1965, p. 277, figure 123, i

APPENDIX J. Pelta palettes, types after Petrie,1920, pl. XLIV (examples with known provenance only)

(1) Abadiya, tomb B 51 [SD 40? = Nagada IIA-IIC?]. Type 100 D, 2 examples (not located), Petrie and Mace, 1901, plate VI, II, 35-36

(2) Abadiya, tomb B 109 [SD 44? = Nagada IIA-IIC?]. Type 100, 2 examples (not located), Petrie and Mace, 1901, plate VI, XII, 37-38

(3) Abusir el-Meleq, tomb 4 d 2 [Nagada IIC-III]. Type (not located), Scharff, 1926, p. 51, tafel 51, 315

(4) Akhmim [date unknown]. Type 100 (Strasbourg?), Forrer, 1901, p. 33

(5) el-Amra [date unknown]. Type 100 D, 2 examples (St. Germain-en-Laye, 77.705p1), Cleyet-Merle and Vallet, 1982, p. 116

(6) Armant, cem. 1400 [date unknown]. Type 100 D (Oxford, Queens College Loan 1226), Mond and Myers, 1937, plate XV, 2; Payne, 1993, p. 239, no. 1981, figure 82

(7) Badari, tomb 1967 [date unknown]. Type 101 G (not located), Brunton and Caton-Thompson, 1928, plate LII, 21

(8) Badari, tomb 3844 [Nagada IIA]. Type 100 D, 2 examples (not located), Brunton and Caton-Thompson, 1928, plate XXXIV, 4, LII, 20

(9) Badari, Town group 3167 [date unknown]. Type 101 S (not located), Brunton and Caton-Thompson, 1928, plate XLVII, 5

(10) Ballas / Zawaida [date unknown]. Type 101 F (St. Germain-en-Laye, 77.709w), Cleyet-Merle and Vallet, 1982, p. 148

(11) Ballas / Zawaida [date unknown]. Type 100 D (St. Germain-en-Laye, 77.709w), Cleyet-Merle and Vallet, 1982, p. 148

(12) Gebel et-Tarif [date unknown]. Type 101 (Cairo CG 14148), Quibell, 1905, p. 223

(13) Gebel et-Tarif [date unknown]. Type 100 D (Cairo CG 14150), Quibell, 1905, p. 223

(14) Gebel et-Tarif [date unknown]. Type 100 (Cairo CG 14151), Quibell, 1905, p. 223

(15) Khozam [date unknown]. Type 100 D (not located), Chantre, 1907, p. 231, figure 4

(16) Khozam [date unknown]. Type 100 D (not located), Chantre, 1907, p. 231, figure 5

(17) Mahasna [date unknown]. Type 100 D (Cairo CG 14149), Quibell, 1905, p. 223

(18) Matmar, tomb 2644 [Nagada IIA]. Type 100 D, 2 examples (not located), Brunton, 1948, plate XV, 36

(19) Matmar, tomb 2720 [Nagada IIA]. Type 100 with double bird amulet (London BM EA.63415), Brunton, 1948, plate XV, 37

(20) Matmar, tomb 3123 [Nagada IIB]. Type 101 H, 2 examples (Cairo JdE 57431A-B), Brunton, 1948, plate XV, 35

(21) Mustagedda, cem. 1800 [date unknown]. Type 100 D (not located), Brunton, 1937, plate XLIII, 11

(22) Nagada, tomb 8 (?) [date unknown]. Type 101 T (London UC.4414), Petrie, 1920, plate XLIV, (not in Baumgartel, 1970)

(23) Nagada, tomb 10 (?) [Nagada IID2]. Type 101 R (London UC.4518), Petrie, 1920, plate XLIV (not in Baumgartel, 1970)

(24) Nagada, tomb 185 [Nagada IIC]. Type 101 S (not located), Petrie, 1920, plate XLIV (not in Baumgartel, 1970)

(25) Nagada, tomb 325 [date unknown]. Type 101 h (Manchester 2378), Baumgartel, 1970, p. XIV

(26) Nagada, tomb 325 [date unknown]. Type 101 h (Berlin 12874), Scharff, 1929, pp. 91-92, no. 127, tafel 24

(27) Nagada, tomb 461 [date unknown]. Type 101 f (London UC.5469A), Baumgartel, 1970, p. XIX

(28) Nagada, tomb 1419 [Nagada IIB]. Type 101 H (Oxford 1895.864), Petrie, 1896, plate XLIX, 64; Payne, 1993, p. 239, no. 1983

(29) Nagada, tomb 1419 [Nagada IIB]. Type 101 H (Oxford 1895.864), Petrie, 1896: plate XLIX, 64; Payne 1993, p. 239, n° 1983

(30) Nagada, tomb 1419 [Nagada IIB]. Type 101 H (London UC.5368), Petrie, 1896, plate XLIX, 64, 1920, plate XLIV, XLV, 21

(31) Nagada, tomb 1865 [Nagada IIB]. Type 101 G (London UC.4345), Petrie, 1914, plate XLII, 251 f, 1920, plate XLIV

(32) Nagada, tomb 1870 [date unknown]. Type 101 (Bonn, Ägyptologisches Seminar 329), Regner, 1996, p. 66 (not in Baumgartel, 1970)

(33) Nagada, tomb unknown [date unknown]. Type 101 L (not located), Petrie, 1896, plate XLIX, 62

(34) Nagada. tomb unknown [date unknown]. Type 101 G (not located), Petrie, 1896, plate XLIX, 63

(35) Nagada. tomb unknown [date unknown]. Type 101 H (not located), Petrie, 1896, plate XLIX, 66

(36) Nagada. tomb unknown [date unknown]. Type 101 (not located), Petrie, 1896, plate XLIX, 67

(37) Nagada, tomb unknown [date unknown]. Type 101 (not located), Petrie, 1896, plate XLIX, 68

(38) Naqa ed-Deir 7008 [Nagada IIA?]. Type 101, 2 examples (Berkeley?), Lythgoe and Dunham, 1965, p. 5, figure 1, o

(39) Naqa ed-Deir 7509 [Nagada IIB?]. Type 101, 2 examples (Berkeley?), Lythgoe and Dunham, 1965, p. 323, figure 143, k

(40) Qaw el-Kebir, tomb 136 [Nagada IIA]. Type 101 G, 2 examples (not located), Brunton and Caton-Thompson, 1928, plate LII, 22-23

APPENDIX K. Rhomboidal palettes of type 91 T-U (Petrie, 1920, plate XLIV, 1921, plate LVIII; examples with known provenance only)

(1) Abadiya, tomb B 101 [Nagada IIA]. Type 91 (not located) Petrie and Mace, 1901, plate V

(2) Abadiya, tomb B 102 [SD 33-41 = Nagada IC-IIB?]. Type 91 (Cairo JdE 34220), Petrie and Mace, 1901, plate V. It is not clear if this palette is an example identical to the previous or if there is an error in the publication and that in reality there is only one palette. It should be noted that this palette is furthermore decorated in carved relief with the symbol of the two horns

(3) el-Amra, tomb a 97 [date unknown]. Type 91 U (Cambridge, University Museum), Baumgartel, 1960, plate VI, 1

(4) el-Amra, tomb a 101 [date unknown]. Type 91 U (not located), Petrie, 1921, plate LVIII

(5) Badari, tomb 3829 [SD 41-48 = Nagada IIA-IIC?]. Type 91 T (not located), Brunton and Caton-Thompson, 1928, plates XXXIII, XXXIV, 3

(6) Dakka, cemetery 99, tomb 3 [Nagada IIA-B?]. Type 91 (not located), Firth, 1915, p. 47, figure 17

(7) Dehmit, tomb 22 [Nagada IIC-D?]. Type 91 (not located), Reisner, 1910, I, p. 248, II, plate 63, b, 15

(8) Mahasna, tomb unknown [SD 41-48 = Nagada IIA-IIC?]. Type 91 T (not located), Petrie, 1921, plate LVIII

(9) el-Mamariya, tomb 35 [date unknown]. Type 91 T (Brooklyn 07.447.600), Needler, 1984, pp. 99, 320-321

(10) Mustagedda, tomb 1825 [Nagada IIA]. Type 91 U (London BM EA.63066), Brunton, 1937, plate XLIII, 3

(11) Mustagedda tomb 1832 [date unknown]. Type 91 (not located), Brunton, 1937, plate XLIII, 4

(12) Nagada, tomb 1440 [SD 41-51 = Nagada IIA-IIC?]. Type 91 U. (not located), Petrie, 1921, plate LVIII (not mentioned in Baumgartel, 1970)

(13) Nagada, tomb 1497 [Nagada IC]. Type 91 U (Oxford 1895.854), Petrie, 1921, plate LVIII; Payne, 1993, p. 227, no. 1867

(14) Nagada, tomb 1904 [date unknown]. Type 91 T (Manchester 2375), Baumgartel, 1970, p. LXI
(15) Nagada [date unknown]. Type 91 U (London UC.6026), Petrie, 1920, plate XLIV
(16) Nagada [date unknown]. Type 91 T (London UC.6025), Petrie, 1920, plate XLIV

APPENDIX L. Objects with horns in one unbroken curve

(1) Abadiya, tomb B 414 [date unknown]. Greywacke amulet (not located), Petrie and Mace, 1901, plate XII, 41. In this tomb, an amulet with horns with curved ends has also been found
(2) Abusir el-Meleq, tomb 60 d 7 [date unknown]. Bone spoon (Berlin 18664), Scharff, 1926, p. 53, no. 339
(3) Ballas, tomb Q 23 [date unknown]. Bone pendant (Oxford 1895.894), Petrie, 1896, plate LVIII, Q23; Payne, 1993, p. 208, no. 1699, figure 72
(4) Ballas, tomb Q 132 [Nagada IC]. Ivory (?) comb (not located), Petrie, 1896, plate LXIII, 57
(5) Ballas, tomb Q 709 [date unknown]. Two stone amulets (not located), Petrie, 1896, plate LVIII, Q709/7, 9
(6) Hu, tomb U 119 [SD 37 = Nagada IC?]. Ivory (?) amulet (not located), Petrie and Mace, 1901, plate X, 12
(7) Minshat Abu Omar, tomb 882 [Nagada IID2?]. Bone amulet, 2 examples (Cairo), Kroeper and Wildung, 2000, p. 121, tafel 36, 6
(8) Mustagedda, tomb 320 [Nagada IC-IIA?]. Bone comb (Berlin 22982), Brunton, 1937, plate XLII, 47; Scharff, 1931, p. 263, no. 854
(9) Nagada, tomb 3 [Nagada IID2]. Copper (?) needle (not located), Petrie, 1896, plate LXV, 22
(10) Nagada, tomb 10 [Nagada IID2]. Bone amulet (Oxford 1895.919), Petrie, 1896, plate LXIV, 80; Payne, 1993, p. 206, no. 1683, figure 72
(11) Nagada, tomb 149 [date unknown]. Ivory amulet (Oxford 1895.922), Petrie, 1896, plate LXII, 37; Payne, 1993, p. 241, no. 1997, figure 83
(12) Nagada, tomb 259 [SD 51 = Nagada IIC-IID?]. Ivory (?) comb (London UC.4400 or 4401), Petrie, 1896, plate LXIV, 70
(13) Nagada, tomb 632 [Nagada IIC]. Noble serpentine pendant (London UC.5100), Petrie, 1920, plate XXIII, 6
(14) Nagada, tomb 1417 [SD 35-41 = Nagada IC-IIB?]. Bone comb (Berlin 12854), Petrie, 1896, plate LXIII, 57; Scharff, 1929, pp. 140-141, no. 270
(15) Nagada, tomb 1517 [date unknown]. Bone (?) pendant (Oxford 1895.918), Petrie, 1896, plate LXIV, 79; Payne, 1993, p. 207, no. 1690, figure 72
(16) Nagada, tomb 1774. [Nagada IC] Bone (?) hairpin (Oxford 1895.952), Petrie, 1896, plate LXIV, 82; Payne, 1993, p. 229, no. 1888, figure 77
(17) Nagada, tomb 1852. [Nagada IIC] Bone comb (Oxford 1895.938), Petrie, 1896, plate LXIII, 57A; Payne, 1993, p. 231, no. 1908, figure 78
(18) Provenance not recorded [date unknown]. Palette type 104 D. (London UC.16272), Petrie, 1920, plate XLIV, XLVI, 18

(19) Provenance not recorded (the inscription 'a 26' which can be seen on the published photograph could refer to el-Amra, tomb a 26) [date unknown]. Palette type 104 G (not located), Petrie, 1920, plates XLIV, XLVI, 19

(20) Provenance not recorded [date unknown]. Palette type 104 L (London UC.16273), Petrie, 1920, plates XLIV, XLVI, 20

(21) Provenance not recorded [date unknown]. Ivory (?) imitation dagger (London UC.16274), Petrie, 1920, plate XLVI, 21

(22) Provenance not recorded [date unknown]. Unidentified bone and metal object (London UC.16281), Petrie, 1920, plate XLVI, 37

(23) Provenance not recorded [date unknown]. Unidentified flint object (London UC.15173), Petrie, 1920, plate VII, 11

(24) Provenance not recorded [date unknown]. Greywacke hairpin (?) (Brussels E.2187), Unpublished (Figure 16.11)

(25) Provenance not recorded [date unknown]. Greywacke palette (Brussels E.7129), Unpublished (Figure 16.12)

(26) Provenance not recorded [date unknown]. Two palettes type 104 L (Berlin 14294), Scharff, 1929, p. 91, no. 126

APPENDIX M. Objects with incurved horns (examples with known provenance only)

(1) Abadiya, tomb B 51 [SD 40? = Nagada IIA-IIC?]. Ivory (?) comb (not located), Petrie and Mace, 1901, plate VI

(2) Abadiya, tomb B 102 [SD 33-41 = Nagada IC-IIB?]. Bone comb (Oxford E.1011), Petrie and Mace, 1901, plate V, IX, 22; Payne 1993, p. 232, no. 1918, figure 78

(3) Abadiya, tomb B 106 [date unknown]. Greywacke amulet type 103, 2 examples (not located), Petrie and Mace, 1901, plate XII, 39-40

(4) Abadiya, tomb B 109 [SD 44? = Nagada IIA-IIC?]. Greywacke amulet type 4 m, 2 examples (not located), Petrie and Mace,1901, plate VI, XI, 2-3

(5) Abadiya, tomb B 414 [date unknown]. Greywacke amulet type 103 (not located), Petrie and Mace, 1901, plate XII, 42

(6) el-Amra, tomb A 88 [SD 36-39 = Nagada IB-IC?]. Greywacke amulet type 103, 2 examples (not located), McIver and Mace, 1902, p. 16, plate VII, 2

(7) el-Amra [date unknown]. Amulet type 103, 2 examples (St. Germain-en-Laye 77.705 p3-p4), Cleyet-Merle and Vallet, 1982, p. 116

(8) Badari, cem. 1600 [date unknown]. Limestone amulet (not located), Brunton and Caton-Thompson, 1928, plate LIII, 50

(9) Badari, tomb 3759 [Nagada IIC]. Ivory amulet, 3 examples (not located), Brunton and Caton-Thompson, 1928, plate XXXIV, 4, LIII, 49

(10) Badari, tomb 3844 [Nagada IIA]. Bone comb (not located), Brunton and Caton-Thompson, 1928, plate XXXIV, 4, LIII, 31

(11) Ballas / Zawaida [date unknown]. Greywacke amulet type 103 (Brussels E.6188b), Unpublished (Figure 16.13 c)

(12) Ballas / Zawaida [date unknown]. Greywacke amulet type 103 (Brussels E.6188c), Unpublished (Figure 16.14 c)

(13) Hu, tomb U 104 [SD 43 = Nagada IIA-IIC?]. Ivory (?) amulet (not located), Petrie and Mace, 1901, plate X, 11

(14) Hu, tomb U 167 [date unknown]. Ivory (?) comb (not located), Petrie and Mace, 1901, plate X, 3

(15) Hu, tomb U 167 [date unknown]. Ivory (?) needle (not located), Petrie and Mace, 1901, plate X, 10

(16) Hu, tomb U 284 [SD 43 = Nagada IIA-IIC?]. Ivory (?) comb (not located), Petrie and Mace, 1901, plate X, 3

(17) el-Mahasna, tomb H 49 [Nagada IC]. Ivory (?) comb (not located), Ayrton and Loat, 1911, plate XVIII, 4

(18) el-Mahasna, cemetery L (= Beith Allam / Alawniya), tomb 229 [SD 36-43 = Nagada IB-IIC?]. Bone amulet (not located), Garstang, 1903, p. 5, plate IV

(19) el-Mamariya [date unknown]. Greywacke amulet (Brooklyn 07.447.614), Needler, 1984, pp. 98-99, no. 63

(20) Matmar, tomb 2622 [Nagada IIB]. Bone comb, 2 examples (not located), Brunton, 1948, plate XVI, 3, plate XVII, 41

(21) Matmar, tomb 2626 [Nagada IIB]. Bone comb (not located), Brunton, 1948, plate XVI, 5, plate XVII, 35

(22) Matmar, tomb 2640 [Nagada IIA]. Bone needle (not located), Brunton, 1948, plate XVI, 2, plate XVII, 28

(23) Matmar, tomb 2646 [Nagada IIA]. Ivory amulet (not located), Brunton, 1948, plate XVI, 12, plate XVII, 54

(24) Matmar, tomb 3092 [date unknown]. Bone needle (not located), Brunton, 1948, plate XVI, 4

(25) Matmar, tomb 3113 [Nagada IC-IIB]. Bone comb (not located), Brunton, 1948, plate XVI, 4

(26) Matmar, tomb 3133 [date unknown]. Greywacke amulet type 103, 2 examples (Oxford 1932.896-897), Brunton, 1948, plate XVI, 10-11; Payne, 1993, p. 240, no. 1988-1989, figure 82

(27) Mustagedda, cem. 1800 [date unknown]. Bone comb (not located), Brunton, 1937, plate XLII, 51

(28) Mustagedda, tomb 1867 [date unknown]. Bone comb (not located), Brunton, 1937, plate XLII, 54

(29) Mustagedda, tomb 11741 [SD 37? = Nagada IC-IIC?]. Greywacke amulet type 103 Q (not located), Brunton, 1937, plate XLIII, 14

(30) Nagada, tomb 149 [date unknown]. Hippopotamus ivory (?) amulet, 2 examples (Oxford 1895.920-921), Petrie, 1896, plate LXII, 38; Payne, 1993, p. 240, no. 1986-1987, figure 82

(31) Nagada, tomb 293 [SD 61-72 = Nagada IID2-IIIA2?]. Ivory comb (London UC.4523), Petrie, 1896, plate LXIII, 56; Baumgartel,1970, p. XIII

(32) Nagada, tomb 884 [date unknown]. Greywacke amulet type 103 N (London UC.4738), Baumgartel, 1970, p. XXXII

(33) Nagada, tomb 1251 [SD 40 = Nagada IIA-IIC]. Greywacke amulet type 103, 2 examples (Oxford 1895.868-869), Petrie, 1896, plate LXII, 42; Payne, 1993, p. 240, no. 1990-1991, figure 82

(34) Nagada, tomb 1348 [Nagada IIB]. Greywacke amulet type 103 F (not located), Petrie, 1896, plate LXII, 40, 1920, plate XLIV, XLV, 41

(35) Nagada, tomb 1468 [Nagada IC]. Greywacke amulet type 103 D/H, 2 examples (Berlin 12853, 12852), Scharff, 1929, p. 90, no. 123-124, tafel 24

(36) Nagada, tomb 1480 [Nagada IC]. Ivory (?) comb (not located), Petrie, 1896, plate LXIII, 58 (not in Baumgartel, 1970)

(37) Nagada, tomb 1497 [Nagada IC]. Ivory (?) comb (not located), Petrie, 1896, plate LXIII, 56 (not in Baumgartel, 1970)

(38) Nagada, tomb 1503 [Nagada IC]. Ivory comb (London UC.4180), Petrie, 1896, plate LXIII, 56, 1920, plate XXIX, 12; Baumgartel, 1970, p. XLVII

(39) Nagada, tomb 1503 [Nagada IC]. Ivory (?) comb (London UC.4178), Petrie, 1896, plate LXIV, 86, 1920, plate XXIX, 7; Baumgartel, 1970, p. XLVII

(40) Nagada, tomb 1586 [Nagada IB]. Bone comb (Oxford 1895.937), Petrie, 1896, plate LXIII, 58; Payne, 1993, p. 232, no. 1917, figure 78

(41) Nagada, tomb 1621 (?) [Nagada IA]. Greywacke amulet type 103 J (London UC.4495), Petrie, 1920, plate XLIV, XLV, 2 (not in Baumgartel, 1970)

(42) Nagada, tomb 1646 [Nagada IC]. Greywacke, amulet type 103 T, 2 examples (London UC.4125-4126), Petrie, 1896, plate LXIV, 89, Petrie, 1920, plate XLIV, XLV, 3

(43) Nagada, tomb 1675 [Nagada IC]. Greywacke amulet type 103 (Oxford 1895.867), Petrie, 1896, plate LXII, 43; Payne, 1993, p. 240, no. 1992, figure 82

(44) Nagada, tomb 1678 [SD 31-56 = Nagada IA-IID?]. Bone hairpin (Oxford 1895.958), Petrie, 1896, plate LXIV, 74; Payne, 1993, p. 229, no. 1885, figure 77

(45) Nagada, tomb 1757 [date unknown]. Greywacke amulet type 103, 2 examples (London UC.5451-5452), Petrie, 1896, plate LXII, 42; Baumgartel, 1970, p. LVI

(46) Nagada, tomb 1871 [Nagada IIA]. Greywacke amulet type 103 J, 2 examples (London UC.4495-4496), Baumgartel, 1970, p. LX

(47) Nagada, tomb 1871 [Nagada IIA]. Greywacke amulet type 103 J (Brussels E.1231), Hendrickx, 1986, p. 40, figure 14 d

(48) Nagada, tomb B 72 [date unknown]. Greywacke amulet type 103 (London UC.4739), Baumgartel, 1970, p. LXIV

(49) Nagada, tomb T 24 [SD 42-47 = Nagada IIB-IIC?]. Bone amulet (Oxford 1895.917), Petrie, 1896, plate LIX, 9; Payne, 1993, p. 238, no. 1963, figure 81

(50) Naqa ed-Deir 7150 [Nagada IIC?]. Ivory comb (Berkeley?), Lythgoe and Dunham, 1965, p. 87, figure 35, g

(51) Naqa ed-Deir 7634 [Nagada IC?]. Greywacke amulets type 103, 2 examples (Berkeley?), Lythgoe and Dunham, 1965, p. 417, figure 188, b

(52) Qaw el-Kebir, tomb 136 [Nagada IIA]. Ivory hairpin (not located), Brunton and Caton-Thompson, 1928, plate LIII, 24

(53) Saghel el-Baghliya [date unknown]. Ivory comb (Cairo CG 14478), de Morgan, 1896, p. 148, figure 343; Quibell, 1905, p. 272

ABBREVIATIONS

Berkeley	Phoebe Apperson Hearst Museum of Anthropology and Archaeology
Berlin	Ägyptisches Museum
Brooklyn	Brooklyn Museum
Brussels	Royal Museums for Art and History
Boston	Museum of Fine Arts
Cairo	Egyptian Museum
London BM	British Museum
London UC	Petrie Museum of Egyptian Archaeology, University College London
Manchester	Manchester Museum
München	Staatliche Sammlung Ägyptischer Kunst
New York	Metropolitan Museum
Oxford	Ashmolean Museum
St. Germain-en-Laye	Musée des Antiquités nationales
Strasbourg	Institut d'Egyptologie

Drought is the most important factor limiting livestock production in Africa..., a drought lasting two or three seasons exhausts the available food sources, decimating herds. In Botswana, for example, drought reduced cattle numbers from 1.35 million to 900,000 in the dry years of 1964-67.

M. Chenje and P. Johnston, *State of the Environment in Southern Africa*, IUCN, The World Conservation Union, Harare, Zimbabwe, 1994.

17

CONCLUSION: ECOLOGICAL CHANGES AND FOOD SECURITY IN THE LATER PREHISTORY OF NORTH AFRICA: LOOKING FORWARD

F. A. Hassan

No one can afford to ignore the current and potential impact of climatic variability on our contemporary human affairs, and it would be irresponsible not to seek in the past for insights into how climatic change has influenced food security and the course of change in our human condition. Africa, while not the only country to suffer from the onslaught of recent droughts, has been in the forefront of our concern, not only because the droughts have plighted poor nations that have just recently emerged from under the yoke of imperialism, but also because the livelihood and stability of its political systems depend on the swings of the Intertropical Conversion Zone. This zone is responsible for the majority of rainfall in Africa and many other parts of the world, and hence an examination of climatic changes and their impact on cultural developments, in response to droughts undermining the food security of African societies in the past, is most relevant to the clarification of the role of climate in the contemporary world.

The contributions in this volume reveal that many of the cultural changes appear to have been related to severe, short-term *cold* climatic episodes (Table 17.1), which are mostly identified in the record of reduced sea surface temperature, given in this volume by Rohling and his colleagues. It is important to note here several issues:

- Reliability and precision of radiocarbon dating
- Sensitivity of certain environments and microenvironments to global climatic events, as well as differences in the imprint of climatic signal, possibilities of post-depositional dilution or erasure of signal, and time delay between climatic event and environmental response.
- Geographic and latitudinal differences (e.g., proximity to large bodies of water, or location relative to longitudinal and latitudinal range of climatic belts).
- Local differences in catchment area and depositional environment.

Table 17.1. Chronology of climatic and cultural events discussed in this work

	Calibrated ¹⁴C years kyr cal BP	Uncalibrated ¹⁴C years kyr bp	Climatic Events	Cultural events
Phase I	18-16	15-13.5	**Reduced SST**	
	16-13.4	13.5-11.5	Transition to wetter conditions in North Africa and replenishment of water table.	
	13.4-11	11.5-10	Reduced SST: Younger Dryas (11-10 kyr bp).	Plant domestication in Southwest Asia.
Phase II	11-10	10-9	Lakes and ponds cover a very large surface in Sahara and Sahel.	
	10-9.5	9.8.6	**Reduced SST**	Cattle domestication in Africa.
	9.5-8.6	8.6-7.8	Continuation of wet conditions in Sahara and Sahel. Climate in general is warm with predominantly monsoon related rain supporting vegetation with Sahelian elements as far north as 27°N.	
	8.6-7.9	**7.8-7.0**	**Reduced SST:** Short arid episode inferred by Vernet, this volume, at ca. 8-7.9 kyr cal BP in Sahara and Sahel.	Rapid dispersal of sheep/goat from Southwest Asia and cattle from Eastern Sahara into Northeast Africa (Nile Valley Eastern Sahara and Central Sahara) and emergence of new settlement strategies, tribal chiefdoms, and intensive utilization of cereals. These events coincide with stress and abandonment of settlement in the southern Levant, and migrations to outlying areas.

Phase III	7.9-6.7	<7-5.9	Moist but with pronounced climatic oscillations in Eastern Sahara characterized by cold spells and thundershowers.	Early village communities established in the Delta and the Fayum, Egypt.
	6.7-5.2	5.9-4.5	Reduced SST: Major arid crisis in Libya at 5.7 kyr cal BP: lakes became sabkhas, shelters collapsed, and wind erosion intensified. Occasional spring and autumn rain in Eastern Sahara ca. 6.7 kyr cal BP.	Agriculture established all along the Nile Valley in the Delta and Upper Egypt, establishment of residential villages, rapid political evolution from village chiefdoms to a state society, especially from 5.6 to 5.3 kyr bp following the emergence of petty states. Depopulation of the Acacus with concentration of human groups in proximity of localized oases as well as long-range pastoralism.
Phase IV	5.2-4.4	4.5-4.0	Desert conditions established in North Africa ca. 5.2-5 kyr cal BP (4.5 kyr bp). Vernet, this volume, identifies an arid crisis ca. 4.7-4.45 kyr cal BP separating the Sahara, which became desert, from the Sahel, which maintained Sahelian climate for another millennium.	Cattle in East Africa and West Africa ca. 4.4 cal BP. In the Sahel people survived where water was available at or close to the surface. Sandy plains were excluded. In Egypt, political developments led to the emergence of a unified nation state ca. 4.5 kyr bp (3300-3100 cal BC).
	4.4-4.1	4.0-3.5	Dry conditions in the Sahara, but Sahel south of 22/21°N dominated by summer monsoon	Concentration of settlements and dense occupations in the Sahel.
Phase V	4.1	3.8/3.7	Severe aridity and intensive dust activity.	Disintegration of unified Egyptian nation by the end of the 'Old Kingdom'.
	3.8-3.0	3.5-3.0	Reduced SST: Retreat of Lake Chad between 3.8 and 3.5 cal BP. Hyperarid conditions established in North Africa by 3 kyr cal BP, but Sahel remained moist.	Cultivating *Pennisetum americanum* in Mauritania 3.8-3.7 cal BP, in the Nigerian Chad Basin around 3.25 kyr cal BP (3 kyr bp or 1200 cal BC) and 2.9 cal BP (ca. 1000 cal BC) north of Burkina Faso.

In general, terrestrial evidence for surface water resources, such as rivers and lakes in the Sahara and Sahel, do not become pronounced following post-glacial global warming until the water table is replenished after the prolonged droughts that accompanied the Last Glacial Maximum. The greening of the desert thus did not occur until after the Younger Dryas, commencing ca. 11,000 yr bp. The question of dating is extremely important in the discussion of the link between climatic and cultural events. If certain climatic events are to be considered as causal factors in culture change, it must be shown that the climatic events are indeed prior to the cultural changes presumed to have been linked to the climatic event under consideration. This fundamental issue is often over-looked, which undermines the credibility of climatic explanations of culture change. The issue is complicated because: (1) authors may recourse to single radiocarbon age determinations; (2) some radiocarbon age determinations are associated with a large magnitude of error; and (3) the association of the dated material and the climatic or cultural events may be questionable. One of the common problems that frustrates comparisons and correlations is the lack of consensus about the abbreviations used to report radiocarbon age measurements and their calibrated equivalent. For this reason, we have consistently used as a bench mark the calibrated radiocarbon dates before present as cal BP, while uncalibrated radiocarbon years are referred to as bp.

The reliability of the age estimates of climatic and cultural events reported in this volume is enhanced by comparing the calibrated radiocarbon age determinations in different regions and the attempt to use, as much as possible, at least three radiocarbon age determinations for each event. Luckily, recent investigations not only provide numerous dates for each event, but the dates are reported with a relatively small margin of error.

Undoubtedly, the climatic conditions influenced the type and distribution of the palaeovegetation in Northeastern Africa, but changes in vegetation were certainly linked to the changing parameters of the local landscape (e.g., hydrography, relief, slope processes and sedimentary environments). It is thus important not only to distinguish the Mediterranean zone from the inner Saharan belt and the Sahel fringe to the south, but also to recognize the particular characteristics of the range and basin of massifs and hills bordering the Eastern Sahara, including Gebel Uweinat and Gilf El-Kebir, the open desert plateaus (e.g., Abu Tartur Plateau), and the oases depressions (e.g. Kharga, Dakhla, Farafra and Baharia). The wadis associated with the depressions and the palaeochannels in the southern area of the Eastern Sahara, for example the Selima sand sheet area, must also be regarded as key features in the palaeolandscape of Northeastern Africa.

Geographically, the distinction between north and south has already been mentioned; areas to the north becoming drier, on a millennial scale, earlier than areas to the south. In addition, there is a clinal change from west to east, with eastern areas becoming drier earlier than West Africa. This explains in part the advance of cattle-keeping westward as conditions became progressively drier. The topographic differences between the highland massifs of the Sahara and the sandy plains also explain the dispersal of cattle-keeping via the range and basin topographic lanes across the Sahara. As conditions became drier, depressions, where ground water is close to the surface, supporting vegetation in isolated oases, became loci of habitation and subsistence activity. This also applied to wadis fed by large catchment areas in the Sahara.

In light of the considerations of dating, geographic and environmental setting, the response to climatic change may be assessed not only in terms of the long-term, millennial, changes in climatic regimes, but also in the light of short-term, abrupt, climatic events, which cause rapid and dramatic disruption of prevailing climatic regimes (cf. Hassan, 1996, 1997).

In considering the differences between various environments in capturing climatic signals we note that, unlike marine records, terrestrial evidence is often patchy and discontinuous. However, with data from lake cores and from piecing together data from playa sediments, springs, caves, rock shelters, and wadi deposits, a composite palaeoclimatic record for the last 10,000 years can be constructed. Again we are fortunate that, after four decades of geoarchaeological investigations, there is sufficient evidence to clearly detect four arbitrary major phases of climatic change:

> 11-7.9 kyr cal BP (10-7.0 kyr bp)
> Warm, wet, dominated by monsoon-related rainfall, supporting vegetation of Sahelian elements as far as 27°N.
> 7.9-5.2 kyr cal BP (7.0-4.5 kyr bp)
> Warm with occasional cold spells associated with spring and autumn rainfall.
> 5.2-4.1 kyr cal BP (4.5-3.7 kyr bp)
> Desert conditions prevalent in North Africa leading to the separation of Sahara and Sahel, with persistence of moist conditions as far north as 22°N.
> 4.1 kyr cal BP to present
> Prevalence of present day desert conditions in North Africa.

In the past, recognition of abrupt, short-term climatic events nested in long-term climatic variations was hampered by the lack of high resolution dating and by the paucity of data. Luckily, this is now changing and several global abrupt climatic events have now been recognized. Contributions in this volume emphasize the impact of the Younger Dryas on the emergence of food production in Southwest Asia (see also Hassan, 2000b; Hillman, 1996), the impact of the abrupt events at 7.9 kyr cal BP (7 kyr bp) on the rapid dispersal of sheep and goats from Southwest Asia to the Nile Valley, and the spread of domesticated cattle into the Central Sahara. The event ca. 7 kyr bp is well documented in a variety of records from North, East and Equatorial Africa (Gasse and Van Campo, 1994; Roberts et al., 1993), and is clearly shown in the record from Lake Malha in the Sudan (Mees et al., 1991). On a global basis, cooling by as much as 6 ± 2° at 7.3 kyr bp (8.2 kyr cal BP) is indicated in the north Atlantic region (Alley et al., 1997). This cooling event is also recognized as a major oxygen isotope excursion in the GRIP and GISP Greenland Ice cores (Snowball et al., 1999). A similar excursion has been reported from the sediments of Lake Ammersee in southern Germany. Also, detailed palaeomagnetic investigations on a 900-year record of varved lake sediments from Lake Sarsjon in northern Sweden clearly point to a climatic excursion at 7.3-6.9 kyr bp (8.2-7.7 kyr cal BP). The palaeomagnetic anomaly coincides with increased glacial activity in northern Sweden at 7 kyr bp (7.9 kyr cal BP) and central and southern Norway, as well as a major abrupt cooling event in China, dated to 7.3 kyr bp (Fang *et al.*, 1993), and to a marked reduction in sea surface temperature at 7 kyr bp (Rohling *et al.*, 1997).

Abrupt climatic events coinciding with the reduction of sea surface temperature from 6.7 to 5.8 kyr cal BP (5.9 to 5.1 kyr bp) culminated at 5.7 kyr cal BP with an arid crisis well documented in Libya, where lakes and swamps became sabkhas, roofs of shelters collapsed, and wind erosion intensified. The event was accompanied with the depopulation of the Acacus and the concentration of settlements in the proximity of localized oases. It also led to the emergence of long-range transhumant pastoralism, and further dispersal of cattle-keeping in West Africa. The period from 6.7 to 5.7 kyr cal BP (4700-3700 cal BC), was also a period of major social and political transformations in Egypt. Sedentary agricultural villages and a shift from chiefdoms of the Badarian (ca. 4400 cal BC) to petty states occurred in Egypt ca. 5.8-5.7 kyr cal BP (3800-3700 cal BC), at the beginning of the ceramic phase called Nagada II in Upper Egypt.

The recognition of abrupt climatic events in the Sahara from 4.5 kyr bp (5.3-5 kyr cal BP) is problematic because of the prevalence of aridity and erosion since that time, leading to a paucity of records. For example, an abrupt cooling event at 4.5-4.4 kyr bp (5.3-5 kyr cal BP) is very well documented in the lake records of lakes Kashiru, Nudurumu, and Kuruyange in Burundi, and Lake Rudolf in Turkana, as well as at Tigalmamine in Morocco, at Bougouma in the Sahel and Malha Carter Lake in the Sudan (Gasse and Van Campo, 1994; Lamb *et al.*, 1995; Bonnefille *et al.*, 1990, 1991) This event in Africa is matched by an abrupt cooling event in Russia (Kremenetski and Patyk-Kara, 1997) and advance of glaciers in Southeast Scotland (Rose *et al.*, 1997).

The abrupt cooling event of ca. 4.5 kyr bp (5.3-5.0 kyr cal BP) signalled a major transformation in climate to cooler conditions in the world and aridification in North Africa. The severity of droughts in the Sahara led to the movement of cattle-keepers following the retreating rainfront westwards. By 4.4 cal BP they reached Mali and Ghana, and Mauritania by 3.8 kyr cal BP. Cattle also reached East Africa by 4.4 cal BP (Hassan, 2000a; Gifford-Gonzalez, 2000; Marshall, 1998).

Another major abrupt climatic event has recently been well documented from a high-resolution study by Street Perrot *et al.* (2000), of dust deposition and ostracods in a 5500-year record in the West African Sahel at Kajemarum Oasis, northeastern Nigeria. The work revealed a pronounced shift in both terrestrial and aquatic ecosystems around 4.1 kyr cal BP (3.8-3.7 kyr bp). This event has previously been recognized in Burundi, Rwanda, the highlands of Uganda, Lake Victoria and Ethiopia (Jolly and Bonnefille, 1992; Taylor, 1992; Kendall, 1969; Bonnefille and Hamilton, 1986).

The effects of this episode of droughts was most felt in the Sahel at a time when the Sahara was already remarkably arid. This would explain the impetus for the emergence of cultivation in Mauritania to supplement other subsistence activities. It was also the event that would have led to a pronounced attenuation and break-up of the Sahelian belt, allowing cattle-keepers to penetrate farther south into Central Africa.

The cultural adaptations in response to climatic changes may thus be examined in terms of the long-term millennial patterns that endow a region with specific parameters that span generations and appear at any particular time as immutable and stable. These climatic parameters shape the overall features of the land, its vegetation, animals, and plants, as well as its depositional and sedimentary processes, thus contributing to the establishment of a geological landscape. The millennial patterns also set the mode and range of seasonal, annual, and interannual variability. However, this seeming stability is a function of variation within a range that can be, and is often, interrupted by freak events that could last for years, undermining the ability to predict and make decisions. The freak events may include changes in the timing, frequency, spacing, or intensity of rain, and may introduce states (conditions) of environmental responses beyond the expected range. Such events can have a pronounced effect on the quality, amount, distribution, interannual variability and spatial unpredictability of water and food resources. Although certain events were localized, the impact of a change in one region inevitably spread, triggering a range of cultural developments on the whole continent. The social impact of a climatic-environmental event depended not only on the amplitude, frequency, duration, and rapidity of climatic signals and the modifications of the landscape, but also on the potential for perceiving environmental change, decision-making and group action.

The preservation and amplification of certain societal actions as cultural modes and traditions depends, in general, on the perceived utility and social matrix of prior actions and the opportunities presented by novel social or environmental situations. Like other cultures, those that emerged in Africa were flexible and responsive to change. The transition from hunting-gathering to pastoralism and farming, and subsequently the emergence of urban centers, complex political entities, from tribal chiefdoms to empires, constituted a sequence of developments chosen by certain communities as a means to overcome food insecurity.

In Africa, the emergence of food production at a much later date than in Southwest Asia, and the prevalence of modes of food extraction and food production that were not a replication of the course of events in Southwest Asia are issues deserving serious consideration. In this volume, Breunig and Neumann note that recent systematic excavations in Africa have produced enough data to demonstrate that models of agricultural origins in Southwest Asia and Europe are not applicable in an African context. Moreover, regional sequences show the interregional variability, which necessitates different models of cultural and economic transition from hunting and gathering to food-producing communities.

The delayed emergence of food production in Africa by comparison to Southwest Asia poses an interesting question. Should we *a priori* assume that the time, tempo and direction of change will be the same everywhere regardless of local circumstances? Clearly, the question is not only that of differences in suitable domesticates or potential for stress (relative to sufficiency) but we must also recognize differences in cultural attitudes, tendencies and values. At the same time, we must emphasize that in many situations the number of options at any particular 'mode' of culture (mostly a function of the type of political organization and socioeconomic conditions) are limited, and that people, even in different physical and social situations, may converge on a 'best' or 'optimal' choice, perhaps one that minimizes the risk of great loss which can lead to unbearable suffering and regret while aiming to gain as much as possible under the circumstances. Such criteria of risk vary depending on one's position in society, the amount of accumulated capital at risk, and the magnitude of gain relative to expenditure. The choices are not made only in terms of monetary returns or costs, but in terms of a host of variables loaded with social, ideological and psychological notions of fear, honor, fame, power, and happiness.

In the Nile Valley, where cultivation of cereals is the earliest in Africa, the cultigens were introduced from Southwest Asia several millennia after their appearance in the southern Levant, within a few weeks distance and with no insurmountable physical barriers. The reasons for this delayed effect, now that it does not appear that there is any presence of cultivated grain in Egypt before 7000 cal BP, must be due to both the lack of motivation to adopt farming by the inhabitants of the Nile Valley, and the lack of a desire by Levantine populations to settle in the Nile Valley, in spite of the presence of cultural connections and exchanges. It appears that this situation changed by ca. 8.4 kyr cal BP, when adverse climatic events precipitated the collapse of late Pre-Pottery Neolithic B. The collapse was not a local event; it affected the western slopes of the Mediterranean ranges in Palestine and Lebanon as much as it affected the eastern side of the Rift Valley (Bar-Yosef, 1998). The climatic events that caused the collapse were most probably associated with arid spells linked with the reduced sea surface temperature between 8.6 and 7.9 cal BP (Rohling et al., this volume). As a result of this collapse, some individuals and families among the mobile hunter-gatherers or sheep/goat herders who lived in symbiosis with farming villages in the southern Levant found it advantageous to disperse to areas where water was plentiful. One of the destinations was the Nile Delta. Another destination was the wadis along the Red Sea Coast.

In the Nile Valley, where cultivating plants proved to be ultimately far more rewarding than herding, farming villages and settled communities were established, leading eventually to the development of an evolutionary trajectory that culminated in the rise of the state. This development in the Lower Nile Valley may be attributed in part to the geomorphic and hydrographic setting of the Nile Valley; the thickness, extent and conditions of its soil, its water resources, including the channel, the swamps and marshes, the seepage of water from the channel to the floodplain, the seasonality of flooding, the layout of basins and levees as well as the changeable character of local conditions from year to year depending on the vagaries of flooding, its timing, level, and duration. We must also include weather conditions in Egypt allowing for the success of certain crops. The cultivation of wheat, barley and pulses from Southwest Asia on the banks of the Nile proved to be a threshold for a major cultural transformation. To simplify a complex course of events, it appears that people aimed to ensure survival by focusing on these crops rather than on herding. This marked a switch to a settled agrarian mode of life, which in turn created social conditions that were favorable to further social and political transformation that were adopted perhaps because they ensured great food security. The role cattle played in Egyptian ideology, as examined by Hendrickx in this volume, seems to indicate that cattle iconography was selected, probably by a group that became politically dominant during the formative stages of Egyptian civilization.

Cattle keeping, as Gautier repeats in this volume, was, given the current state of evidence, most probably initiated in the Eastern Sahara in the area of Nabta Playa and Bir Kiseiba. From this nuclear area, cattle keeping spread to the Central Sahara, apparently at the same time as sheep and goat were penetrating the Eastern Sahara, the Red Sea Coast, and the Nile Valley. The expansion of desert conditions and intensification of aridity in North Africa appear to have led occupants of desert regions to emphasize sheep and goats relative to the indigenous cattle. However, judging from rock art in the Central Sahara and East Africa, cattle appear to have played a major role in African ideology.

The adoption of cattle keeping, culminating in favorable habitats, as in the range and basin topography of the Saharan massifs and in West Africa, was also probably in part related to the difficulty of surviving on rain-fed agrarian products in a habitat with great interannual variability. In this volume, di Lernia remarks that recent droughts from 1961 to 1997, which affected all of Africa, seem to have been much more devastating for agricultural yields than animal products. He states that, although there are differences according to regional variability and the specific breeds of domestic livestock, it is interesting, because animals are less susceptible to the effects of climatic fluctuations than cultivated plants, that pastoralism emerged as a more successful long-term economic strategy under the progressive aridity of the later part of the Holocene. From this perspective, it would be rewarding to explore the deep cultural relationships and contexts at the core of this important mode of food production, and to attempt to trace the persistence and modification of the management of livestock from the start of the Holocene.

As di Lernia notes in this volume, pastoralism is undoubtedly one of the most important economic pursuits in Africa today. The success of pastoralism by comparison to cultivation, which appears no earlier than 1800 cal BC in West Africa, as much as three millennia after its appearance in the Nile Valley, and with indigenous African plants, suggests that cultivation was not initially the primary objective of the human groups in North Africa, outside the Nile Valley, during the early and mid Holocene. This may be attributed, in part, to the possibility that hunters and gathers during the early Holocene were reluctant to change their traditional methods and that they cautiously accepted keeping cattle, and subsequently sheep and goat, because it was compatible with their nomadic mode of life, and that the transition to cultivation was undertaken only by groups in favorable habitats in the Sahel that found themselves, as a result of increasing desert conditions, incapable of continuing their traditional subsistence pursuits. The impact of cattle keeping and herding, which arrived shortly before the emergence of plant cultivation in West Africa, was perhaps a contributing factor in destabilizing the fragile ecology of the Saharan fringe of the Sahel, enhancing the potential for moving to the practice of plant cultivation.

Choices were thus apparently made on the basis of ecological potentials, historical antecedents, and the relative economic and social pay-off of different options. As certain options were pursued, e.g., the addition of cattle-keeping to hunting-gathering, further options became potentially feasible and may thus have been considered and eventually accepted. One track was to engage more fully in cattle-keeping at the expense of hunting. Subsequently, specialized pastoralism would have been feasible. Alternatively, the adoption of plant cultivation within a broad spectrum of hunting and gathering could have eventually led to the adoption of specialized farming. Settled agrarian communities might have then entered into symbiotic relationships with pastoralists. Economic changes, however, were not simply that. Any economic change is a social change and it is erroneous to assume just because we use the words 'social' and 'economic', mostly for heuristic purposes and to refer to predominantly behavioral relationships in the first and material transactions in the other, that the two terms belong to unrelated categories. This tendency of cognitive dissociation of social phenomena is enhanced by the palpability of material evidence and the intangible aspects, especially in the archaeological record, of ideas and behavioral actions. Indeed it may be asserted that economic decisions are based on the scope and range of socially acceptable notions of values, actions, and pursuits. The appearance of discontinuities in culture is not a matter of a total re-creation of social conditions and economic behavior. Instead, it denotes relatively rapid episodes of social change in which fundamental elements of social relationships are quickly altered, leading to a domino effect and a multiplication of such effects in a short time. In the later prehistory of Africa, the rise of pastoralism and settled agrarian communities was accompanied with the rise of hierarchical and eventually militarized societies to secure territorial ranges, enlarge the labor force, integrate regional resources, or extract revenues and tribute. This social and political development marked a threshold of cultural evolution that aggravated the impact on the landscape and engendered modes of social display and consumption that precipitated gender and social inequalities, magnified population size, and triggered, in many cases, hostile inter-group contacts. At present, the cultural impact of climatic-environmental events cannot be evaluated without due consideration being given to the current socio-political organization of African societies and their global context.

The record clearly shows that cattle as well as sheep/goat herding were important economic pursuits in the Nile Valley and the Central Sahara over the last 7000 years. There is also plentiful evidence both of long-term and short-term climatic variations. In the past, archaeologists tended to focus on long-term adaptive strategies. However, more recent investigation have begun to examine the possible role of abrupt, brief climatic oscillations which might have severely affected food security and related economic regimes. The impact of such abrupt events (from decades to centuries) must have been particularly grave in the fragile and impoverished environments of North Africa. Human groups in such habitats must have developed a variety of strategies to cope with frequent and unpredictable oscillations.

The emergence of cattle-keeping and herding during the early Holocene was perhaps a response to droughts that interrupted the wet phase that began following the Last Glacial Maximum.

Looking beyond what we know now, and realizing the deficiencies in our knowledge and the need to re-orient our archaeological and environmental research toward a better understanding of the the relationship between climatic change and food security, I join my colleagues in charting the following avenues as some of the most urgent requirements for future research:

- Develop and encourage studies that provide high-resolution environmental and archaeological data.
- Promote accurate and precise dating of climatic, environmental, and cultural events and develop a standard, uniform nomenclature of uncalibrated and calibrated radiocarbon dates, as well as dates obtained by other methods.
- Focus on terrestrial sediments and palaeolandforms to enhance our understanding of landscape responses to climatic change.
- Promote efforts to develop high-resolution climatic models of regional utility.
- Encourage palaeoclimatic investigations of changes in the seasonality and type of rainfall.
- Examine the complex circulation features influencing rainfall in Africa and the possibility of different climatic regimes other than the those that explain palaeoclimatic variability in Africa in the simplistic terms of the 'monsoons'.
- Support archaeological investigations that focus on the ecodynamics of social change and the means by which environmental signals are perceived and assessed, and the processes of decision making and action.
- Develop methodologies that integrate ethnoarchaeological and ethnohistorical data with archaeological models of land-use, diet, resource management and environmental anthropogenic factors based on archaeobotanical, archaeozoological and other archaeological investigations.
- Promote archaeological investigations of the impact of climatic change on food security and political stability of past societies with different modes of social organization.
- Focus on the role of animal husbandry and pastoralism in the cultural transformations of Africa over the last 8000 years, with special attention to the impact of such activities on the African landscape and inter-group social interactions.
- Examine the role of the scale of social and ecological processes, especially those in the order of decades and 100-400 years.
- Focus on the identity, temporal pattern, amplitude, rapidity, and duration of abrupt, short-term intervals, especially those that are suspected to have occurred at 9.6 kyr cal BP (8.6 kyr bp), 8.75-8.45 (7.9-7.7 kyr bp), 8 kyr cal BP and 7.85 cal BP (7.26 bp and 7 kyr bp), 5.2-5 kyr cal BP (4.5-4 kyr bp), 4.1 kyr cal BP (3.7 kyr bp), 3.6-3.1 kyr cal BP or 1600-1300 BC (3.3-3 kyr bp), and 780-1290 AD.

- Integrate existing data from archaeological, palaeoclimatological, palaeo-environmental, and palaeoecological studies in a single database.
- Implementation of multiregional approach of interdisciplinary scope to clarify the role of local versus megaregional and global climatic-ecological events, as well as to disentangle anthropogenic factors from climatic signals.
- Endorse the recommendations of the INQUA-PAGES and PAGES-START workshops on past global changes in Africa, especially the urgent need for the provision of technical facilities in Africa, and for the development of an adequate infrastructure and a viable research environment for African scholars, which includes opportunities for training in Africa and abroad, provision for funding for local research projects, travel to international meetings, participation in the information network, and participation in international research programs.
- Disseminate information to local communities, the public and policy makers, and seek their participation, contributions, and views on issues of common interest.

REFERENCES

Alley, R. B., Mayewiski, P. A., Sowers, T., Stuiver, TM., Taylor, K. C., and Clark, P. U. (1997). Holocene climatic instability, a prominent widespread event 8200 years ago. *Geology* 25: 483-486.

Bar-Yosef, O. (1998). Jordan prehistory: a view from the west. In Henry, D. O. (ed.), *The Prehistoric Archaeology of Jordan*, British Archaeological Reports International Series 705, Oxford, pp. 162-178.

Bonnefille, R., and Hamilton, A. (1986). Quaternary and late Tertiary history of Ethiopian vegetation. *Acta Universitas Upsalienses, Symbolae Botanicae Upsalienses* 26(2): 48-63.

Bonnefille, R., Roeland, J. C., and Guiot, J. (1990). Temperature and rainfall estimates for the past 40,000 years in equatorial Africa. *Nature* 346: 347-349.

Bonnefille, R., Riollet, G., and Buchet, G. (1991). Nouvelle séquence pollinique d'une tourbière de la crête Zaïre-Nil (Burundi). *Review of Palaeobotany and Palynology* 67: 315-330.

Fang, Z., Thompson, L. G., Mosley-Thompson, E., and Yao, T. (1993). Temporal and spatial variations of climate in China during the last 10,000 years. *The Holocene* 3(2): 174-180.

Gasse, F., and Van Campo, E. (1994). Abrupt post-glacial climate events in West Asia and North Africa monsoon domains. *Earth and Planetary Science Letters* 126: 435-456.

Gifford-Gonzalez, D. (2000). Animal disease challenges to the emergence of pastoralism in Sub-Saharan Africa. *African Archaeological Review* 17(3): 95-139.

Hassan, F. A. (1996). Abrupt Holocene climatic events in Africa. In Pwiti, G., and Soper, R. (eds.), *Aspects of African Archaeology. Papers from the 10th Congress of the Pan African Association for Prehistory and Related Studies*, University of Zimbabwe Publications, Harare, pp. 83-89.

Hassan, F. A. (1997). Holocene Palaeoclimates of Africa. *African Archaeological Review* 14: 213-230.

Hassan, F. A. (2000a). Climate and cattle in North Africa. In Blench, R., and MacDonald, K. C. (eds.), *The Origins and Development of African Livestock: Archaeology, Genetics, Linguistics, and Ethnography*, University College London Press, London, pp. 61-86.

Hassan, F. A. (2000b). Holocene environmental change and the origins and spread of food production in the Middle East. *Adumatu* 1: 7-28.

Hillman, G. (1996). Late Pleistocene changes in wild plant foods available to hunter-gatherers of the Northern Fertile Crescent: possible preludes to cereal cultivation. In Harris, D. (ed.), *The Origins and Spread of Agriculture and Pastoralism in Eurasia*, University College London Press, London, pp. 159-203.

Jolly, D., and Bonnefille, R. (1992). Histoire et dynamique du marécage tropical de Ndurumu (Burundi), données polliniques. *Review of Palaeobotany and Palynology* 75: 133-151.

Kendall, R. L. (1969). An ecological history of Lake Victoria Basin. *Ecological Monographs* 39: 121-176.

Kremenetski, C. V., and Patyk-Kara, N. G. (1997). Holocene vegetation dynamics of the southeast Kola Peninsula, Russia. *The Holocene* 7: 473-479.

Lamb, H. F., Gasse, G., Benkaddour, A., El Hamouti, N., Van der Kaars, S., Perkins, W. T., Pearce, N. J., and Roberts, C. N. (1995). Relations between century-scale Holocene arid intervals in tropical and temperate zones. *Nature* 373: 134-137.

Marshall, D. (1998). Early food production in Africa. *The Review of Archaeology* 19: 47-58.

Mees, F., Verschuren, D., Nijs, R., and Dumont, H. J. (1991). Holocene evolution of crater lake at Malha, Northwest Sudan. *Journal of Palaeolimnology* 5: 227-253.

Roberts, N., Taieb, M., Barker, P. Damnati, B., Icole, M., and Williamson, D. (1993). Timing of the Younger Dryas event in East Africa from lake level changes. *Nature* 366: 146-148.

Rohling, E. J., Jorissen, F. J., and De Stiger, H. C. (1997). 200-year interruption of Holocene saptopel formation in the Adriatic Sea. *Journal of Micropalaeontology* 16: 97-108.

Rose, J., Whiteman, C. A., Lee, J., Branch, N. P., Harkness, D. D., and Walden, J. (1997). Mid- and Late-Holocene vegetation, surface weathering and glaciation, Fjallsjökull, southeast Iceland. *The Holocene* 7: 457-471.

Street-Perrot, F. A., Holmes, J. A., Waller, M. P., Allen, M. J., Barber, N. G. H., Fothergill, P. A., Harkness, D. D., Ivanovitch, M., Kroon, D., and Perrott, R. A. (2000). Drought and dust deposition in the West African Sahel: a 5500-year record from Kajemarum Oasis, northeastern Nigeria. *The Holocene* 10(3): 293-302.

Snowball, I., Sandgren, P., and Petterson, G. (1999). The mineral magnetic properties of an annually laminated lake-sediment sequence in northern Sweden. *The Holocene* 9(3): 353-362.

INDEX

Abeke rock shelter, 261
Abkan, 217
Abrupt climatic change, 56, 209
Abrupt climatic events, 21, 325, 326
Abu Ballas, 114-115
Abu Darbein, 116
Abu Hureyra, 117, 162, 165
Abu Roash, 282, 310
Abu Tartur Plateau, 324
Abusir, 285, 307, 311, 314
Abydos, 276, 278-283, 285, 288, 298, 306,
    307, 309, 310
Acacia, 70, 75, 231, 238
Acacus, 12, 65, 67, 70, 71, 73, 75, 200, 213,
    219-220, 227-232, 234-236, 238-
    243, 245-246, 251, 323, 326
    Adrar Acacus, 12
    Tadrart Acacus, 65-71, 76, 78, 209, 211,
        214, 231
Accra Beach, 98-99
Acheulean, 73, 75
Adi Ainawalid, 173-185
Adi Gudem, 173, 176, 178
Adrar, 61
    Adrar Acacus, 12
    Adrar Bous, 13-14, 17, 20, 210, 213, 215,
        237, 252, 255
    Adrar Ifoghas, 54, 61
    Adrar Tiouyine, 14
Adriatic Sea, 35, 158, 165
Afrogyrus oasiensis, 73
Agean Sea, 35, 39, 42-44
Agordat, 172
Agriculture, 51-52, 57, 60, 113, 123, 126,
    130-131, 143, 171, 182, 184, 259;
    see also Crops, Cultivation *and*
    Domestication
    ethnographic observations of, 173-182
    origins in Africa, 19-20, 23, 57, 60, 62,
        102, 111-119, 148-150, 157, 160,
        164-165, 166, 171-173, 275, 323
    origins in Southwest Asia, 117-118, 161-
        164, 166
Agropastoral communities, 131, 136, 143,
    145-146
A-Group, 20, 219; *see also* Kerma
Ahaggar, 12, 52, 61, 148, 252
Aïr, 16, 36, 45, 216, 236
Akameru cave, 263
Akjoujt, 60
Aksum, 172, 182

Alchemilla, 85, 88-89
Algeria, 16
    Algerian Sahara, 14, 52, 54, 231
Amekni, 13, 20
AMS dating, 36-37, 84-85, 130, 158, 160,
    162, 165, 172, 253; *see also*
    Radiocarbon dating
Anatolia, 19, 22
Aneibis, 116
Anezrouft, 14
Angola, 257, 262
Animal fodder, 6, 89, 118, 143, 177, 180
Animal husbandry, 16, 210, 331; *see also*
    Cattle *or* Pastoralism
Animal protein, 263, 267
Anthemidae, 111
Apiaceae, 85
Arabia, 149, 160-161, 203
    Arabian Peninsula, 148, 171
    Saudi-Arabia, 148
Arab peoples, 22, 60-61
Arad, 163, 165
Archaeobotany
    assemblages, 117-118, 172
    studies, 111, 113, 115, 117, 130, 140,
        148 149, 261
Archaeozoology, 195-204
    assemblages, 140, 234-235, 240, 252-253,
        261, 263
    studies, 137, 228, 331
Areschima, 213
Aridification, 97, 326
Aristida, 116
Arkinian, 210
Arlit, 252, 255
    Arlit-Somaïr, 252, 255
Armant-Qurna region, 215
Arsi Mountains, 83-91
Artemisia, 42, 70, 158-159, 165, 238
    Artemisia herba-alba, 77
Asfun el-Mata'na, 277
Asia, 164, 201, 204
    Asian Levant, 200
    Southwest Asia, 111, 117-119, 157-167,
        171-172, 182, 322, 325, 327-328
Aswad, 162,
Aswan, 43, 111, 199
Aterian, 66, 68, 71, 76
Atlantic, 7, 18, 40, 42, 48, 50, 61, 160, 201,
    235
Atlas, 8, 50, 52

Auenat, 220
Aurès, 219
Aurochs, 196, 198, 200, 228
Avena abyssinic, 116
Azaouak Basin, 253
Azawad, 55, 60
Azawagh, 52, 55, 60-61, 147

Bab edh-Dhra, 163
Badari, 118, 215, 219, 285, 287, 311, 313,
    315
Badarian culture, 217, 219, 276, 287, 326
Bafounda Swamp, 96, 99
Baharia Oasis, 324
Bahr el Ghazal, 57
Bale Mountains, 83-91
Bama-Konduga group, 135
Bama Ridge, 132, 135, 137
Bamenda Highlands, 97, 102
Bamileke Plateau, 96
Banana cultivation, 261
Bandiagara, 52
Bantu-speaking peoples, 100, 261-262
Barbary sheep, 196, 201, 202-203
Bardagué, 13, 14, 220
Barley, 16, 20, 162-163
    in Egypt, 16, 118, 164-165, 166, 328
    in Ethiopia, 172
        today, 175-178, 181
    in Southwest Asia, 117, 119, 157, 161,
        163, 166
    in the Levant, 16, 162-163
Bashendi, 202
Basil, 181
Basins, 8, 55, 85, 99, 123, 133, 324, 328
    agriculture in, 143, 148, 323
    archaeological sites in, 73-75, 132, 135,
        137-138, 140, 143, 146-147, 150,
        162, 202, 253, 258
    cattle in, 253, 260, 329
    formation of, 73-75
Basketry, 183
Bateké Plateau, 96, 100
Beer, 181
Beidha, 163, 165
Benedougu, 21
Berber peoples, 51, 60-61
Beta Giyorgis, 172
Bilanko Depression, 96
Bioanthropology, 21
Biomphalaria pfeifferi, 73
Bir Atrun, 231
Bir Kiseiba, 8, 12, 113, 198-200, 202, 203,
        210, 217, 329
Bir Moghreïn, 60
Birds, 101, 111, 132, 140, 176, 202
    in Egyptian iconography, 277, 289-292,
        294, 297, 311-312
Birim River Valley, 98

Boar, 204
Bölling-Allerød, 158-160, 166,
Bone, 140, 172, 217, 253
    beads, 145
    carved, 178, 193, 278, 292, 306, 314-317
    dating of, 129, 138, 254-255
    of cattle, 13, 131, 140, 235, 275
        domestic, 13, 113, 172, 202
    of domestic livestock, 116, 196
    of fish, 111, 113, 265
    of ovicaprids, 131, 202-203, 260-261, 265
    of wild animals, 111, 202, 204
    preservation of, 200, 264, 269
    reconstructing age profiles from, 267-268
    tools, 143, 145
    weight calculation of, 264
Boraginaceae, 238
Borkou, 54, 146, 220
Borno, 138, 146
Bos, 198, 202, 210
Bos sp., 196, 197, 198-201, 260
Bougdouma, 231, 238
Bougdouma-Oyo event, 231
Bourgou, 116
Bovidian, 213, 219-220
Bozo peoples, 263
Brachiaria, 20, 115, 215
Brachiaria deflexa, 116
Bread, 178, 181
Bread wheat, 175
Bronze Age, 160, 163
Broomrape, 176
Bucrania, 280, 286, 287
Buffalo, 198, 258, 260
Bukarkurari, 138, 259
Bulls, 17
    in Egyptian iconography, 276-289, 291-
        298, 307, 310
Burials, 126; *see also* Funerary practices
    dating of, 128
    horn cores as grave markers, 275
    monumental, 17, 215; *see also* Funerary
        architecture
    of cattle, 17, 213, 217, 276, 280
    tumuli, 245
    variation in, 236
    with dwarf goat, 262
Burkina Faso, 18, 56, 123, 125-127, 130-131,
        147, 149-150, 253, 255, 258, 269,
        323
Burundi, 238, 263, 326
Butare, 263
Buto, 275-276

Calendar circles, 217
Calligonum, 238
Camel, 180, 183
Cameroon, 14, 18, 23, 96-101, 256, 260-261;
        see also Grassfields

Canada, 40
Canoe, 132
Cape, 14, 23
Capharnaum, 160
Capparaceae, 70
Caravan routes, 51, 61, 76
Caryophyllaceae, 85, 238
Catch-cropping, 174, 178, 183
Cattle
    and environmental change, 51, 55, 213,
        228-229, 237, 240-241, 245, 324
    burials, 213, 217, 252-253, 275
    domestic, 140, 172, 198-201, 245, 275
    domestication, 8, 60, 210, 228, 322
    emergence of cattle keeping, 12-15, 113,
        119, 198, 204, 263, 275, 329
    genealogy of, 196-197
    in East Africa, 323, 326
    in iconography, 275-300, 329
    in ideology, 245, 275, 329
    in varying economies, 12-13, 235-236
    in West and Central Africa, 251-270, 323,
        324-325
    meat yield, 267-268
    secondary products of, 116, 210, 235, 267
    spread of cattle keeping, 13, 16-18, 146,
        150, 216-217, 219, 229, 237, 251,
        322, 324-327, 329
    today, 6, 180, 183, 262
    watering, 234, 240, 244
    wild, 111, 148, 198-201, 228
Cave sites, 65-66, 69, 78, 172, 203, 230, 238-
    239, 258, 261-263, 269, 325
Celtis, 116-117
Cenchrus, 115-116, 213
Central Africa, 95-103, 200, 251-270, 327
Central African Republic, 256, 262
Ceramic artifacts, 150, 210, 215, 217, 219,
    227, 237, 253, 275; *see also*
    Pottery
    decoration, 210, 219-220
    plant impressions in, 140, 149
Cereals, 5, 16, 20, 116, 162-166, 172, 210,
    215, 225-226, 232, 322
    cultivation of, 16, 20, 117-118, 140, 161-
    162, 175-182, 328
C-Group, 220; *see also* Kerma
Chad, 14, 60-61, 146, 210, 231, 258, 267
Chaff, 172, 177
Chaîne de Gobnangou, 253, 258
Chami, 18
Charcoal, 69, 101, 113-114, 117, 126, 128-
    129, 131, 138, 212, 254-257
Charred plant remains, 217
Chat, 172
Chemical indicators, 78
Cheno-Amaranthaceae, 231, 238
Chenopodiaceae, 42, 158-159
Chicken, 180

Chickpea, 20, 117, 162, 175-178
Chiefdoms, 11, 17, 23, 215, 322-323, 326-
    327
Children, 116, 176, 181, 309
Chili, 181
Chin Tafidet, 17, 252-253, 254
China, 1, 4, 326
Chronology, 5, 62, 88, 92, 160, 210, 285; *see*
    also Radiocarbon dating
    of agriculture, 20
    of domestication, 12, 51, 195, 198, 200,
    203, 251
    of the Gagiganna complex, 137-138, 140
    radiocarbon, 16, 127, 134
Cities, 22, 61, 160
Climatic change, 4, 35-44, 47-62, 65-79, 89,
    92, 157-167, 209-221, 225
Climatic events, 2, 5, 7-8, 11, 19, 21, 49, 95-
    103, 150, 225-247, 251, 321-322,
    324-326, 328
Climatic oscillation, 12, 18, 66, 83-93, 210,
    215, 323, 330
Cocorba, 126, 128
Coffea Arabica, 20
Colorado Plateau, 4
Complex societies, 6, 11
Compositae, 111, 238
Conflict, 4, 21, 131, 244; *see also* Warfare
    and Weapons
Congo, 95-96, 262
Congo-Brazzaville, 96-101, 257
Congo-Kinshasa, 257
Copper, 21, 60
Coprolites, 68, 72, 234
Côte d'Ivoire, 98
Cotton, 172, 183
Couscous, 116
Crocodiles, 54, 276
Crops; *see also* Agriculture, Cultivation *and*
    Domestication
    cultivation of, 20, 119, 164, 166, 173, 176
    domesticated, 118, 148, 162
    failure of, 164, 179, 182, 225
    founder crops, 117-119, 157, 161-162,
    164
    harvesting of, 6, 177
    indigenous African, 20
    Near Eastern, 20
    production of, 178-179
    rotation of, 175
    sowing of, 119, 175-176
    spread of, 165
    storage of, 6, 179-180
    wild, 148
Culling, 197
Cultigens, 118, 328
Cultivars, 118

Cultivation, 13, 101-102, 131, 261, 329; *see also* Agriculture, Crops *and* Domestication
    origins in Africa, 8, 19-20, 22-23, 119, 124, 148, 150, 161-162, 166, 171-173, 219, 246, 326, 328-330
    origins in Southwest Asia, 117-118, 163-164
    spread, 16, 165
    today, 124, 176, 180, 181, 183
    tools for, 243-244
Cultural evolution, 11, 16, 21, 61-62, 323, 328, 330
Cumin, 181
Cushite peoples, 171, 220
Cyinkomane, 263, 267
Cyperaceae, 70, 85-87, 89-91, 233
Cyrenaica, 16

Dactyloctenium aegyptium, 116
Daima, 143, 256, 258
Dakhla Oasis, 202, 215, 217, 220, 324
Damascus, 160, 162
Dams, 51
Danakil, 171, 183
Darfur, 238
Dead Sea, 163-164
Deforestation, 6, 102
Dega Sala, 83-91
Déguesse mound, 256, 260
Délébo, 13, 14
Dentition, 196, 202, 253-254, 258, 263, 267,
Depressions, 51, 53, 60, 74, 96, 97, 114, 125, 132, 137, 119, 171, 174, 253, 259, 324
Desertification, 16-18, 47, 51, 55, 253
Dhar Tichitt, 14, 18, 20, 50-51, 55, 60, 124, 149, 253, 265, 267
Digitaria, 115, 116, 213
Dikwa, 145
Dimba, 257, 262
Disease, 54, 183, 263
    trypanosomiasis, 260, 262
Dja'de, 162
Djabarona, 55
Djalo, 61
Djebel Ichkeul, 198
Djed, 279, 286, 288
Djenne, 60-61
DNA, 12, 118, 201
Dog, 17, 196, 197, 203, 258, 262
Dogomboulo, 213
Dogon peoples, 52
Dom palm, 111
Domesticated
    animals, 16, 118, 140, 143, 203-204, 259, 262, 265
    cattle, 16, 111, 172, 202, 228, 244, 245, 275, 325

    ovicaprids, 16, 202, 204
    plants, 20, 101, 111, 117-119, 130-131, 140, 143, 148-149, 157, 161-163, 165-166,
Domestication
    of animals, 195-204
    of cattle, 198-201, 228-229, 322
    of ovicaprids, 165, 201-203
    of plants, 19, 116-117, 119, 143, 148-150, 322, 327-328
Dongodien, 14
Donkey, 180, 183
Dori, 126, 128-129, 258
Drainage, 125, 174, 183
Dromedary, 60
Droughts, 1, 4-8, 12, 16-18, 21-23, 47, 49, 52, 60-61, 89, 92, 118-119, 146, 158, 160, 181, 182-184, 225-226, 240, 321, 324, 326, 329, 331
Dufuna, 132, 135
    canoe, 132
Dunes, 8, 50, 55, 65-66, 68, 73-76, 98-99, 125-126, 128-131, 135, 150, 228-229, 233, 238, 240, 253
Dung, 70, 72, 175, 179, 182, 203, 239, 241, 256, 260
Durum wheat, 178, 183
Dwarf goat, 18, 197, 260-262
Dwarf sheep, 18

East Africa, 13, 18, 20, 22-23, 92, 263, 269, 323, 326, 329
Eastern Desert, 203
Eastpans, 114-115
Echinochloa, 115-116, 213
Echium, 238
Edeyen of Murzuq, 65, 72-75, 213, 231, 238, 240
Eghazer Basin, 252
Egypt, 17, 199
    agriculture in, 16-17, 20, 111, 118-119, 148, 164-167, 171-172, 219, 323, 328
    cattle in, 17, 22, 198, 202, 204, 210, 216-217
    civilization of, 16-18, 21, 51, 323, 326, 329
    climate of, 12, 18, 50, 52, 78, 328
    iconography of, 275-298, 329
    ovicaprids in, 201-202, 204
    Western Desert, 50, 52, 54, 199, 200, 202, 204, 209-210, 215-217, 219, 275
Einkorn, 117, 162-163, 165
El-Amra, 276, 305, 311, 313, 315
El Damer, 116
El Ghorab, 210, 217
El Niño Southerly Oscillation (ENSO), 184
El Omari, 118

Elephant, 50, 54, 229, 285-286, 308
Elkab, 278, 286
Elmenteitan, 263
Emigration, 5, 16, 147, 261
Emmer, 20, 117-118, 161-165, 172, 178
Ennedi, 12, 146-147, 220
Ensete ventricosum, 171
Environmental determinism, 4-5
Eocene, 101, 164
Epipalaeolithic, 73-75, 112, 117
Equatorial zone, 17, 56, 100, 157, 238, 252, 261-262, 325
Eragrostis spp., 116
Eragrostis tef, 172, 175
Erg Admer, 219
Erg Ine Sakane, 265
Ericaceae, 88-89
Ericaceous belt, 85, 89
Eritrea, 172-173
Erosion, 12, 56, 72-73, 99, 137, 173-174, 182, 183, 184, 238, 323, 326
Esh Shaheinab, 203, 220
Ethiopia, 17, 57, 78, 83-93
    agriculture in, 20, 116, 171-185
    climate of, 83-93, 182-184, 326
Ethnographic studies, 115-116, 171-185, 331
Eucalyptus, 182
Euphrates, 162, 165
Europe, 21-22, 47, 116, 118, 123, 132, 157, 164, 327
Exchange, 2, 5, 21, 176, 179, 180, 183, 210, 217, 219, 241, 328

Faba bean, 172, 176, 178
Faboura, 254, 260
Fachi, 50, 61
Fadar, 128
Fallow, 101, 130, 175
Famine, 5, 83-93, 182-183
Farafra Oasis, 114, 202, 215, 217, 219-220, 324
Farming, 5, 6, 11, 13, 16, 18, 20, 21, 119, 131, 164, 173-174, 178, 182-184, 330, 327-328
Faunal analysis
    age profiles, 267-268
    bone weight, 264
    MNI, 264
    NISP, 264
    total weight, 266-268
Faunal remains, 13, 37-40, 111, 113, 126, 195-204, 210, 213, 228, 234, 253, 258-265, 269-270
Fava bean, 20
Fayum, 21, 39, 52, 118-119, 164, 166, 171, 200, 203, 219, 238, 275, 323
Fenugreek, 175, 178, 181
Fertile Crescent, 161
Fezzan, 61, 65-79, 198, 199

Figurines, 129, 145, 172, 217, 275-277, 283, 291, 293
Final Stone Age, 126-131, 135
Finger millet, 172, 175, 178, 181
Fireplaces, 113, 233, 240; *see also* Hearths
Firki plains, 132, 136, 146, 148, 150, 259-260, 265, 267
Fish, 21, 50, 112, 264-265, 277
    bones, 111, 113, 126, 132, 140, 259, 263-267
Fishing, 5, 7, 20, 52, 60, 111, 113, 119, 131, 140, 145, 150, 215, 219, 251-253, 260-264, 266-269
Flax, 20, 117-118, 172, 176
Fluvial activity, 8, 75-76, 98-100, 244
Fonio, 140
Food
    production, 4, 11, 20-21, 95, 111, 119, 123-124, 135, 150, 157, 161-167, 173, 178-179, 182, 225, 227-228, 325, 327, 329
    security, 1, 4, 8, 11, 21, 119, 225, 237, 240, 246, 251, 270, 321, 327, 329-331
    shortage, 5, 16, 21, 179, 183
    storage, 16, 179-180, 183, 210
Foraging, 13, 20, 162, 166
Forest, 4, 18, 20, 55-56, 60, 85, 88-89, 97-103, 131, 160, 252, 260-262, 267, 269; *see also* Rainforest
Fouta, 54
Fox, 202
Fozzigiaren, 236
Fruit, 12, 101, 111, 114
Fuel, 102, 114, 181-182, 183,
Funerary architecture, 18, 215; *see also* Burals, monumental
Funerary practices, 236-237; *see also* Burials
Furnace, 263
Furniture, 281-282

Gabon, 256, 262
Gajiganna, 14, 18, 132, 135-147, 149-150, 255-256, 259-260, 265, 267
Game, 50, 140, 204, 229, 267; *see also* Hunting
Gao, 61
Garlic, 181
Gas chromatography, 115
Gathering, 7, 11, 52, 95, 111, 115-116, 119, 123, 131, 150, 161, 162, 166, 219, 327, 330
Gazelle, 50, 54, 111, 165, 200, 202, 203, 210, 279
Gebel Faradi, 286
Gebel Tarif, 276, 305, 312
Gebel Tjauti, 283
Gebel Uweinat, 324
Genetic studies, 21

animals, 195-196
plants, 163
Geochemical data, 37
Geographic determinism, 246
Geomorphology, 8, 65, 97, 132, 137, 164, 262-263
Gerza, 280, 308, 309, 311
Ghab Valley, 159-160, 163, 165
Ghana, 14, 50, 57, 96, 98-102, 255, 260, 326
Ghat Oasis, 75
Ghat-Tahala-Serdeles system, 242
Ghoraïte, 162
Giba Plateau, 173
Gilf El-Kebir, 324
Gilgal, 163
Giraffe, 55
Gisagara, 257, 263
Global warming, 4, 324
Globigerinella, 39
Globigerinoides, 36, 39, 43
Globorotalia, 40
Globoturborotalita, 39
Goats
    in Central Sahara, 213, 228-229, 239, 240, 244, 324
    in East and southern Africa, 18
    in Ethiopia, 180
    in Northeast Africa, 16, 51, 119, 201-203, 215, 219, 323, 325, 328-329
    in the Sahel, 148
    in West and Central Africa, 18, 95, 260-263
Gobedra, 172
Gogo Falls, 263
Gonder, 181
Gorgol, 55
Gourma, 253, 265, 269
Grains, 17, 20, 114-117, 128, 130-131, 148, 158, 183, 215, 328
    storage, 179-180
Graminaceae, 56
Graminae, 111
Grape, 172
Grasses, 8, 12, 16, 20, 50, 69, 97-100, 102, 115-117, 119, 140, 149-150, 165, 213
Grassfields, 100-102, 261-262
Grasspea, 177, 178, 181, 183
Grave goods, 131, 237
Grave markers, 275
Great Western Erg, 54
Greece, 158
Greenland, 40
    ice core, 78, 158, 165, 325
Grinding tools, 5, 111, 113, 115, 126, 143, 232, 233
Grotte Capéletti, 13, 16, 238
Guinea, 56, 95
Gulf of Lions, 40

H. Aponensis duveyrieri, 231
Hagenia, 85, 89
Halula, 162
Hanfetse, 177, 178
Haoussas peoples, 61
Hare, 111, 200, 202, 210
Hartebeest, 111, 200, 202, 279, 306
Harvesting, 5, 6, 115-117, 119, 150, 163, 174, 177-180, 183
Hassi-el-Aboid, 265
Hastigerina pelagica, 39
Haua Fteah, 16
Hearths, 67, 72, 113-114, 202, 213, 217
Helichrysum, 85
Helwan, 17, 282, 309
Hemamieh, 172
Herding, 13, 20-21, 51-52, 57, 60, 113, 119, 165, 203, 210, 215, 219, 227, 229, 232, 237, 244, 246, 264, 266, 268, 328-331
Hidden Valley, 114-115, 202, 215
Hierakonpolis, 277-282, 285, 298, 308-310
Hippopotamus, 50, 54, 229, 254, 276-277, 280, 298, 308, 316
Historical records, 1, 7, 83, 92, 172, 247
Hodh, 61
Hoes, 5, 171, 243
Hoggar, 146-147
Holocene, 35-44, 65-79, 113-116, 157-167, 209-221
    Climate Optimum, 54-55, 95, 97, 158, 160, 164-165
    early, 7, 42, 67-68, 76, 78, 83, 113, 115, 119, 125, 132, 158-161, 163-166, 209-210, 215, 229, 236, 329, 331
    Interglacial, 78
    late, 7, 47-62, 83-93, 95-103, 113, 116, 126-131, 211, 217-218, 251-270
    middle, 51, 54, 65, 68, 72-73, 78, 95, 97, 113, 116, 124, 126, 132, 149, 211-217, 219, 225-247
Hor-Aha (Egypt), 278, 283
Horncores, 196, 202, 204, 260, 275
Horse, 60, 196, 211
Hunting, 11, 13, 17, 20, 51-52, 60, 95, 111, 113, 123, 131, 140, 150, 165, 201, 202, 204, 210, 215, 219, 234, 251-253, 258-270, 327, 330
Husbandry, 16, 210, 331
Hydrophytes, 111
Hypericum, 85

Ibex, 201, 203
Ice Age, 92-93, 157
Ice-rafted debris (IRD), 7, 40
Ice sheets, 157-158
Ichthyofauna, 267
Iconography, 17, 197-198, 275-298

Ighazer wan Agades, 60
Ilex mitis, 99
Imenennaden, 236
In Habeter, 66, 245
In Tuduf, 17, 147, 253-254
India, 5, 19, 22, 148
Indian Ocean, 40, 160
Inland Niger Delta, 8, 21
Insects, 54, 176, 179
Intertropical Convergence Zone (ITCZ), 173, 225, 321
Iran, 158, 161
Iraq, 161
Iron, 21, 60, 102, 262
Iron Age, 126, 128-129, 131-132, 135, 146, 258, 261-263
Irrigation, 6, 22, 176
Israel, 158, 165
Iwelen, 61

Jackal, 203
Jebel Rahib, 200
Jebel Tomat, 148
Jenné-jeno, 20-21, 148
Jerf al Ahmar, 162
Jericho, 162-164
Jordan, 160, 162-165
Juniperus, 85, 89

Kabbashi al-Haita, 116
Kadada, 220
Kadero, 116
Kajemarum Oasis, 326
Kamabanga, 257, 262
Kanem, 50
Kanemis, 61
Kanouris, 61
Karakoro, 55
Kariari, 138, 265
Karkarichinkat, 14, 18, 253-254
Kashiru, 238, 326
Kawar, 50, 61
Kayes, 55
Kelumeri, 138
Kenya, 13-14, 18, 89, 263
Kerma, 20, 55-56, 172
Kharga, 8, 324
Khartoum, 164, 219-220, 237
Khartoum Neolithic, 202, 219-220
Khatt Lemaïteg, 14, 18
Khattara, 164
Kintampo, 14, 18, 102, 255, 260-261, 269
Kissi, 128-129, 131
Kitala, 257, 262
Kob, 126
Kobadi, 200, 253-254
Koidu River Basin, 98-100
Kolel, 129
Kolima Sud, 253-254, 265, 267

Konduga, 132, 135
Koumbi Saleh, 61
Kreb, 116, 140
Kufra, 12
Kursakata, 150, 256, 260
Kuruyange, 238, 326

Labe, 138
Lacustrine sediments, 12, 73-75, 160, 229, 231, 233
Lake Afar, 78
Lake Barombi, 96-100
Lake Besaka, 172
Lake Bosumtwi, 57, 96, 98-102
Lake Chad, 8, 18, 48, 50, 54-56, 61, 148-150, 259, 265, 269, 323
    Chad Basin, 57, 123, 132-133, 135, 136-137, 140, 143, 146-148, 150, 158, 260, 323
    Mega Chad, 18, 132, 135, 137, 258
Lake Hula, 158-160, 163
Lake Kitina, 96, 98-100
Lake Kivu, 262
Lake levels, 10, 57, 73, 75, 76, 78, 83, 89, 92, 98-99, 160, 230, 231, 233, 238
Lake Malha, 57, 238, 325-326
Lake Mundafan, 160
Lake Njupi, 96
Lake Ossa, 96, 98-100
Lake Rudolf, 238, 326
Lake Sinnda, 96, 98, 101
Lake Turkana, 89
Lake Victoria, 326
Lalibela, 172
Landraces, 178, 179, 183
Languages, 123, 146, 171-172, 220, 261
    Afro-Asiatic, 220
    Bantu, 100
    Chadic, 146
    Cushitic, 171
    Niger-Congo, 146
    Nilo-Saharan, 200, 210, 220
    Semitic, 171
Last Glacial Maximum, 43, 326, 331
Late Stone Age, 95
Lathyrism, 181
Latin names, 196, 197
Lebanon, 328
Legumes, 175-176
Leiterband pottery, 220
Lentil, 20, 117-118, 162-165, 172, 175-176, 178
Les Saras, Mayumbe, 257
Levant, 16, 22, 117-118, 157, 159, 161-163, 165, 200, 322, 328
Levantine Sea, 43

Libya, 14, 50-52, 54, 65-79, 125-147, 149, 199, 203, 209-212, 214, 219, 323, 326
Liliaceae, 111
Limnological data, 83, 97
Lingbangbo, 256, 262, 265, 267
Linguistic evidence; *see* Languages
Linseed, 172, 175, 181
Lion, 54, 277, 282
Lithics, 12, 143, 172, 231, 227, 232, 240, 259, 261, 275; *see also* Stone tools
Lithology, 35-37
Locust, 176, 183
Lowlands, 4, 101, 171-172, 232, 234, 237, 239, 240
Lymnaea natalensis, 73

Maadaga, 255, 258
Maadi, 275-276
Macedonia, 165
Maghreb, 198
Magnetic susceptibility, 101
Maize, 176
Malakal, 160
Malha event, 238
Mali, 8, 14, 18, 52, 54-55, 61, 78, 125, 131, 150, 231, 252-254, 265, 326
Manga Grasslands, 146, 150
Mare d'Oursi, 125
Marine records, 35, 158, 163, 325
Markets, 178-180, 183
Mashita Mbanza, 257, 262
Massifs, 8, 237, 324, 329
Material culture, 126, 130-131, 135, 143, 147, 172, 213, 220, 229, 233, 240, 247, 259, 261
Mathendusc cave, 70
Maures peoples, 51
Mauritania, 14, 19, 50-51, 60-61, 149, 253, 323, 326
Mbi, 256, 261
Mboandong, 96, 98-100
Medieval, 132
Mediterranean, 8, 16, 35-44, 48, 123, 157-161, 163-166, 199, 204, 220, 237, 324, 328
Megalithic architecture, 17, 21, 217, 244-245
Mege, 150, 256, 260, 265, 267
Meibob Hills, 78
Mekelle, 173-185
Méma, 253, 265, 267, 269
Menaka, 253
Merimde Beni Salama, 13-14, 16, 118, 159, 164-166, 171, 203, 204, 219, 275-276
Meroe, 117
Mesolithic, 116-117
Mesopotamia, 19, 22

Messak Settafet Plateau, 65-71, 220, 240, 245
Metal, 12, 21, 50, 52, 60, 62, 261, 315
Methane, 78
Middens, 262, 267
Migration, 5, 16, 60, 135, 143, 146-147, 150, 215, 232, 237, 251-252, 261, 322; *see also* Population movement
Millet, 16, 19-20, 60, 118-119, 125, 131, 143, 145, 149, 172, 175, 178, 181, 219
Miocene, 101
Mobility, 111, 143, 150, 210, 217, 219, 229, 237, 241, 247, 259, 328
Molluscs, 37, 72-75, 140, 231, 261, 264, 267
Moltkiopsis, 238
Mondongué, 256, 262
Monkey, 101
Monocropping, 6, 176, 183
Monsoons, 8, 43, 78, 92, 111, 160-161, 173, 225, 237, 322-323, 325, 331
Morocco, 238, 326
Morphology, 70, 115, 143, 148, 161, 196, 211, 260
Mossi peoples, 61
Mount Kenya, 89
Mountains, 6, 47-48, 50, 52, 54, 60, 62, 65-67, 70, 73, 76, 78, 83-92, 149, 201, 213, 219, 227-228, 232, 234, 237, 239, 240, 241, 243, 245, 246, 252, 279
Mousterian, 76
Mules, 180
Mureybit, 162
Murzuq, 65, 72-75, 213, 231, 238, 240
Muturu cattle, 260
Myrtaceae, 102

Nabta Playa, 8, 12-13, 16-17, 113-115, 198-201, 202, 203, 209-210, 217, 219, 275, 329
Nagada
    period, 276-283, 285, 287-289, 299, 291-292, 297, 298, 305-318, 326
    site, 17, 118, 164, 276, 279, 282-284, 287, 299, 305-314, 316, 317
Naqa ed-Deir, 279, 281, 305, 308, 310, 311, 317
Natchabiet, 172
Natufian, 162, 166
Ndama cattle, 260
Ndiamon-Badat, 254, 260, 265, 267, 268
Near East, 20, 123, 160, 166
Neguev, 163, 165
Neogloboquadrina pachyderma, 40
Neolithic, 50-52, 54, 57-58, 60, 62, 113-117, 123, 146-147, 150, 157, 161-164, 166, 198-204, 210, 213, 215, 217, 219-220, 228-229, 263, 275, 328

Netiv Hagdud, 162
Ngamalaka Pond, 96, 98, 100
Ngamuriak, 13
Ngovo, 257, 262
Niger, 14, 18, 50, 60-61, 125, 146-147, 150,
    209-210, 213, 217, 219, 231, 237,
    252, 254
Niger River, 8, 20, 54, 61-62, 147, 263
    Delta, 8, 54, 56-57, 96, 98-100, 253
Nigeria, 14, 17-18, 21, 61, 96, 98-102, 123-
    124, 132-147, 149-150, 255, 258,
    265, 323, 326
Nile, 8, 17, 51, 56-57, 61, 111, 118, 164, 166,
    172, 201, 203, 204, 210, 217,
    219-220, 238, 253, 263-264, 283,
    328
    Delta, 16, 50, 61, 164, 203, 328
    floods, 7, 16, 43, 52, 92, 165
    Valley, 16-17, 22-23, 50-51, 56, 112-113,
        116, 118-119, 157, 164, 166, 199-
        200, 201, 203, 204, 210, 217-220,
        229, 232, 237, 238, 241, 247,
        322-323, 325, 328-330
Nkag, 23
Nkang, 18, 256, 261, 265, 269
Nkile, 262, 265, 268
Nomadism, 16-17, 21, 50, 60-61, 219-220,
    229, 240-247, 329
Nomenclature, 12, 196, 197, 331
Noog, 172
Ntereso, 19, 255, 260, 269
Nubia, 17, 51, 78, 172, 219, 278
Nudurumu, 238, 326
Nuer peoples, 145, 263

Oak, 158, 261
Oases, 60-61, 75, 113-114, 157, 202, 215,
    217, 219, 240-245, 247, 323-326
Oat, 116, 172
Obobogo tradition, 261
Oil palm, 20, 97-103
Olea sp., 97-99
Ona Nagast, 172
Onion, 181
Oral traditions, 6
Orbulina universa, 39
Orgoba, 83-91
Oryza sp., 116, 140
Osaru Pond, 101
OSL dating, 66
Ostrakon, 279
Ostrich, 50, 54, 202, 289, 298
Ostrich egg shell, 47, 129, 279
Oudalan Province, 123, 253, 258
Oursi, 125-126, 128-129, 150, 258
Oveng, 256-257, 262
Ovicaprids, 18, 23, 55, 56, 131, 195, 198,
    201-203, 213, 228-229, 234, 237,

    244, 251, 253, 258-262, 265-266,
        269
Oxen, 171-172, 174, 177, 180
Oxygen Isotope data, 36-37, 41-43, 325
Oyo, 231

Palaeoclimatic data, 4-5, 7, 65, 124, 227,
    325, 331-332
Palaeoecological data, 5, 94, 100, 124, 132,
    332
Palaeoenvironmental reconstruction, 48, 59,
    92, 265
Palaeomagnetism, 325-326
Palestine, 285, 328
Palettes, 17, 277-280, 282, 286, 289-294,
    296-298, 305-306, 310, 311, 313-
    315
Palm, 20, 97-103, 111
Palm nut, 255-257
Palynological data, 78, 97, 101, 103, 126,
    213, 230-231, 241, 263
Paniceae, 140, 149
Panicoideae, 213
Panicum sp., 71, 111, 116, 213, 215, 231
Parasites, 54
Paronnychia, 238
Paspalum, 116
Pastoralism, 5, 11, 12-17, 118, 327, 329-331
    in East Africa, 20, 23, 171
    in Egypt and the Sahara, 17-19, 22, 51,
        68, 73-74, 76, 111, 119, 148, 198-
        204, 209-221, 225-247, 323
    in West and Central Africa, 18-19, 125,
        131, 136, 140, 143, 145-146, 148,
        150, 251-260, 262-265, 267, 326
    nomadic pastoralism, 240-247
    seasonal pastoralism, 232-239
    vertical pastoralism, 232, 237, 240
Pea, 20, 162-164, 117-118, 172, 175-176,
    178
Pearl millet, 19-20; see also Millet
Pedological data, 65, 68
Pennisetum, 115, 131, 140, 143, 148-150
Pennisetum americanum, 128, 130, 148, 323
Pénténga, 255, 258
Petrographic data, 241
Peul peoples, 263
Pig, 195, 197, 203, 204, 262
Pistacia, 42
Pits, 16, 114, 116, 128, 126, 179-180, 233,
    261, 265
Plains, 8, 47, 54, 57, 60, 76, 111, 125, 132,
    135, 137, 143, 145, 148, 173, 213,
    219, 323-324
Plankton, 36-37, 39-40, 43
Plant macro remains, 97, 111, 116
Plateaus, 4, 50, 60, 65, 96, 99-100, 160, 172-
    173, 220, 227, 240, 245, 324

Playas, 114, 119, 202, 215, 325; *see also* Nabta Playa
Pleistocene, 47, 65-78, 100, 111-113, 125, 137
Plowing, 5, 22, 171, 174-176, 178, 180
Poaceae, 85, 140
Podocarpus, 85, 97-99, 102
Political stability, 4, 18, 331
Pollen, 20, 42, 56, 70, 78, 83-84, 86-92, 95-97, 101-103, 124, 126, 130-131, 150, 158-160, 163, 165, 230-231
Population movement, 2, 12, 60-62, 118, 149-150, 232, 236, 251, 261-262, 269, 323, 326, 328; *see also* Migration
Post-processual theory, 246
Potato, 176
Pottery, 172; *see also* Ceramic artefacts
    in Egypt, 276-277, 281, 283, 291, 305
    figurines, 275, 276, 279, 307
    in the Sahara, 210, 213-214, 220, 232, 237, 240-241
    in West and Central Africa, 123, 128-129, 135, 137, 139, 140, 143, 146-147, 253, 258-259, 261-262
    seed impressions in, 20, 116-117, 140, 142-143, 149-150, 215
Preboreal, 158
Precipitation, 4, 8, 43, 49, 55, 57, 61, 65, 67-70, 75, 78, 158, 163, 182
Pre-Pottery Neolithic, 117, 163, 166, 328
Primulacaea, 85
Processual theory, 246
Prolonged Drift, 13
Prosopis africana, 55
Pulicaria, 238
Pulses, 117, 157, 162-163, 165-166, 175-181, 183, 328
Punt/Gash, 20

Qaat, 20
Qadan, 210, 275
Qasr Ibrim, 118, 148
Quaternary, 83, 163, 226, 229
Queen Icheti (Egypt), 165
Quercus, 159

Rabak, 13, 14
Rachis, 161-165
Radiocarbon dating, 2, 7, 13, 16, 23, 36-37, 49, 55, 57, 70, 83-84, 137, 147, 158, 160, 172, 229, 240, 252, 259, 321, 324, 331
    presentation of dates, 14, 17, 37, 39, 41, 53, 58, 70-71, 73-76, 79, 92, 111, 127-129, 134, 137-138, 157-160, 162-165, 209-210, 212, 215, 217, 219-220, 229-232, 236-239, 254-258, 262

Rainforest, 95, 98, 100, 101-103, 261-262, 269
Ramad, 162
Raphia, 99, 102
Red Sea, 16, 164, 166, 201, 204, 217, 328-329
Reedbuck, 126
Regional approach (to pastoral economies), 213, 226, 228
Remera, 257, 263
Reptiles, 140
Reservoirs, 4, 174, 183, 215
Rhinoceros, 50
Rhizomes, 111, 114
Rice, 20-21, 60-62, 140
Rift Valley, 85, 89, 328
Rinderpest, 183
Ritual practices, 2, 11, 13, 17, 233, 237, 245, 262
Rivers, 17, 43, 48, 50, 54, 56, 60-62, 98-100, 112, 119, 135, 147, 166, 174, 176, 247, 324
Rocher Toubeau, 213
Rock art, 17, 50, 54, 197-198, 199, 203, 215, 228, 237, 245, 277-278, 283, 329
Rock dassie, 203
Rock shelters, 65-68, 70-72, 78, 172, 211, 229, 230, 234, 238, 239, 258, 260-261, 269, 323, 325-326
Rodents, 164, 176, 179
Roman period, 160, 198
Rub Al Khali, 160
Ruhengeri, 263
Rwanda, 257, 263, 326
Rye, 117, 162

Sagai, 116
Sahara, 211
    agriculture in, 20, 116, 124, 148-150
        Central Sahara, 149
        Eastern Sahara, 16, 115, 119
    cattle in, 16, 146, 148, 150, 199, 210, 219-220, 237, 240-241, 251-253, 259, 324
        Central Sahara, 16-17, 22, 51, 119, 146, 217, 251, 322, 325, 329-330
        Eastern Sahara, 113, 119, 146, 225-247, 251, 263, 322, 329
    ceramics of, 146, 150, 217, 219
        Central Sahara, 135, 146
    climate of, 7-8, 12, 18, 41-44, 47-62, 78, 113, 118, 124, 146, 149-150, 160, 209, 219, 226, 229, 231, 247, 322-326
        Central Sahara, 18, 65-79, 226-227, 238-239
        Eastern Sahara, 113, 116, 160, 323
    contacts with Nile Valley, 210, 217-219, 237

contacts with West Africa, 21, 147
fishing in, 113
human groups in, 113, 118, 124, 147,
        201, 237, 246-247
    Central Sahara, 115, 232, 247
livestock in, 18
Nilo-Saharan languages, 200, 210, 220
ovicaprids in, 16, 329
rock art of, 197-198
    Central Sahara, 17, 50, 329
stone tools of, 126, 143
tombs of, 17-18, 21
Sahel, 1, 4, 18-19, 95, 160, 232, 238, 251-
        253, 269, 323-324, 326-327, 329
    climate in, 18, 47-62, 92, 322-323, 325-
        326, 329
    West African Sahel, 20, 123-150, 251,
        326
Salt, 21, 183
Sampling, 83,111-112, 149, 230, 264-265,
        267
Sand seas, 65, 73, 227
Sanga, 262
Sanga cattle, 200
Santorini eruption, 35, 39
Saouga, 129
Saqqara, 164, 279, 282, 284, 286
Saudi-Arabia, 148-149
Savannas, 21, 55, 56, 78, 97, 99-100, 103,
        113, 116-117, 119, 123, 149, 158,
        160, 233, 237, 251-252, 258-263,
        269
Scorpion King (Egypt), 283, 288
Sea surface temperature (SST), 7, 39-40, 42,
        44, 323-323, 326, 328
Seasonality, 8, 56, 68, 229, 259, 328, 331
Sedentism, 50-51, 60-61, 76, 119, 123, 131,
        143, 150, 210, 232, 240-247, 259,
        269, 326
Sedges, 97, 99-100, 102, 111, 118-119
Sedimentological data, 65, 70
Seeds, 12, 20, 101, 102, 114-115, 117, 128-
        129, 161-164, 176, 177, 179, 255
Segmentorbis angustus, 73
Selima sand sheet, 324
Semitic-speaking peoples, 171
Senegal, 56, 254, 260
    River, 20, 54, 56, 61
    Valley, 57, 60
Serir Tibesti, 241
Setaria sp., 115-117, 213
Settlement mounds, 126, 128-129, 131-132,
        135-136, 140, 143, 256, 258-260,
        265
Shaqadud, 220
Sheep, 16, 18, 95, 119, 148, 165-166, 180,
        196, 197, 201-203, 213, 215, 219,
        228-229, 232, 234, 235, 240, 244,

        260-263, 306, 322, 325, 328-330;
        see also Ovicaprids
Shell mounds, 260, 262, 267
Shells, 37, 47, 73-75, 217, 231, 254, 256,
        262, 282, 306
Shorthorn cattle, 196
Shum Laka, 96, 98-99, 102, 256, 261
Sierra Leone, 98-100
Sinai, 165-166, 204, 217
Sine-Saloum, 260
Siwa, 8
Slash-and-burn agriculture, 101-102
Slaves, 116
Social organization, 2, 11, 16, 21, 145, 146,
        331
Social stratification, 22, 220, 240, 244-245,
        282, 330
Sodmein Cave, 203
Soil, 8, 48-49, 67-68, 76, 83, 92, 101, 140,
        158, 161, 173-175, 213, 269, 328
Songhaï peoples, 61
Soninke peoples, 61
Sorghum, 16, 19-20, 60, 115-119, 148-149,
        172, 175-179, 181, 215, 219
Sorghum bicolor, 115, 148, 175
South Africa, 14
South Town, 164
Southern Africa, 18, 21
Sowing, 162, 166, 175-176
Speleothems, 160
Spherulites, 234
States, 6, 16-17, 23, 323, 326, 328
Stele, 245
Stone tools, 12, 126, 128-129, 143-144, 213,
        215, 232, 241, 242
    bifacial, 126, 143, 213, 215, 219
    microliths, 126, 128-129, 131, 209, 213,
        215, 258
Storage pits, 16, 179-180
Stratigraphy, 48, 55, 66-68, 70, 72-76, 83,
        92, 210, 231, 264
Striga, 176
Subsistence farming, 182-183
Subsistence strategies, 4, 8, 16-17, 20, 23,
        95, 111, 113, 115-116, 118, 171,
        179, 183, 259, 261, 275, 325, 327,
        329
Sudan, 14, 50, 55, 61, 199, 200, 210, 215,
        220, 237
    agriculture in, 16, 20, 111, 116-118, 118,
        148
    climate, 17, 56-57, 97, 99, 124, 145, 160,
        231, 325-326
    livestock in, 13, 16-18, 22, 56, 119, 203,
        260
Surface water, 8, 50, 54, 60, 143, 323-325
Swidden cultivation, 101
Syria, 158, 161-163, 165

Tadrak, 240
Taferjit, 253
Tagalagal, 210
Tagant, 60-61
Tahala, 75-76, 242, 244
Talak-Timersoï region, 252
Tamarix, 76, 231, 238
Tamaya Mellet, 14
Tanzania, 6
Tarifian, 219
Tarkhan, 282, 309
Tassili n'Ajjaralso, 13
Tassili n'Ajjer, 219
Taurus-Zagros arc, 161
Technological innovation, 2, 5, 11-12
Tefedest, 219
Teff, 20, 172, 176-177, 178, 181
Tegdaoust, 61
Tekrour, 50
Tenaghi Philippon, 165
Ténéré, 50, 219-220
Tenerian culture, 50, 52, 213, 215, 220
Termit, 50, 52, 60
Terracing, 52, 174, 183
Tethering stones, 198
Tetraploid wheats, 20, 161
Thermoluminescence (TL), 47, 66, 220
Threshing, 161-162, 166, 172, 176, 177, 179, 180
Ti-n-Akof, 126, 128-131, 150, 253, 255, 258
Ti-n-Hanaketen, 13
Ti-n-Torha, 71, 149, 209-210, 217, 228, 230-231, 235
Tiberias, 160
Tibesti, 50, 54, 61, 220, 241, 251
Tichitt-Oualata, 253
Tigalmamine, 238, 326
Tigray, 171-185
Tilemsi, 55, 60, 237, 253
Timbuktu, 57
Tomato, 176
Tongo, 14, 257, 262, 267
Tortoise, 277, 307
Touat, 57
Toubou peoples, 51, 61
Toucouleur peoples, 61
Trade, 2, 21-22, 131
Travertine, 67, 69, 76
Tropical roots, 171
Tropical zones, 40, 54, 56, 92, 157, 160, 171
Trypanosomiasis, 260, 262
Tuareg peoples, 51, 61, 115, 241
Tuba Ajuz, 138, 256
Tuba Lawanti, 138
Tubers, 171
Tuduf, 17, 147, 253-254
Tunisia, 198
Turborotalita quinqueloba, 40
Turkana, 89, 238, 326

Turkana peoples, 263
Turkey, 158, 161
Turquoise, 217
Turtle, 258
Tushka, 275
Typha, 70, 213, 233

U/Th dating, 67, 69, 170
Uan Afuda, 12, 66-68, 70, 209, 228
Uan Kasa, 65-72, 73-75, 213, 231, 233, 238, 240, 245
Uan Muhuggiag, 13, 70-72, 211-212, 214-215, 220, 228-229, 231, 236
Uan Tabu, 66, 70-71, 228
Uan Telocat, 70, 240, 251
Uganda, 57, 326
Urban centres, 11, 22, 327
Urewe tradition, 263
Urochloa, 115, 213
Uweinat, 12, 324

V. Nilotica, 231
Valleys, 16-17, 20, 22-23, 50-51, 56-57, 60, 62, 75-76, 78, 85, 87, 98, 112-116, 118-119, 157, 159-160, 162-166, 199-204, 210, 215, 217-220, 227, 229, 231-232, 237, 238, 240, 241, 244, 247, 253, 322-323, 325, 328-330
Vegetation, 4, 49, 55-56, 92, 95-97, 101-102, 113-114, 116-117, 124, 131, 150, 158, 160-161, 165-166, 231, 259, 322, 324-325, 327
    Afro-alpine, 85
    arboreal, 157
    floodplain, 8
    forest, 85, 101, 103
    Mediterranean, 8
    montane, 97
    oasis, 113
    savanna, 78, 116, 158, 237
    swamp, 85, 89
Vetch, 117-118
Village de la Frontière, 18
Villages, 5, 16, 20, 51-52, 60, 128, 131, 135, 143, 173-174, 179, 180, 184, 202, 213, 217, 323, 326, 328

Wadi El Ajal, 242
Wadi Howar, 55, 200, 220
Wadi Kubbaniya, 111
Wadi Shaw, 200, 231
Wadi Tanezzuft, 65, 75-76, 78, 241, 242, 244-246
Wadis, 8, 50-55, 65, 68, 69, 73-76, 78, 111, 200, 220, 231, 242, 244-246, 324-325, 328
Walasa group, 135

Warfare, 21, 51, 245; *see also* Conflict *and*
    Weapons
Water table, 12, 16, 49, 55, 72, 75-76, 89,
    232, 322, 324
Weapons, 51, 131; *see also* Conflict *and*
    Warfare
Weaving, 183
Weeding, 6, 20, 176
Weeds, 20, 117-118, 130, 175-177, 180, 182,
    183
Wells, 16, 51-52, 62, 209, 174
West Africa, 95-103, 123-150, 329
    agriculture in, 22-24, 124, 140, 148, 329
    climate in, 56, 95-103, 324, 326
    livestock in, 18-19, 21, 146, 251-270,
        323, 326
Wheat, 16, 20, 117-119, 157, 161-162, 164-
    166, 172, 175-178, 183, 328
Wild animals, 12-13, 111, 113, 148, 157,
    195-204, 210, 228-229, 234, 253,
    258-259, 261, 263, 265, 267, 279

Wild plants, 16, 101, 111, 115-119, 140, 143,
    148-150, 161-164, 166, 210, 213,
    229, 232
Wind erosion, 12, 72, 173, 323, 326
Windblown chemical indicators, 78
Windé Koroji, 131, 150, 253-254
Wolf, 203
Women, 17, 116, 220, 283-285, 287

Yam, 20
Yedseram River, 135
Younger Dryas, 40-42, 47, 117, 158-160,
    162-166, 322, 324-325

Zaire, 14
Zampia, 253-254
Zarmas peoples, 61
Zebu cattle, 196, 197, 200
Ziway-Shalla basin lakes, 85, 89
Ziziphus sp., 116
Zoser pyramid (Egypt), 164-165

Printed in the United Kingdom
by Lightning Source UK Ltd.
130456UK00007B/136/P